D1226838

Anatomy of the Head, Neck, Face, and Jaws

Anatomy of the Head, Neck, Face, and Jaws

LAWRENCE A. FRIED, D.D.S.

Associate Professor of Anatomy
Department of Dental Hygiene
Shawnee State Technical College
Portsmouth, Ohio

LEA & FEBIGER · 1976 · PHILADELPHIA

Library of Congress Cataloging in Publication Data

Fried, Lawrence A.
Anatomy of the head, neck, face, and jaws.

Includes bibliographical references and index.
1. Head. 2. Neck. 3. Face. 4. Jaws.

I. Title. [DNLM: 1. Head—Anatomy and histology. 2. Neck—Anatomy and histology. WE705 F899a]
QM535.F74 611'.91 75-22105
ISBN 0-8121-0539-7

Published in Great Britain by Henry Kimpton Publishers, London

PRINTED IN THE UNITED STATES OF AMERICA

To my wife, Phyllis

Preface

As members of the dental division of the health professions, we too often lose sight of the fact that our patients are complete organisms, designed to function best when the individual is in good health, totally. Just as a marine biologist knows that the sea is at its best state when all living parts are in balance, we must remember that the human being is in the best state when all parts are without disease. Though we may not treat all the pathologic conditions within the region of the head and neck, we can only do our part well if we have a clear understanding of the whole.

Pathology of dental origin usually affects other portions of the head and neck. Oftimes, the first signs of dental disease manifest themselves in nondental symptoms. For example, a patient with an infected molar may say he has a pain in the ear. This symptom arises because of the nerve supply that the ear has in common with the jaw. It is evident that the component parts of the head and neck, in their healthy state, must be thoroughly understood in order to successfully recognize a pathologic condition.

This text is written not only to supply knowledge sufficient to pass examinations but also to be used in the practitioner's office as a ready reference to applied anatomy in clinical concepts.

After several years of practice and teaching, I have realized that a text reference for my lectures was heretofore not available. Hopefully, this volume will bridge the gap between didactic knowledge and applied science.

Because cadaver laboratories are becoming less readily available, extensive time and work have been devoted to providing clear illustrations to make the material more easily understood and more interesting. The student is encouraged to refer frequently to the illustrations provided.

In addition, with the trend toward increased duties and responsibilities of dental auxiliary personnel, it is mandatory that dental auxiliaries become more knowledgeable in the details of head and neck anatomy.

LAWRENCE A. FRIED, D.D.S.

Portsmouth, Ohio

Acknowledgments

Many of the illustrations are redrawn from various reference texts. The authors and publishers of these texts were most gracious and helpful in obtaining the illustrations from their works. Much of the information in this text was gleaned from the books and works of other authors. Their work and contributions to the literature are duly acknowledged.

The original illustrations were done by my close and dear friend, Kay Deitchel, a superior and most talented artist. Her patience and loyalty remained unshaken throughout our labors.

L.A.F.

Contents

Anatomy of the Head, Neck, Face, and Jaws

Introduction: Nomenclature

The study of anatomy is similar to the study of the makeup of any entity. One must understand the relationships of structures. A good automotive mechanic knows where the parts of an engine are located and what they do. Without that knowledge, he cannot repair a malfunction. The language of anatomy must be understood before the general subject can be learned.

Relationships and position of anatomic structures vary as the body changes position. The lungs are in a somewhat different relation to the heart and chest wall when the body is supine than when the body is erect or flexed at the waist. Therefore, we try to study anatomy and structural positions with the subject in the same position for each description of anatomic relationships.

Descriptions of the body are based on the assumption that the person is in the *anatomic position.* This is standing erect, arms at the sides, face and palms directed forward (Fig. A). The locations of various parts of the body are described in relation to different imaginary planes. A vertical plane passing through the center of the body and dividing the body into equal right and left halves is the *median* (*midsagittal*) *plane* (Fig. B). A structure located closer to the median plane than another is said to be *medial* to the other.

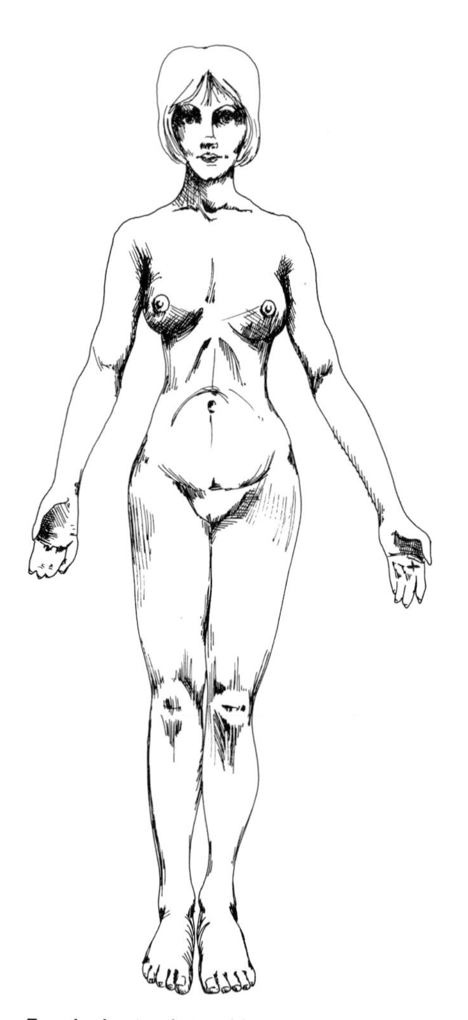

FIG. A. Anatomic position.

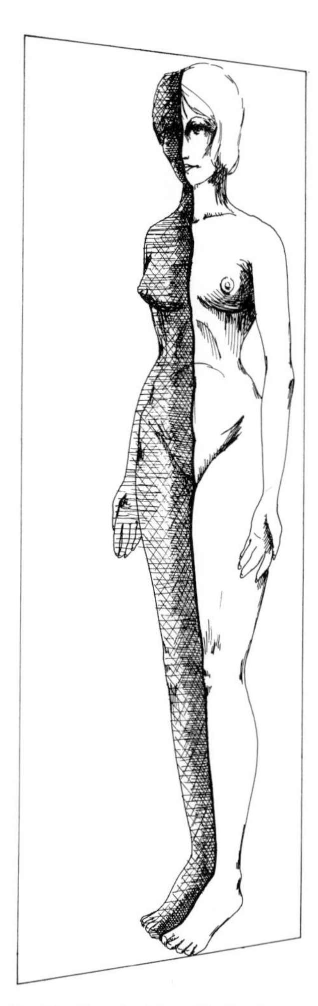

FIG. B. Median (midsagittal) plane.

A structure lying further away from the median plane than another is said to be *lateral* to the other.

A *coronal plane* is also a vertical plane, but it is directed at right angles to the median plane (Fig. C). The *horizontal* (or *transverse*) plane is at right angles to both the coronal plane and the median plane (Fig. D).

The terms *anterior* and *posterior* refer to the front and the back of the body, respectively. For example, one structure is described

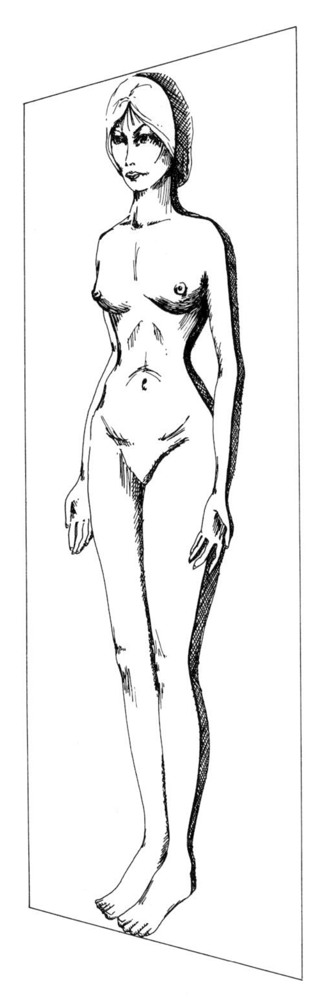

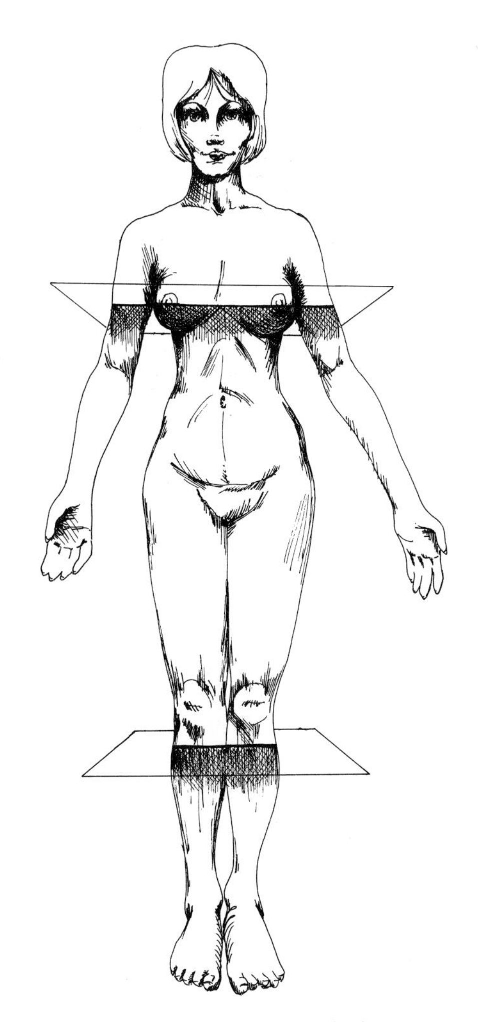

FIG. C. Coronal plane.

FIG. D. Horizontal (transverse) plane.

as anterior or posterior to another structure insofar as it is closer to the anterior or posterior surface of the body.

Proximal and *distal* describe the relative distances from the roots of the limbs or other organs. The wrist, for example, is distal to the elbow and proximal to the fingers. An artery in the lip is distal to the heart.

Superficial and *deep* refer to the relative distances of the structures from the body surface. The brain, therefore, is deep to the bone of the skull. The muscles of the abdomen are superficial to the intestine but deep to the skin.

Superior and *inferior* describe levels with reference to the upper or lower ends of the body. The jaws are superior to the shoulder but inferior to the eyes.

Ipsilateral refers to the same side and *contralateral* refers to the opposite side. The right eye is contralateral to the left molar and ipsilateral to the right ear.

With an understanding of this nomenclature, we can begin to describe the anatomy of the head and neck and relate their various structures to one another. More specific terms will be defined as the need arises.

It is emphasized that the study of anatomy requires more than one text. Therefore, I recommend that an atlas of anatomy or another text or two be consulted occasionally. These may be found in any medical or dental library and should be used, without reluctance, as often as the need dictates. An accepted up-to-date medical dictionary should be purchased and be available at all times during the study of anatomy. It will prove to be an invaluable aid and should be referred to frequently by every student of anatomy.

1

Skull and Spine

Basic to the study, knowledge, and understanding of any structure is the knowledge and understanding of its foundation. Just as any machine or building must first have a framework, so the human body must also. The skeleton of the human body, composed of bone and cartilage, serves as the framework upon which the most beautiful, complex, and intricate machine has been developed (Fig. 1-1). It is therefore axiomatic that the student of anatomy have a thorough knowledge of the skeleton before he can learn the other more complex features of the human organism.

The text here is meant to deal only with anatomy of the head and neck. Therefore, only the skeleton of this area of the human body will be discussed in detail. However, the student must remember that the head and neck are not separate entities but merely arbitrary divisions for the purpose of simplifying study. The human body is a structure that operates at its optimum only if all parts are intact, and oftimes will not function at all if some sections are absent or nonfunctional. With this basic understanding of the unity of the human body, let us begin the study of the framework and foundation of the head and neck portion.

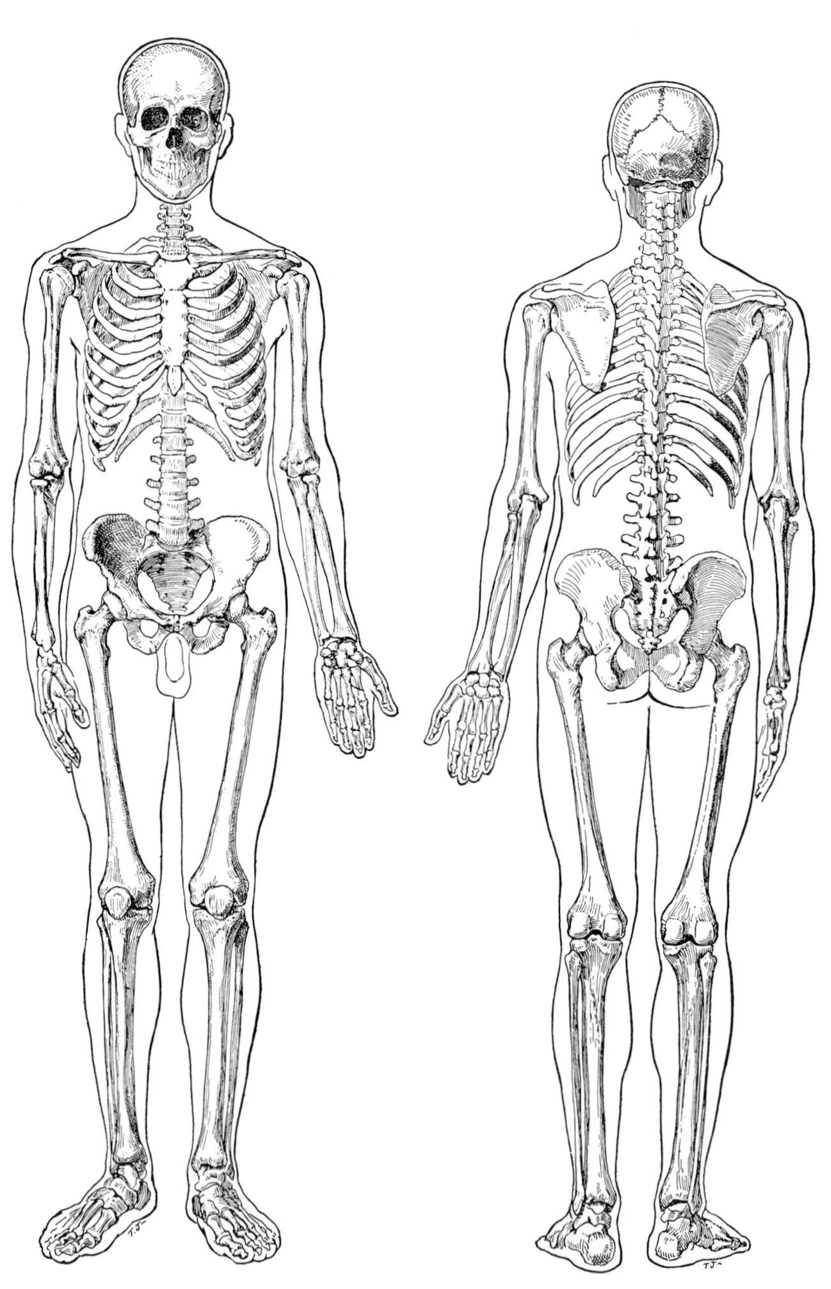

FIG. 1-1. The skeleton as projected on the surface of the body. Ventral and dorsal views. (After Eycleshymer and Jones in Gray's Anatomy of the Human Body, 29th ed. C. M. Goss, editor. Philadelphia, Lea & Febiger, 1973.)

Cervical Vertebral Column

The skeletal portion of the neck is known as the *cervical vertebral column* (Fig. 1-2). There are seven *cervical vertebrae.* The first cervical vertebra is the *atlas* and serves as the connection (articulation) of the vertebral column with the skull (Fig. 1-3). The *occipital bone* of the skull articulates with the atlas. The body of the *axis,* or second cervical vertebra, is fused with that of the atlas. The remaining five cervical vertebrae are joined by movable joints, and the neck is thereby flexible at any one of these cervical joints. The vertebrae are bony, ringlike structures through which passes the cervical portion of the central nervous system, the *spinal cord.* The cervical nerves pass out from the spinal cord, through openings between pairs of vertebrae known as the *intervertebral foramina.*

Fibrocartilagenous pads, the *intervertebral disks,* are interspaced between adjacent surfaces of the bodies of the vertebrae. These act as cushions between the rigid bony articular surfaces and

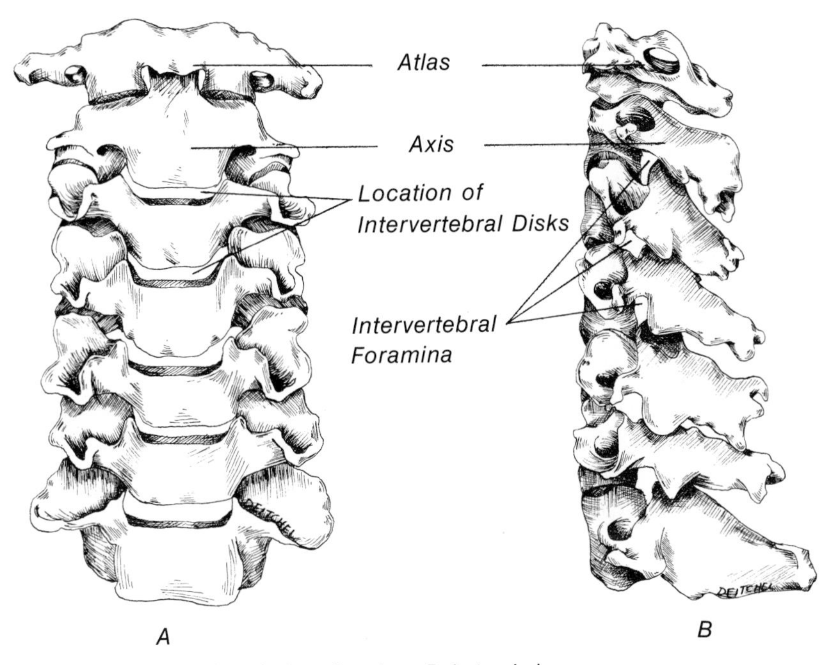

FIG. 1-2. Cervical spine. *A.* Anterior view. *B.* Lateral view.

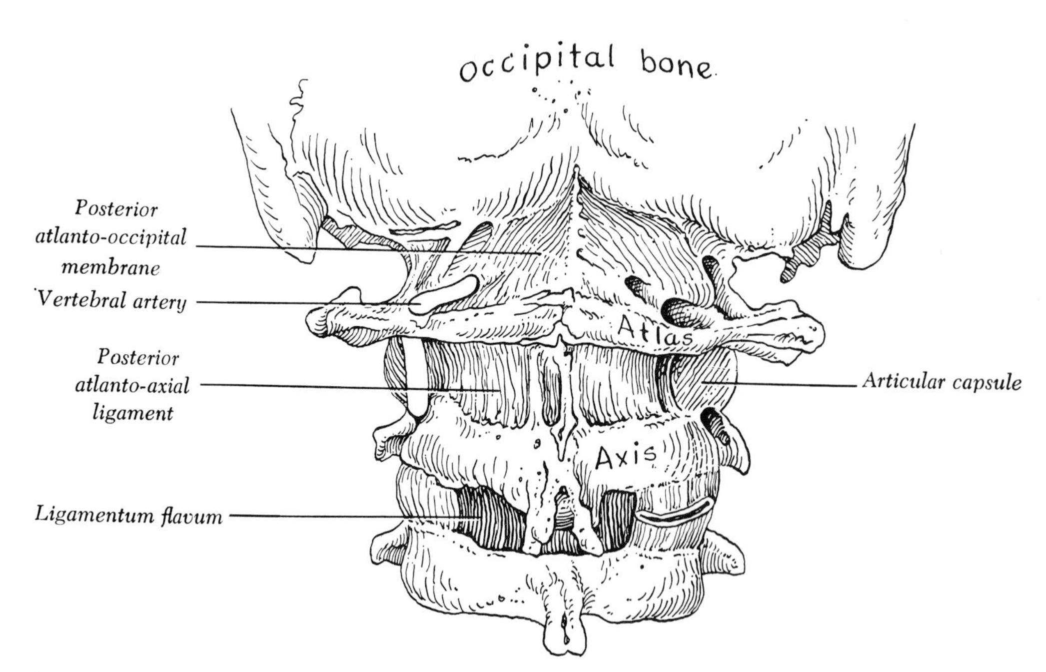

FIG. 1-3. Posterior atlantooccipital membrane and atlantoaxial ligament. (From Gray's Anatomy of the Human Body, 29th ed. C. M. Goss, editor. Philadelphia, Lea & Febiger, 1973.)

maintain the patency of the intervertebral foramina through which the cervical nerves pass.

The seventh or last cervical vertebra articulates with the remainder of the vertebral column.

The Skull

The skull is supported upon the vertebral column and is composed of a series of irregular or flattened bones (Fig. 1-4). All of these bones are immovably joined together with one exception, the *mandible* (lower jaw). These immovable joints of the skeletal bones are known as sutures. The skull is divided into two parts: (1) the *cranium* or brain case, which houses the brain, and (2) the *facial skeleton.* Some bones in the skull are paired and some are not.

The brain case can be further subdivided into the cranial base and the calvaria. The cranial base is the bottom half of the brain case, and the calvaria is the top half. These are purely arbitrary divisions, and several of the bones may contribute to both the cranial base and the calvaria, as well as the facial skeleton.

The facial skeleton is positioned below and anterior to the cranial base. The upper one third of the facial skeleton consists of the *orbits* and *nasal bones.* The middle one third consists of the *nasal cavities* and *maxillae,* (upper jaw), and the lower one third consists of the mandibular region.

Bones of the Calvaria

The bones of the calvaria are the *occipital, frontal,* and *parietal* bones. The parietal bones are paired; the others are not.

The Occipital Bone (Fig. 1-5)

The occipital bone forms the most posterior part of the skull. It also contributes to the cranial base. It articulates with the atlas by way of the *occipital condyles* flanking the opening for the spinal cord, the *foramen magnum.* The outer surface is divided into a large upper part and a small lower part by a rough curvilinear eminence known as the superior nuchal line. A faint linear elevation above and parallel to the superior nuchal line is the *supreme nuchal line.* The various rough markings and eminences on the bone, as on all bones, are dependent on the formation and attachments of various muscles, tendons, and ligaments.

The *basilar* portion is a four-sided plate of bone projecting forward in front of the foramen magnum. The external surface presents a small elevation in the midline, the *pharyngeal tubercle.*

The inner surface is divided into four depressions or *fossae* (Fig. 1-5B). The two inferior fossae are the areas where the hemispheres of the cerebellum of the brain rest. The two superior fossae house the occipital lobes of the cerebrum.

In addition to the foramen magnum, the right and left *hypoglossal canals* for the hypoglossal nerve and the right and left *condylar canals* for transmission of a vein perforate the occipital bone. The condylar canals are just lateral to the foramen magnum, and the larger hypoglossal canals lie in a more anterolateral position.

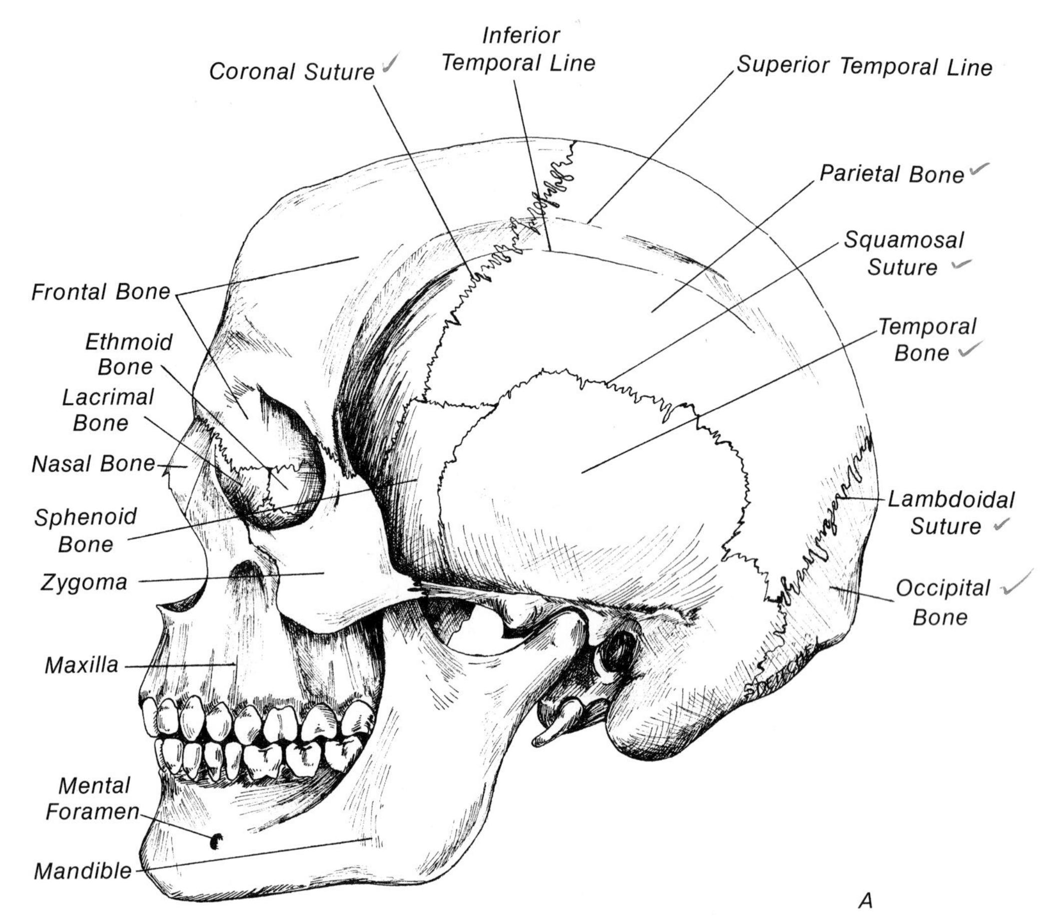

FIG. 1-4. Skull. *A*. Lateral view.

The Frontal Bone (Fig. 1-6)

The large unpaired frontal bone forms the anterior portion of the calvaria. Posteriorly, it articulates with the parietal bones (Fig. 1-4). Posteroinferiorly, it joins the wings of the sphenoid bone bilaterally. Inferiorly, it connects with the ethmoid bone and lacrimal bones. Anteriorly, it articulates with the zygomas, maxillae, and nasal bones (Fig. 1-4).

The curved, smooth area of the frontal bone is known as the *squama.* Anteriorly, two deep curved depressions form the roof of the orbit. Laterally, two projections, the right and left zygomatic

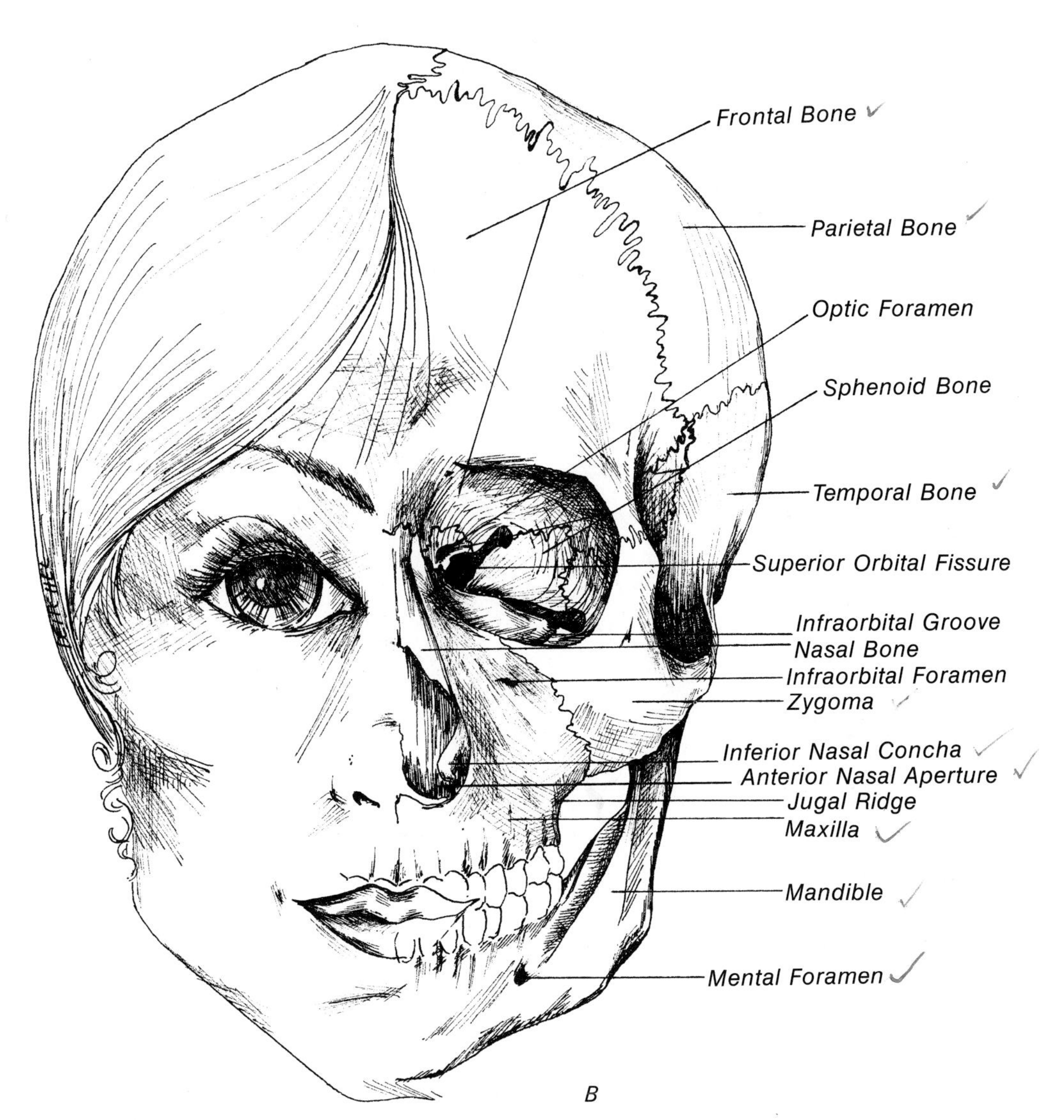

FIG. 1-4. Skull (continued). *B*. Anterior view.

processes, form the lateral wall of the orbit with the zygoma. The supraorbital ridges are curved elevations connecting the midportion of the frontal bone with its zygomatic process. Each is interrupted by its respective supraorbital notch or foramen that transmits the frontal vessels and nerves (Fig. 1-6).

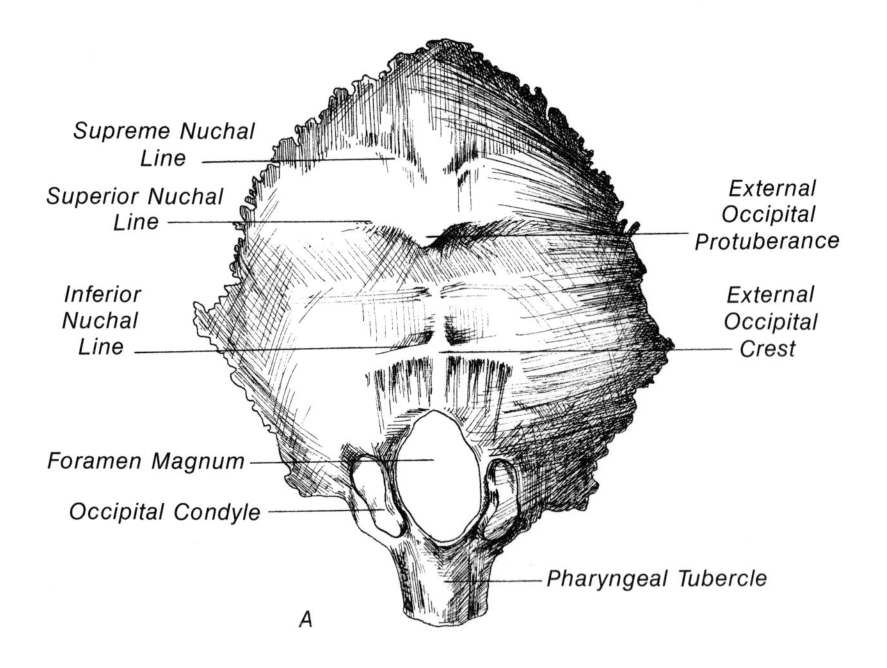

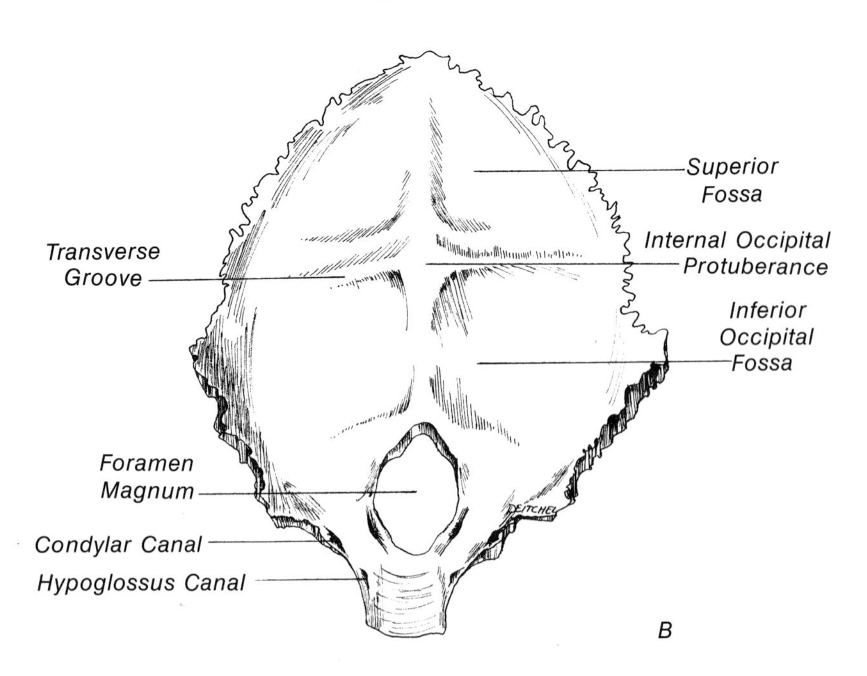

FIG. 1-5. Occipital bone. *A.* Outer surface. *B.* Inner surface.

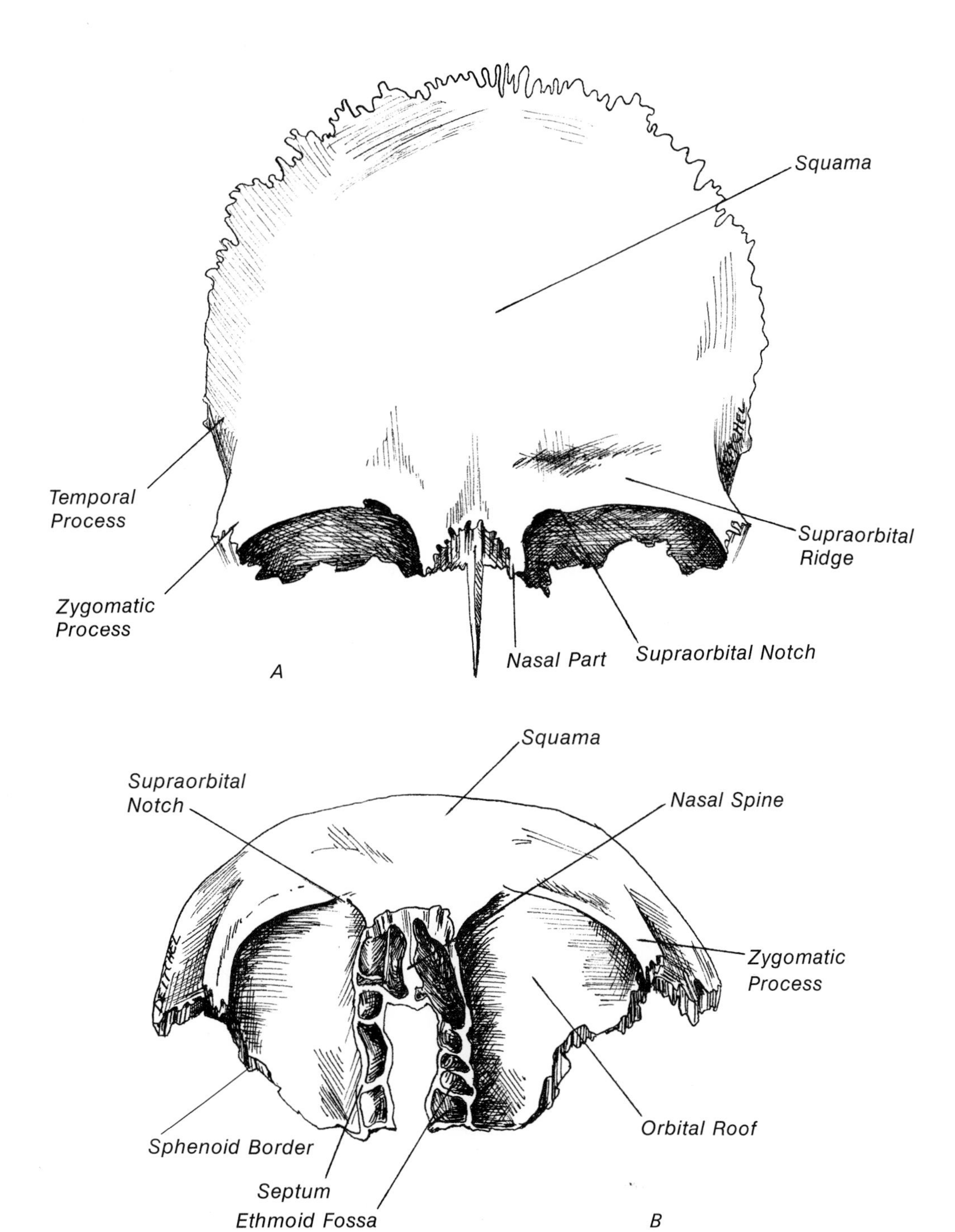

FIG. 1-6. Frontal bone. *A.* Anterior view. *B.* Inferior view.

The frontal sinuses lie in the frontal bone in an area just superior to the articulation with the nasal bones.

Looking at the frontal bone from its inferior aspect, one sees in the midline, the inner and outer plates of the ethmoid portion of the frontal bone. Between these plates are seen fossa separated by multiple transverse ridges or septa. These fossae form the roofs of the upper ethmoid air cells. The most anterior fossa is quite deep and extends superiorly into the depths of the bone to form the frontal sinuses. It is in this region that the nasal bones and frontal processes of the maxillae, together with the ethmoid bone, articulate with the frontal bone (Fig. 1-4).

The Parietal Bones (Fig. 1-7)

The parietal bones are a pair of quadrangular cup-shaped bones that articulate with one another at the midline at the top of the calvaria and with the occipital bone posteriorly. Inferiorly, they meet the temporal bone and the right and left great wings of the sphenoid bone. Anteriorly, they join the frontal bone (Fig. 1-4).

The outer surface is convex. A curving eminence is seen dividing the lower one third from the upper two thirds. This eminence is known as the *inferior temporal line* and serves as the superior limit of the origin of the temporalis muscle. Above this line is a second, but much less distinct, line known as the *superior temporal line.* This serves as the superior boundary of the temporal fossa.

The inner surface of the parietal bone is concave and contains multiple deep branching grooves. They contain the branches of the middle meningeal artery.

Bones of the Cranial Base

The bones of the cranial base are the *temporal bones, sphenoid bone,* and the basilar portion of the *occipital bone* described on page 11. The other two are quite complex and thus quite interesting and intriguing.

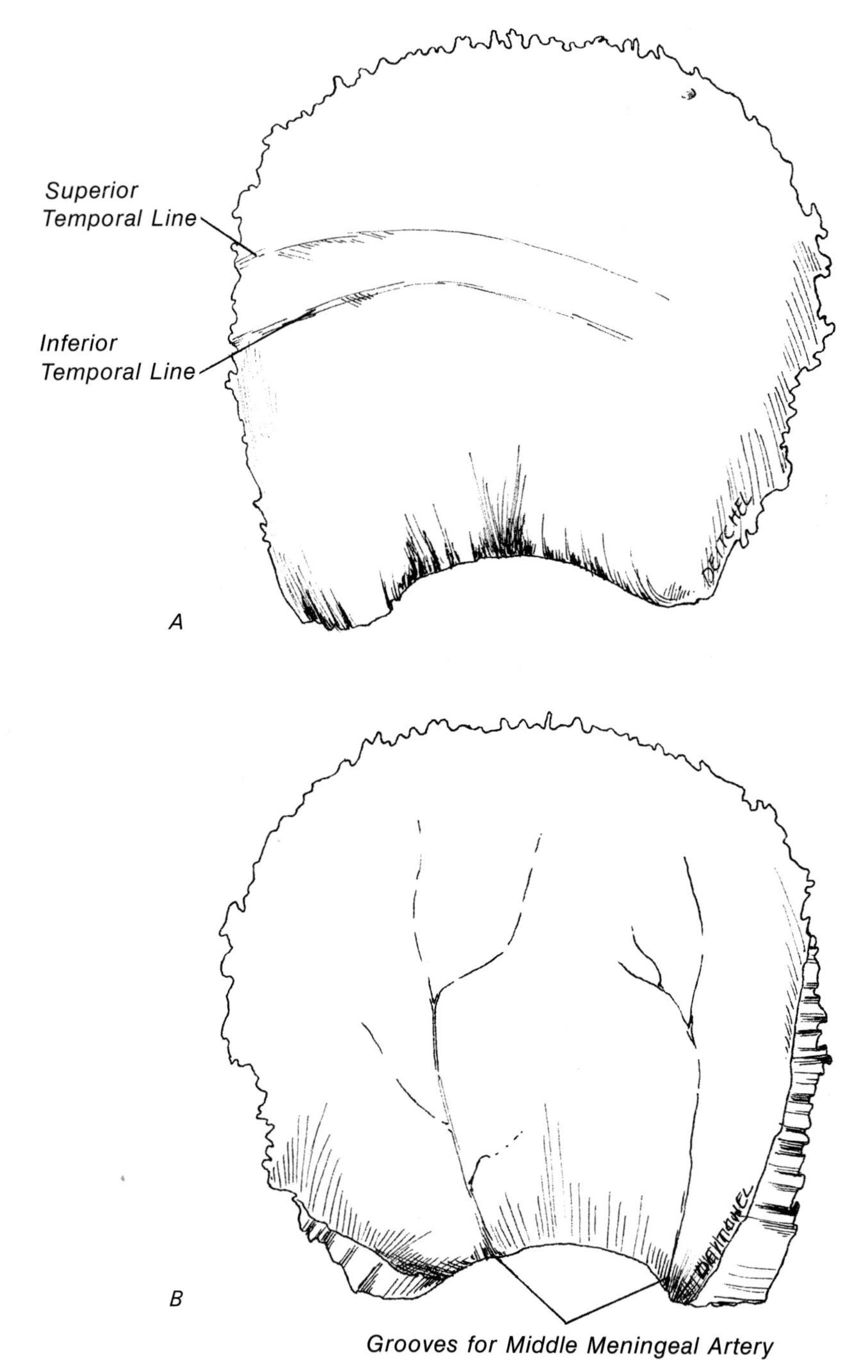

FIG. 1-7. Left parietal bone. *A.* External surface. *B.* Internal surface.

The Temporal Bone (Fig. 1-8)

The temporal bone has a large flat portion known as the *squama.* The squama forms part of the lateral wall of the skull and contains on its inferior surface a depression known as the glenoid fossa, into which the mandible articulates. Just superior to this fossa, is a fingerlike projection, the zygomatic process, which joins with the zygoma anteriorly to form the zygomatic arch.

Immediately posterior to the root of the zygomatic process is a large opening into the depth of the bone. This opening is the entrance to the middle ear and is known as the *external auditory meatus.* Posterior to the meatus is a rounded, rough prominence, the mastoid process. This structure is variably hollowed out by air spaces, the *mastoid air cells,* which communicate with the middle ear.

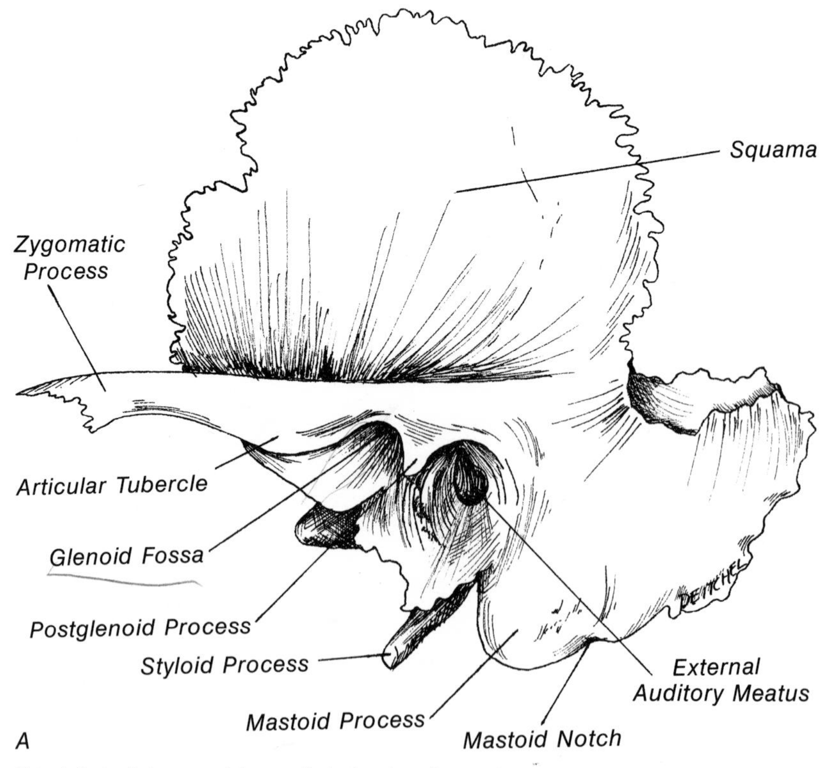

FIG. 1-8. Left temporal bone. *A.* Lateral surface.

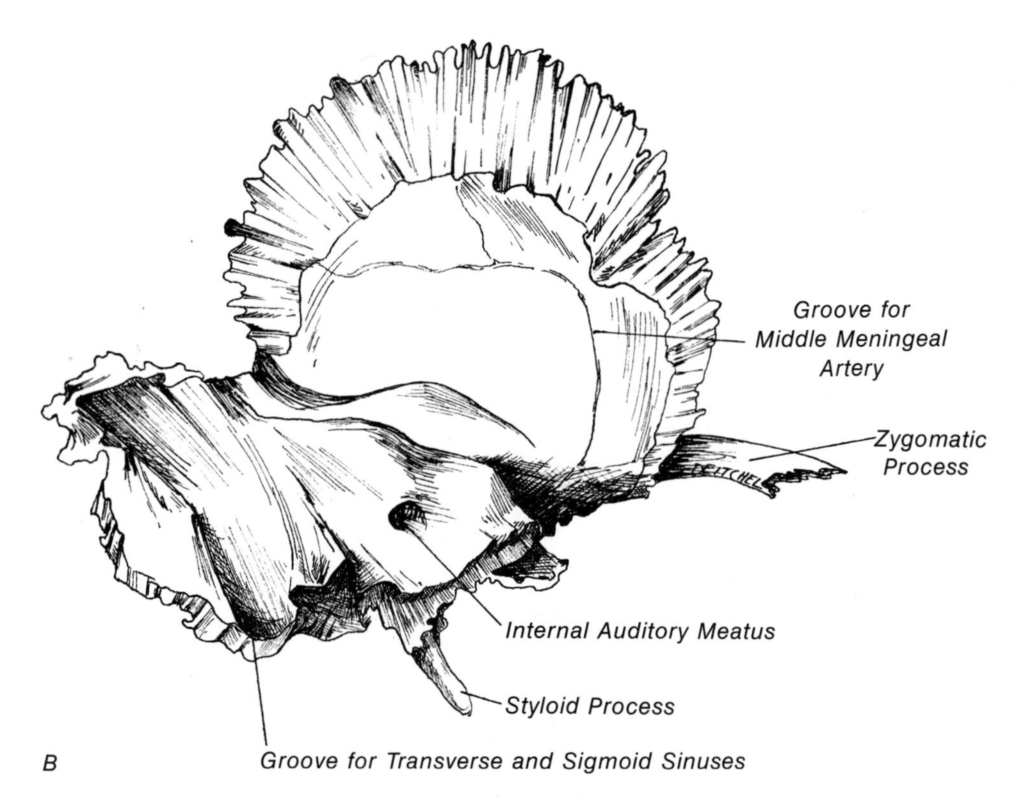

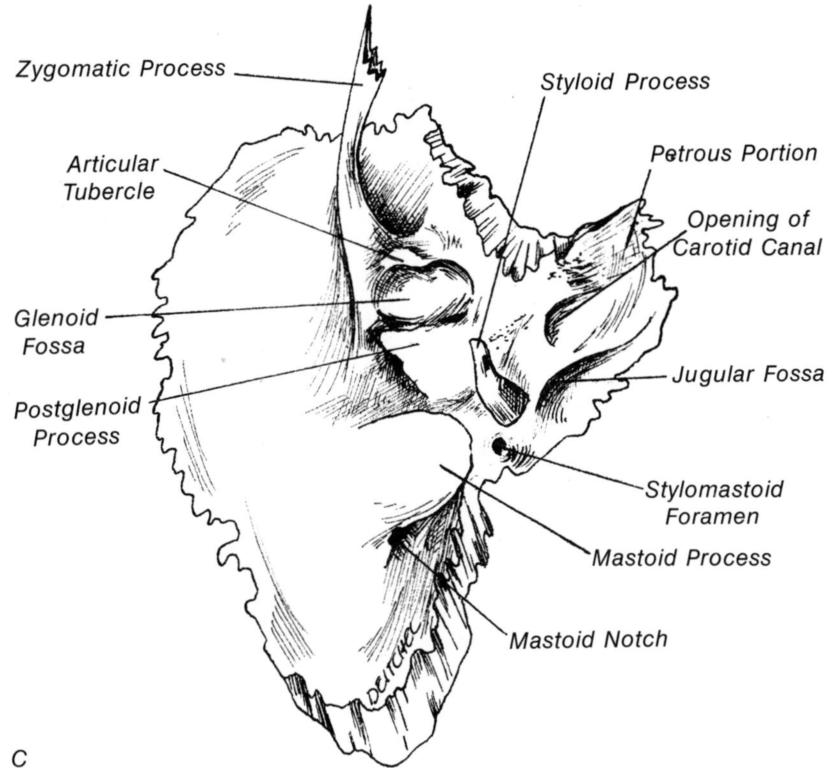

FIG. 1-8 (continued). *B.* Inner surface. *C.* Inferior view.

Inferior and medial to the external auditory meatus is a pointed, bony projection, the *styloid process,* which serves as the attachment of several muscles and ligaments.

The internal surface of the squama is concave. The depressions therein correspond to the convolutions of the temporal lobe of the brain. Also seen are grooves for the branches of the middle meningeal artery. The internal surface of the mastoid portion contains a deep curved groove which houses the transverse and sigmoid sinuses. Just anterior to the groove for the sigmoid sinus is a large opening, the *internal auditory meatus.*

Observing the temporal bone from the inferior aspect, one sees the *articular tubercle* on the anterior lip of the glenoid fossa. The posterior border of the fossa is also elevated by a variably high process, the *postglenoid* process. Between the styloid process and the mastoid process is situated the stylomastoid foramen through which the facial nerve exits from the skull. Medial to the styloid process is the *jugular fossa* which contains a portion of the jugular vein. In front of this fossa lies the opening into the carotid canal. This canal transmits the internal carotid artery and the sympathetic carotid plexus. The *mastoid notch* is the depression that separates the mastoid process from the inferior surface of the temporal bone. It serves as the origin for the posterior belly of the digastric muscle. That portion of the temporal bone wedged between the sphenoid and occipital bones is known as the petrous portion. It contains in its interior the essential parts of the organ of hearing (Fig. 1-9).

The Sphenoid Bone (Fig. 1-10)

The sphenoid bone is a midline bone that articulates with the occipital bone and the temporal bones to complete the cranial base (Fig. 1-9). Anteriorly, it articulates with the maxillae and palatine bones. Superiorly, it joins the parietal bone, and anterosuperiorly, it meets the ethmoid bone and frontal bones (Fig. 1-4).

The sphenoid bone is composed of a body from which arise three paired processes. The *lesser wings* project from the anterior portion of the body, and the *greater wings* project from the posterolateral portion. The pterygoid processes project inferiorly from the root of

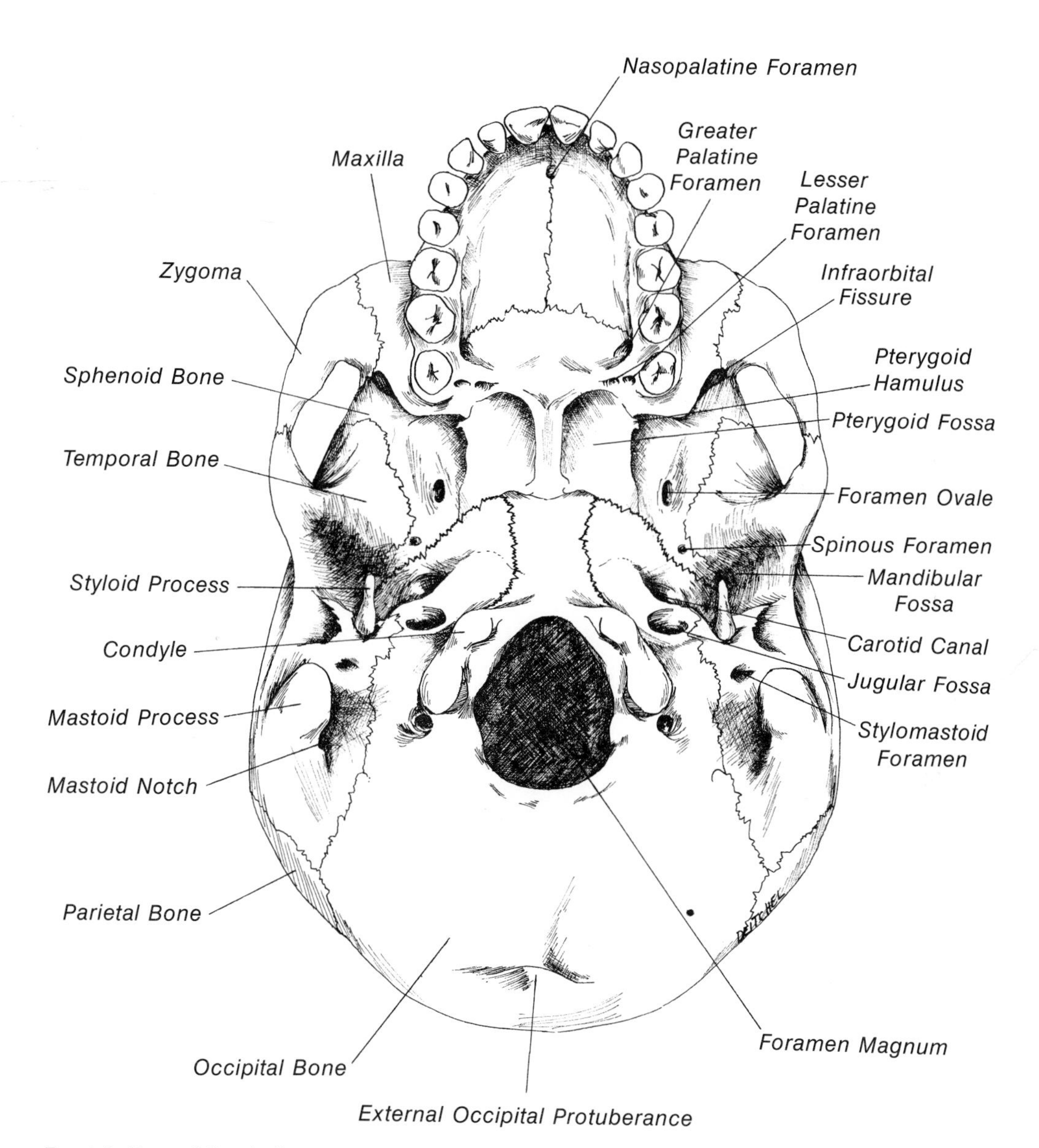

FIG. 1-9. Base of the skull.

the greater wings. A deep depression is seen on the cranial surface of the sphenoid bone in which the pituitary (hypophyseal) gland is housed. This depression is termed the Turkish saddle, or *sella turcica,* because of its saddle-like appearance. The bony midline

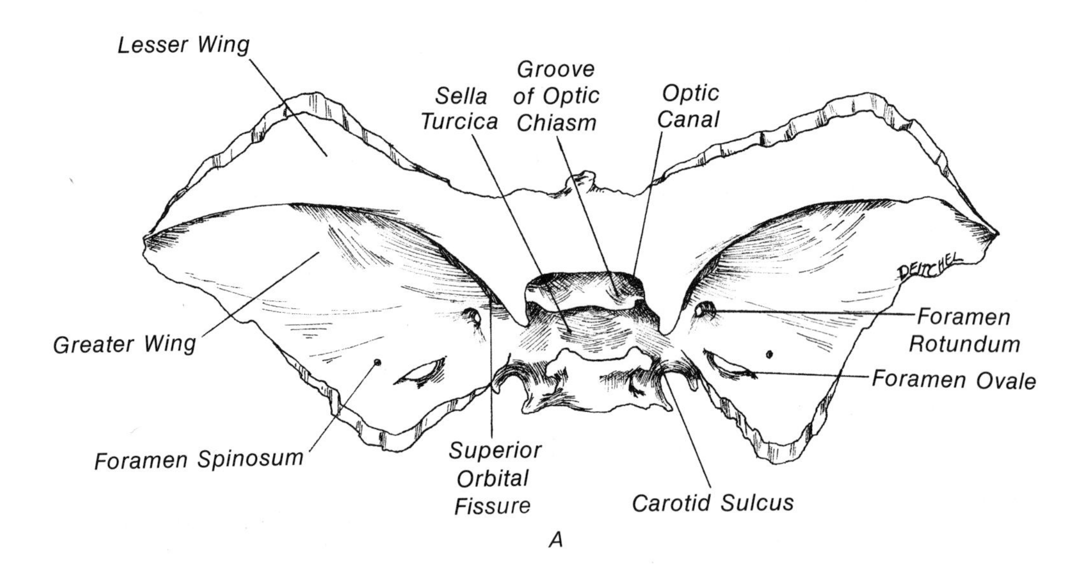

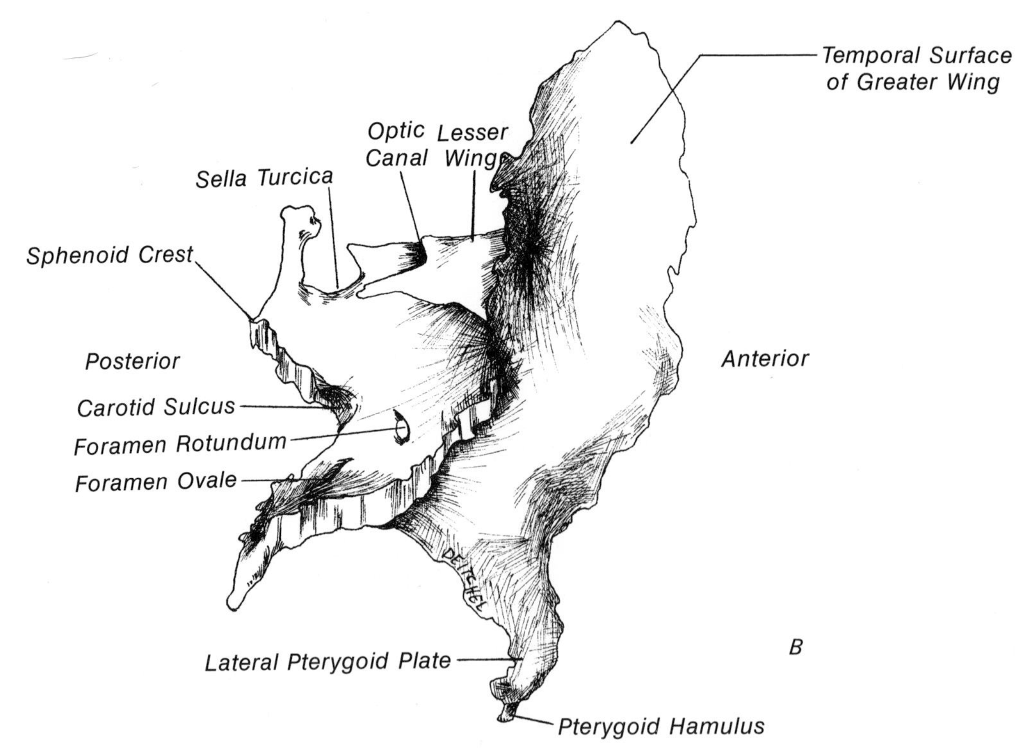

FIG. 1-10. Sphenoid bone. *A.* Superior view. *B.* Lateral view.

prominence on the anteroinferior aspect of the sphenoid bone is the *sphenoid crest* which joins the perpendicular plate of the ethmoid bone and aids, along with the vomer, in forming the bony nasal septum (Fig. 1-11). The body of the sphenoid bone is hollowed out into two air-filled cavities, the sphenoid sinuses, which communicate with the nasal cavity. The depression lateral to the sella turcica is the *carotid sulcus* (groove) and is in direct communication with the intracranial opening of the carotid canal of the temporal bone (Fig. 1-12). Also, the cavernous sinus occupies this area. Anterior to the sella turcica, the two large optic canals can be seen. These structures transmit the optic nerve and ophthalmic artery into the orbit. The lesser wings of the sphenoid bone arise in this area. The superior surface of the wings form part of the anterior cranial fossa, and the inferior surfaces form the most posterior part of the orbital roof and are thus perforated by the optic foramen (Fig. 1-12).

The greater wing arises from the lateral aspect of the body of the sphenoid bone. The anterior surface of this wing is the posterior border of the superior orbital fissure, which opens into the orbit (Fig.

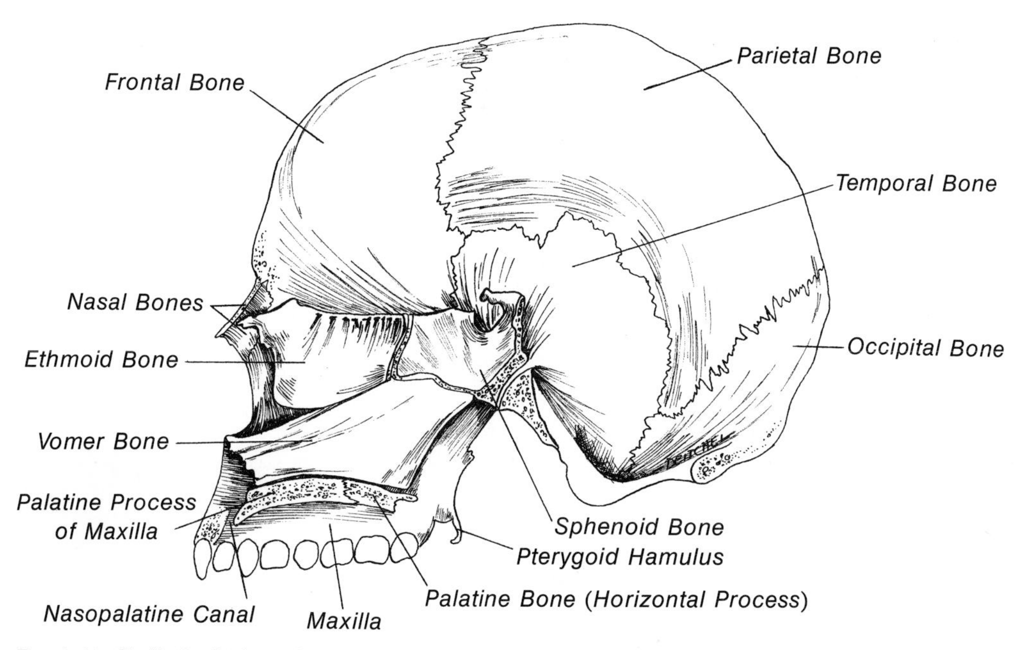

FIG. 1-11. Skull. Sagittal section.

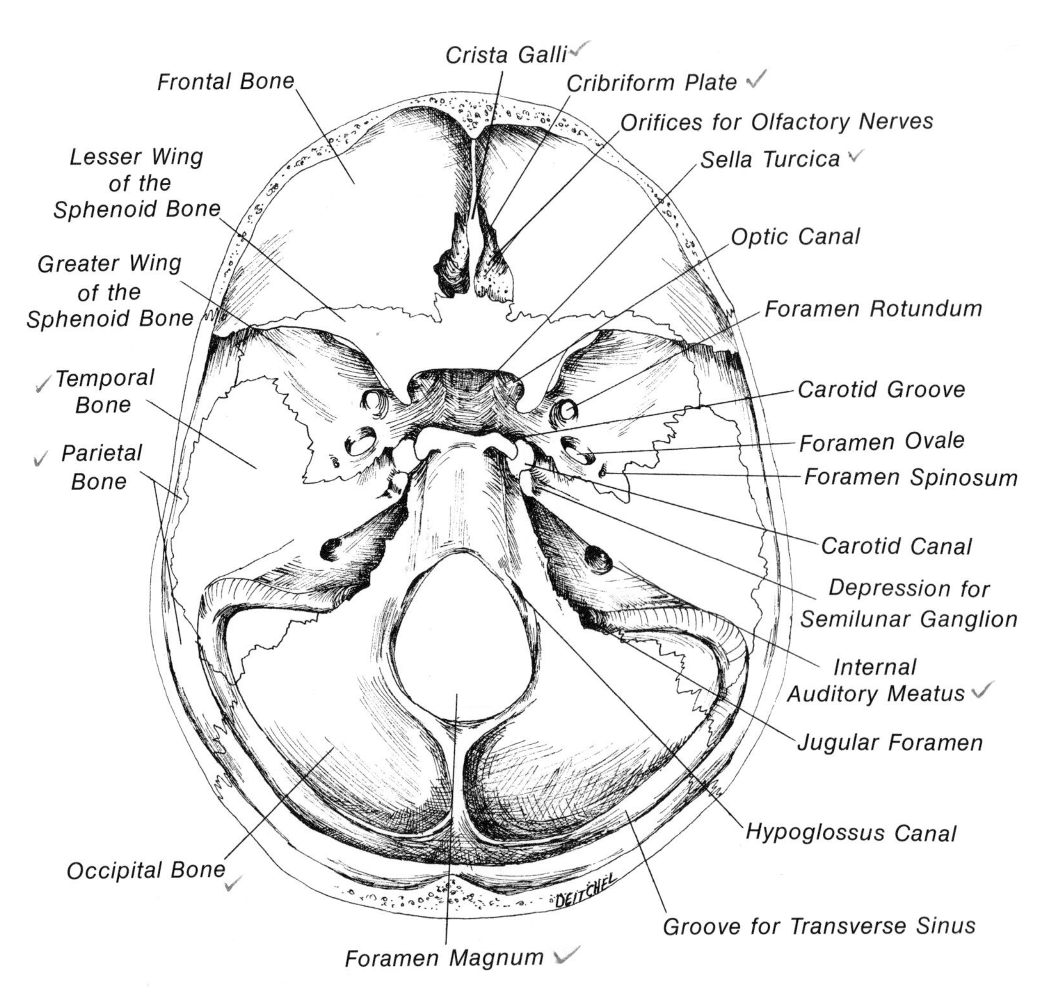

FIG. 1-12. Base of skull from inside.

1-4). Laterally, the greater wing articulates with the frontal bone and the parietal bone; posteriorly, it joins the squama of the temporal bone.

The *foramen rotundum* for the second division of the trigeminal nerve is located lateral to the carotid sulcus and anterior to the larger *oval foramen* (*foramen ovale*) for the third division of the trigeminal nerve (Fig. 1-12). The middle meningeal artery enters the cranial vault through the *foramen spinosum,* a small opening located just lateral to the foramen ovale.

The pterygoid processes arise at the connection of the greater wing and the posterior portion of the body. Each process has two plates separated by a deep fossa (Fig. 1-9). The inferior end of the medial plate ends in a thin curved process, the pterygoid hamulus. This serves in a functional relationship with the tensor palatini muscle.

Bones of the Facial Skeleton

The facial skeleton is composed of the following bones: sphenoid, ethmoid, vomer, maxillae, mandible, lacrimal, nasal, inferior concha, palatine, and zygoma. The sphenoid bone we have discussed (page 20).

The Ethmoid Bone (Fig. 1-13)

The unpaired ethmoid bone fits into the midportion of the anterior medial area of the frontal bone (Fig. 1-11). It contains a midline *perpendicular plate* that crosses a horizontal *cribriform plate.* This cribriform plate is perforated to allow passage of the olfactory nerves between the brain case and the nose. Hanging off the outer lateral edge of the cribriform plate are the *superior and middle nasal conchae,* bilaterally. These each have multiple septa passing laterally to another more lateral paper-thin plate of vertical bone, the *lamina orbitalis* (lamina papyracea). Between the lamina orbitalis and the conchae, separated by these septa, are the ethmoid air cells. The lamina orbitals aids in the formation of the medial orbital wall (Fig. 1-4).

The midline vertical plate aids the vomer and cartilage in formation of the nasal septum. Posterosuperiorly, it articulates with the sphenoid crest of the sphenoid bone, and posteroinferiorly, it meets the vomer (Fig. 1-11).

The cribriform plate articulates with the frontal bone anteriorly and laterally and joins the wing of the sphenoid posteriorly. The lamina orbitalis meets the lacrimal bone anteriorly and the maxilla inferiorly.

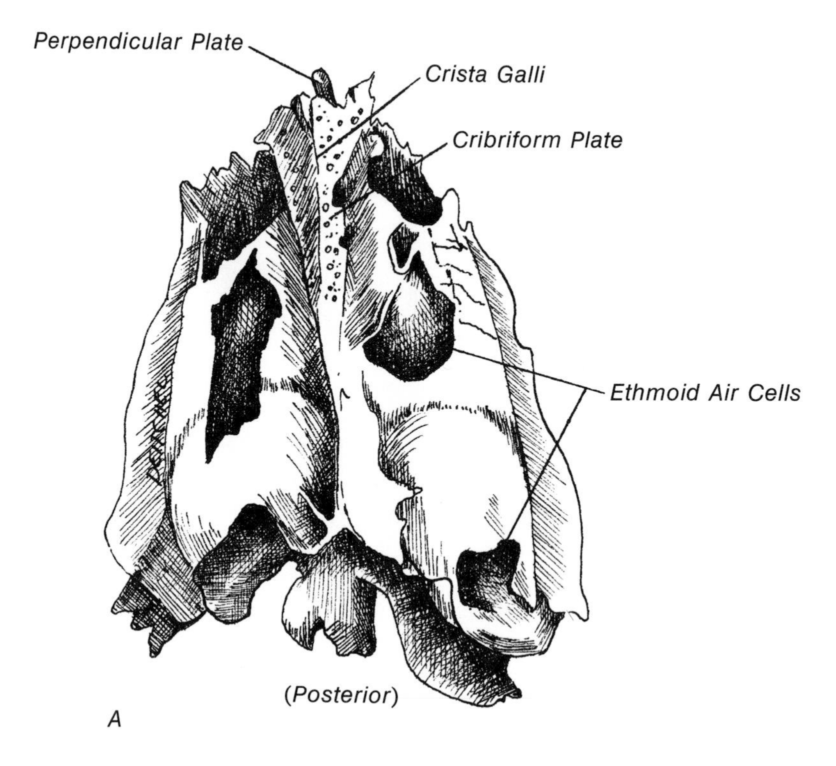

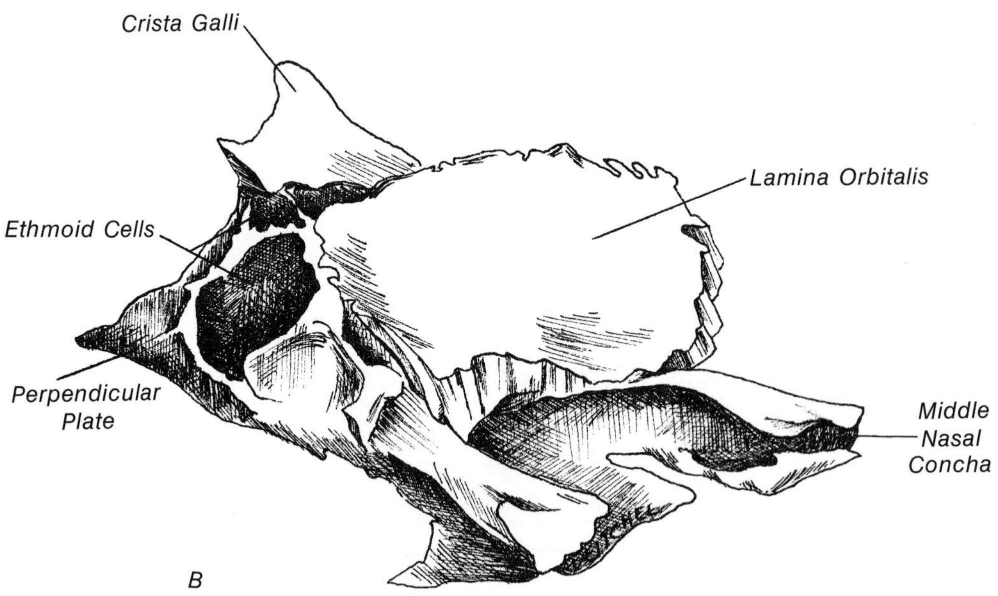

FIG. 1-13. Ethmoid bone. *A*. Superior view. *B*. Lateral view.

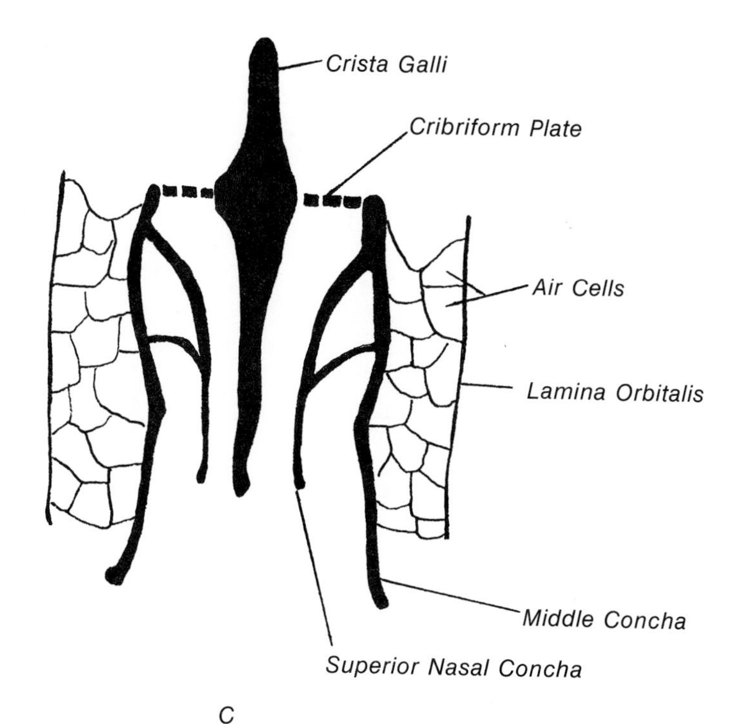

FIG. 1-13 (continued). *C*. Diagram.

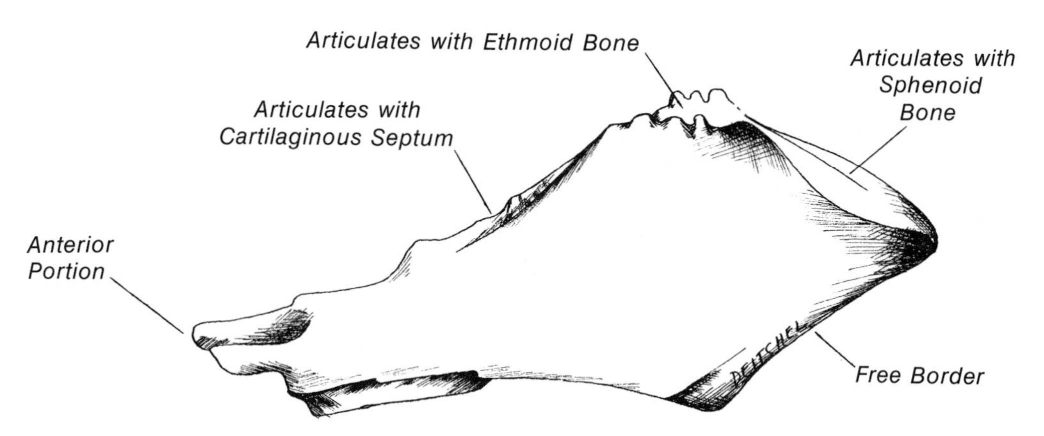

FIG. 1-14. Vomer. Left surface.

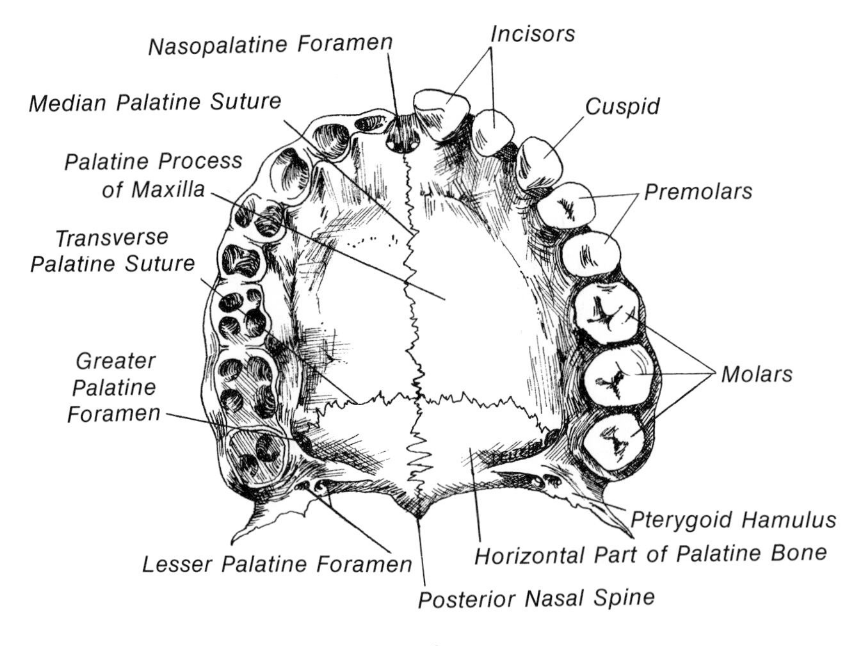

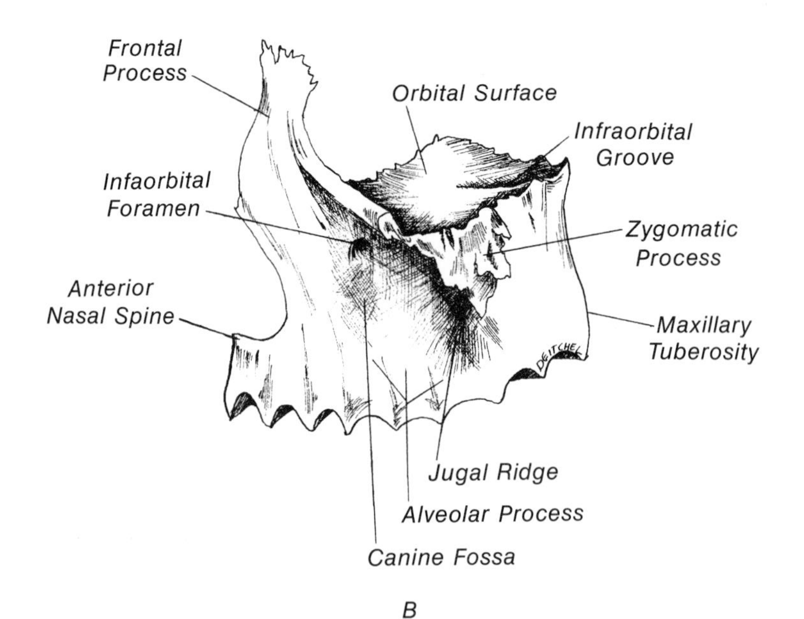

FIG. 1-15. Maxilla. *A*. Bony palate. *B*. Lateral view.

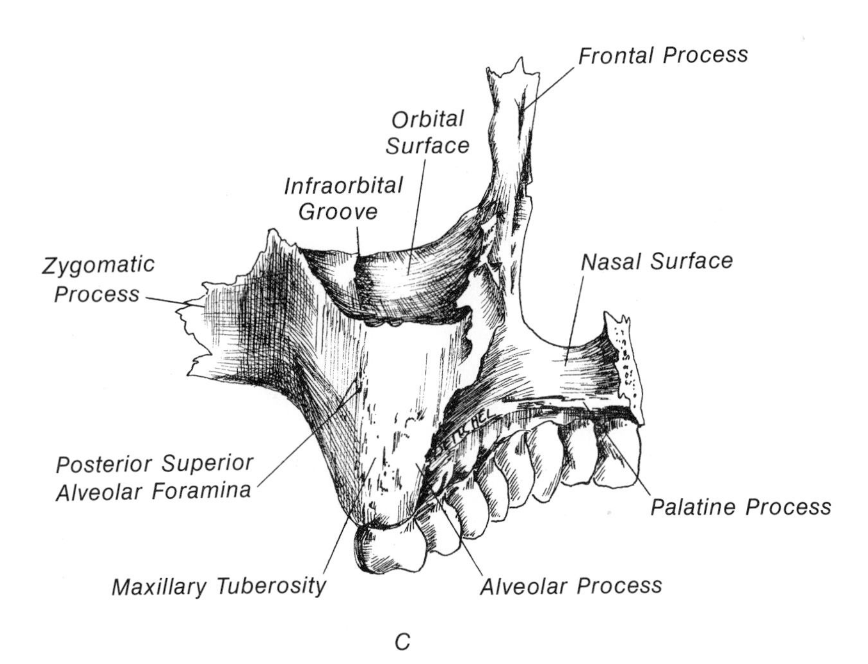

FIG. 1-15 (continued). *C.* Posterior view.

The Vomer (Fig. 1-14)

The vomer forms the posterior portion of the nasal septum. It is situated in the midsagittal plane of the nasal fossa. It articulates inferiorly with the palatine and maxillary bones. The posterior portion of the anterior superior border meets the perpendicular plate of the ethmoid bone. The anterior portion connects with the cartilaginous septum of the nose. The posterior border is free (Fig. 1-11).

The Maxillae (Fig. 1-15)

The maxillae, or the upper jaw, are two of the most important bones to us in the dental profession. This interesting structure consists of a large hollow body and four prominent processes. The *frontal process,* arising from the anteromedial corner of the body, reaches to and connects with the frontal bone and forms the medial orbital rim (Fig. 1-4). Posteriorly, this process articulates with the lacrimal bone which joins the ethmoid bone, the three of which form the medial orbital wall with a portion of the frontal bone (Fig. 1-4). The medial rim of the frontal process fuses with the nasal bone.

A second process of the maxillae arises from the anterolateral corner of the maxillary body. This process, the *zygomatic process,* reaches laterally to the zygoma, and together they form the infraorbital rim and the greatest portion of the orbital floor. This zygomatic process contains the infraorbital foramen on its anterior surface.

The horizontal *palatine process* arises from the lower edge of the medial surface of the body. It joins the process of the other maxilla to form the major part of the hard palate.

Extending downward from the body of the maxillae is the rounded, elongated *alveolar process* which serves to house the teeth. The posterior aspect of this process is known as the *maxillary tuberosity.* This area is somewhat rough and contains several foramina for the superior alveolar nerves and vessels.

The body is pyramidal in shape with four sides. The base faces the nasal cavity and the apex becomes the zygomatic process. The four sides of the pyramid are the orbital surface, forming a large part of the orbital floor; the malar surface, forming, with the zygoma, the cheek; the infratemporal surface turning toward the infratemporal fossa; and the inferior surface covered for the most part by the alveolar process.

The base or nasal surface contains the opening into the maxillary sinus. Behind the opening one can see the roughened edge of the maxillae where the vertical plate of the palatine bone articulates. The lacrimal groove for the lacrimal duct can be seen anterior to the sinus opening.

The orbital surface is concave, with a sharp medial edge that joins the lacrimal bone anteriorly and the lamina orbitalis of the ethmoid bone posteriorly (Fig. 1-4). Laterally, the orbital surface articulates with the zygoma and the frontal bone. It is separated from the sphenoid bone by the *inferior orbital fissure.* A groove in the floor of the orbit, the *infraorbital sulcus,* contains the infraorbital nerves and vessels. This sulcus is covered further forward and becomes the *infraorbital canal* that opens at the infraorbital foramen.

The anterior lateral surface aids the zygoma in the formation of

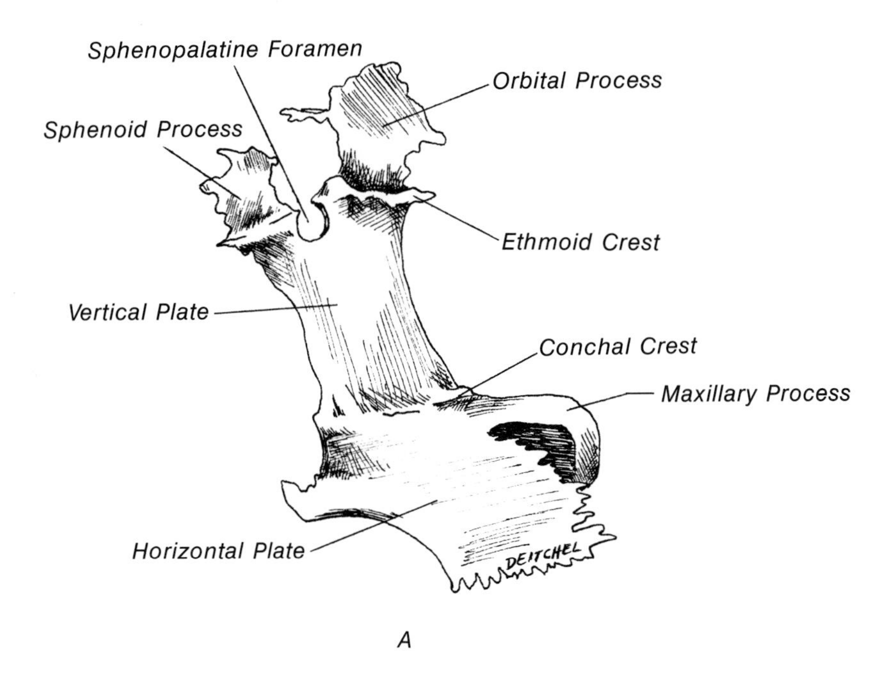

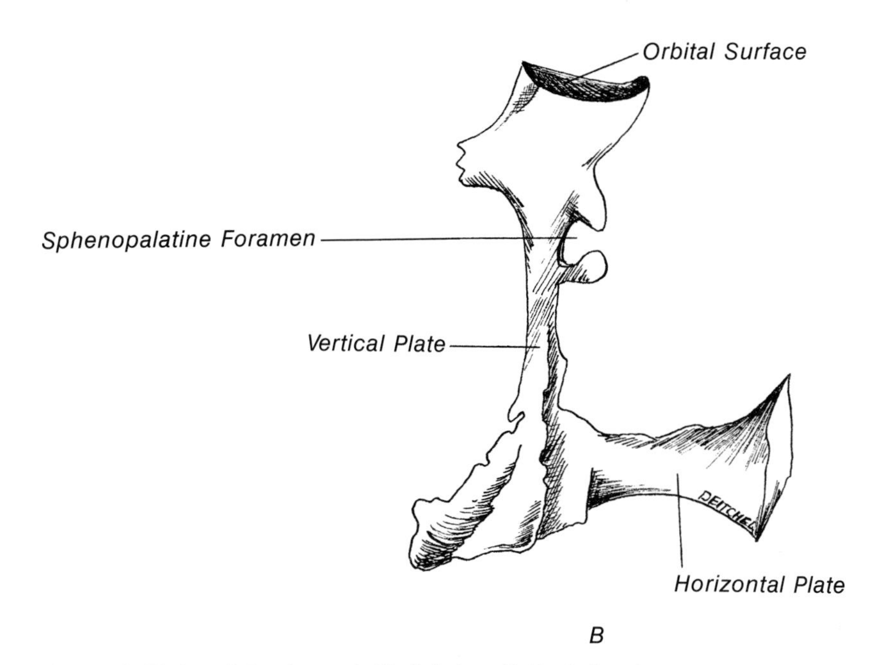

FIG. 1-16. Right palatine bone. *A.* Medial view. *B.* Posterior view.

the cheek. Its boundary posteriorly is the bony crest which begins at the tip of the zygomatic process and courses in a concave arc inferiorly to end at the alveolar process. This ridge is termed the *jugal ridge* or zygomatic-alveolar crest. Just inferior and medial to the infraorbital foramen is a smooth depression on the surface of the maxilla. This is the *canine fossa.*

The posterior lateral surface is that surface behind the jugal ridge. It forms the anterior wall of the infratemporal fossa.

The right and left palatine processes have a rough oral surface, but the nasal surface is smooth. Along the midline between the two processes on the nasal surface is a sharp elevation, the nasal crest. This serves as the attachment for the nasal septum. Two small canals begin on either side of this crest and travel downward anteriorly and medially to exit with one another in the *nasopalatine canal,* which then opens just behind the central incisors. The opening is the *nasopalatine foramen.* On the palatal aspect of the maxilla is a notch. With the corresponding notch of the horizontal plate of the palatine bone it forms the greater palatine foramen. This structure is located at the posterior of the hard palate at a point where the alveolar process joins the palatine process near the second molar region. It is this foramen through which the greater palatine arteries, veins, and nerves pass onto the hard palate.

The suture between the right and left palatine processes is known as the median palatine suture.

Palatine Bone (Fig. 1-16)

The palatine bone is an extremely irregular shaped bone that acts as a link between the maxillae and the sphenoid bone. It consists of two main parts, a horizontal plate and a vertical plate. The horizontal plate forms the posterior of the hard palate. This is the area where the palatine bone articulates with the horizontal or palatine process of the maxilla. The horizontal plate of the palatine bone also articulates with the horizontal plate of the palatine bone of the opposite side to complete the posterior hard palate.

The vertical plate reaches up behind the maxilla to contribute a small lip to the posterior orbital floor. On the posterior aspect of

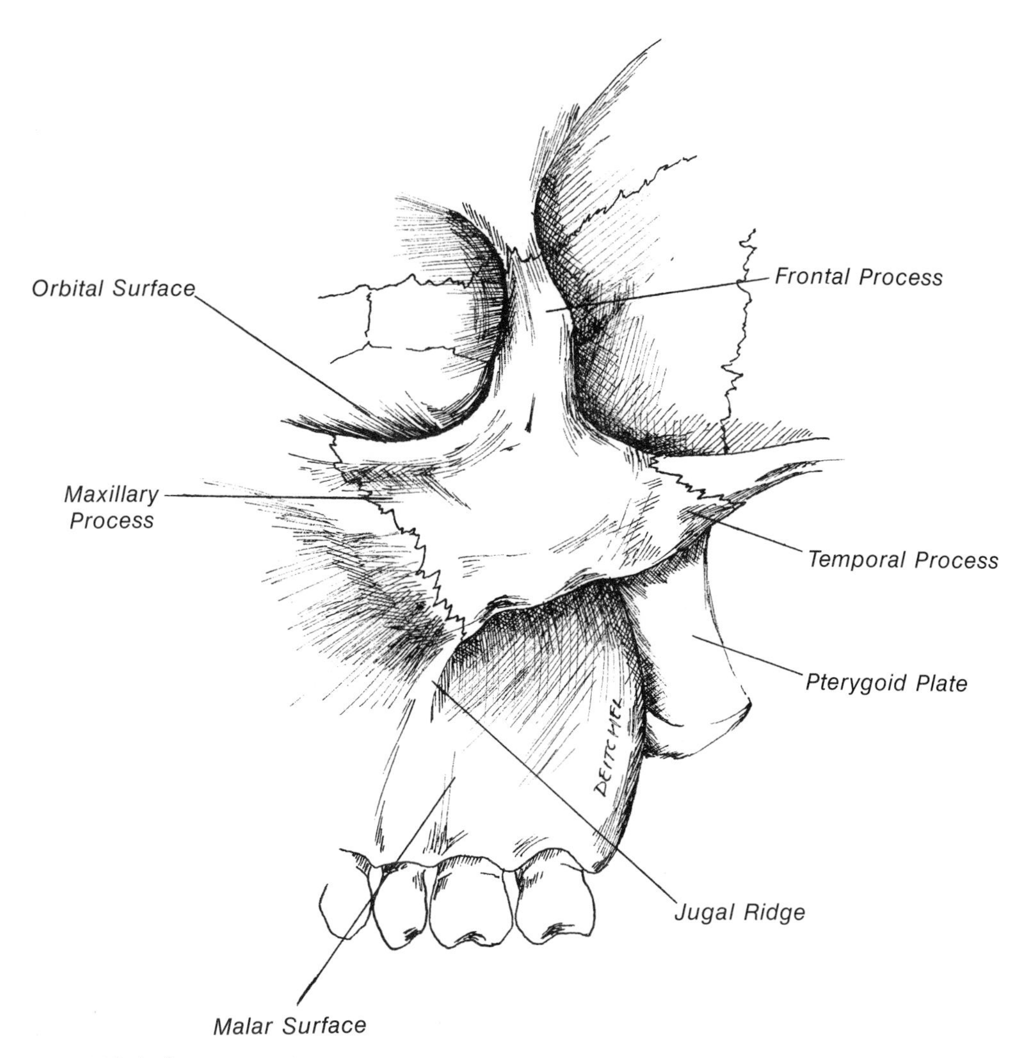

FIG. 1-17. Left zygoma.

the vertical plate is the area of articulation for the lateral pterygoid plate of the sphenoid bone.

The Inferior Nasal Concha (Fig. 1-4B)

The inferior nasal concha is a small oval bone. It lies within the nasal cavity near the floor and articulates to the lateral wall of the nasal cavity formed by the maxilla.

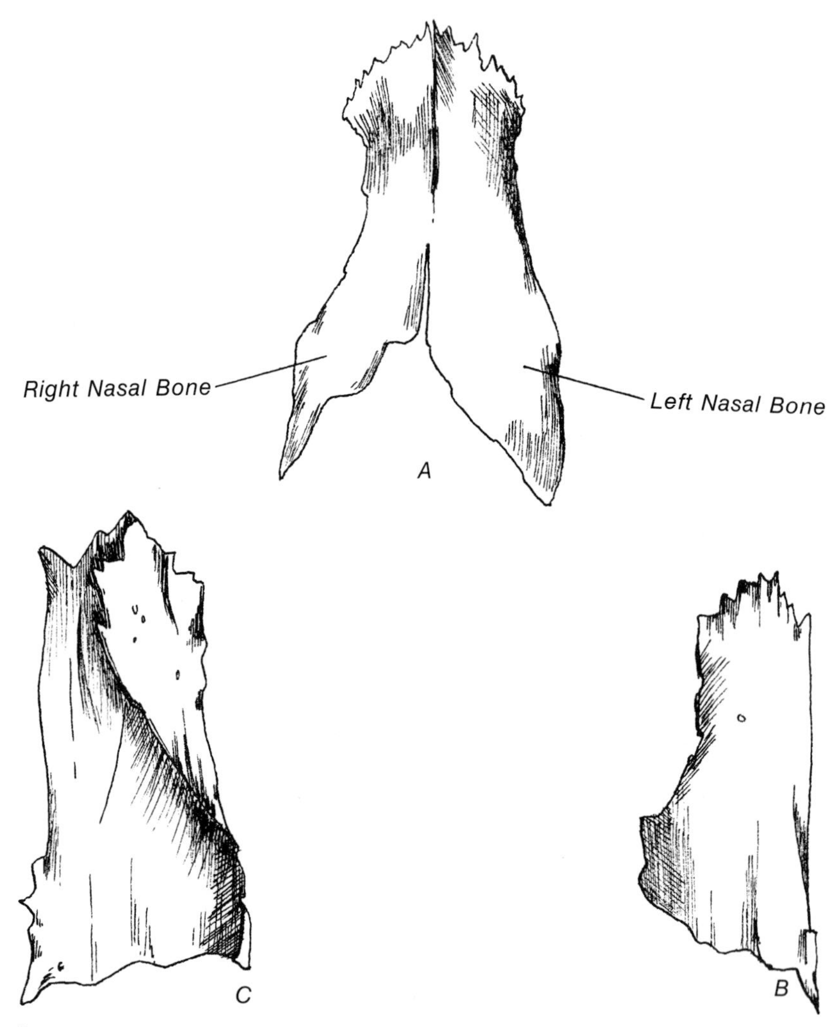

FIG. 1-18. Nasal bones. *A.* Anterior view. *B.* External surface of right nasal bone. *C.* Internal surface of right nasal bone.

The Zygoma (Fig. 1-17)

The zygoma or zygomatic bone is the cheek bone. It is composed of a diamond-shaped body with four processes. The *frontal process,* which forms part of the lateral orbital wall, attaches to the frontal bone. The *temporal process* joins the zygomatic arch with the temporal bone. The *maxillary process,* together with the zygomatic process of the maxilla, forms the infraorbital rim and part of the orbital floor. The fourth process joins the maxilla at the lateral wall of the maxillary sinus in an eminence, the *jugal ridge,* just above the molar region. The articulation of the zygoma with the great wing of the sphenoid occurs on the posterior aspect of the frontal process as it turns to help form part of the posterolateral orbital wall (Fig. 1-4).

Nasal Bones (Fig. 1-18)

The nasal bones are paired and lie in the midline just above the nasal fossae. They fit between the frontal processes of the maxillae and articulate superiorly with the frontal bone (Fig. 1-4). They are somewhat quadrilateral in shape, the outer surface is convex, and the inner surface is concave.

Lacrimal Bone (Fig. 1-4A)

The lacrimal bone is an irregular thin plate of bone. It lies in the anteromedial aspect of the orbit and articulates with the lamina orbitalis of the ethmoid bone posteriorly, the maxilla inferiorly and anteriorly, and the frontal bone superiorly.

Mandible (Fig. 1-19)

The mandible is a strong, horseshoe-shaped bone that is the heaviest and strongest bone of the facial skeleton. The horizontal portion is known as the *body* and the right and left vertical sections are known as the *rami.*

The rami each have two terminal processes: one called the *condyle* is blunt and somewhat rounded; the other, serving as the insertion of the temporalis muscle, is the *coronoid process.* The deep curved notch between the two processes is the *sigmoid notch.* The condyle has a smooth rounded *head* that is attached to the ramus by a thinner elongated *neck.* The posterior border of the ramus meets the inferior border of the body at the *mandibular angle.* The right and left bodies meet at the chin point, *the symphysis,* on which is a variably elevated area, *the mental protuberance.*

On the lateral surface of the body, one sees the *mental foramen*

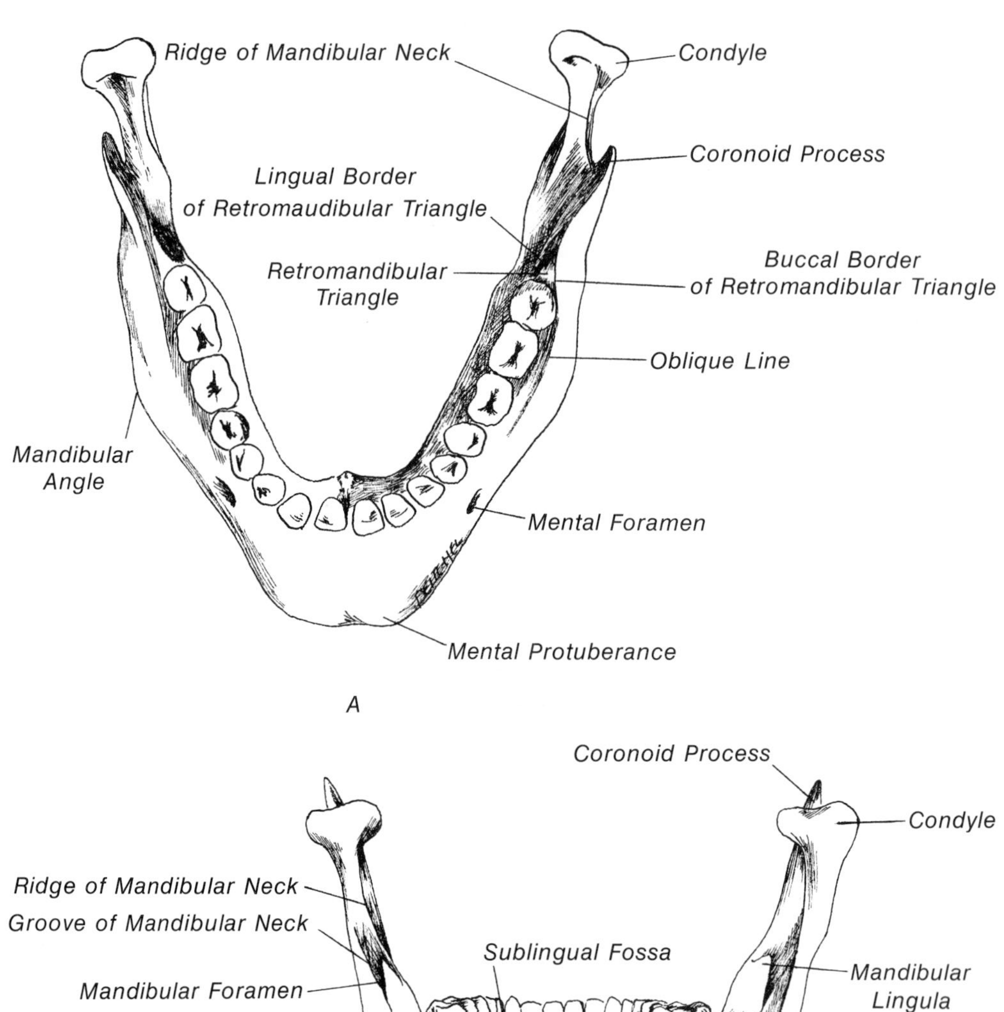

FIG. 1-19. Mandible. *A.* Superior view. *B.* Posterior view. *C.* Anterior view. *D.* Three-quarters view.

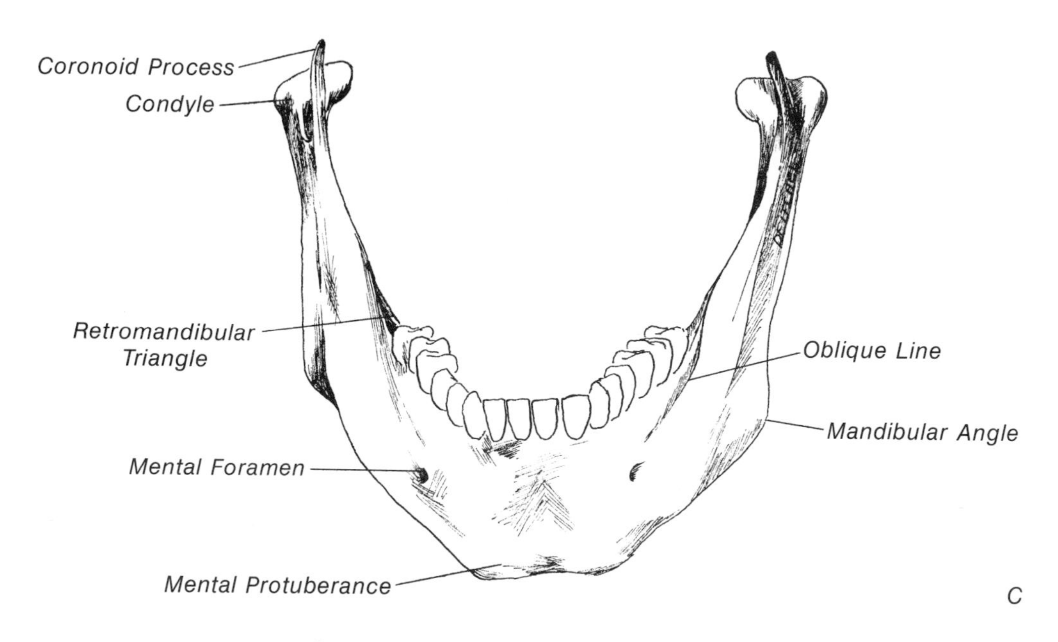
Coronoid Process
Condyle
Retromandibular Triangle
Oblique Line
Mandibular Angle
Mental Foramen
Mental Protuberance

C

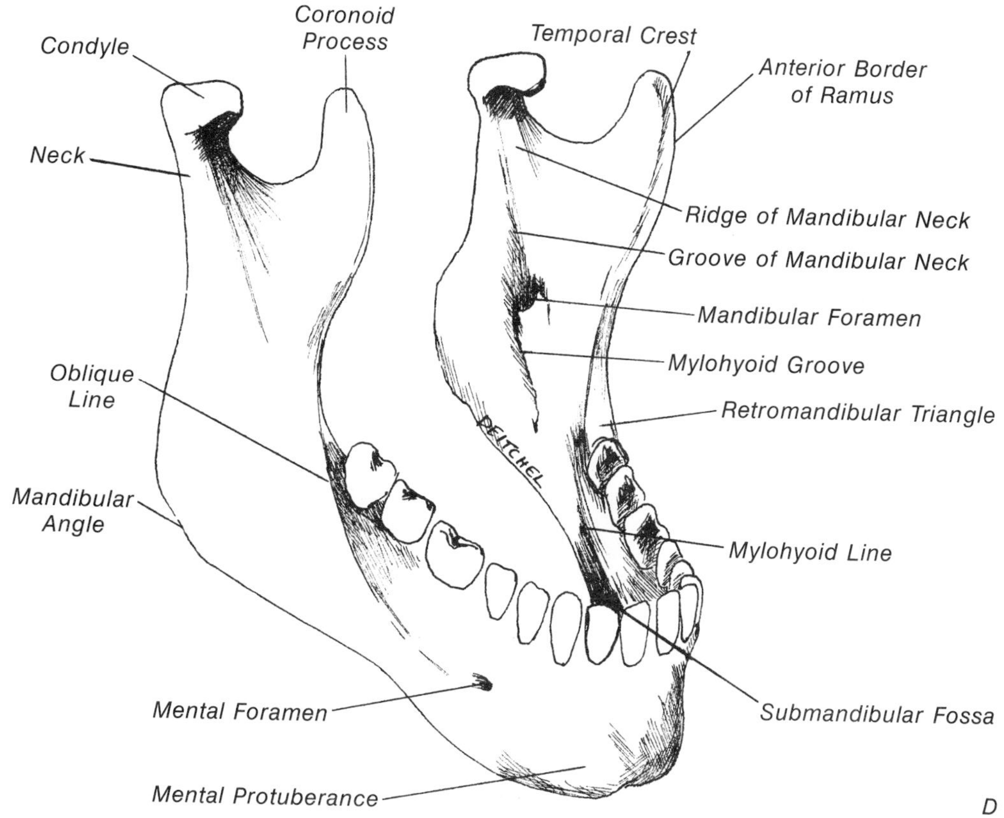
Condyle
Coronoid Process
Temporal Crest
Anterior Border of Ramus
Neck
Ridge of Mandibular Neck
Groove of Mandibular Neck
Mandibular Foramen
Mylohyoid Groove
Oblique Line
Retromandibular Triangle
Mandibular Angle
Mylohyoid Line
Submandibular Fossa
Mental Foramen
Mental Protuberance

D

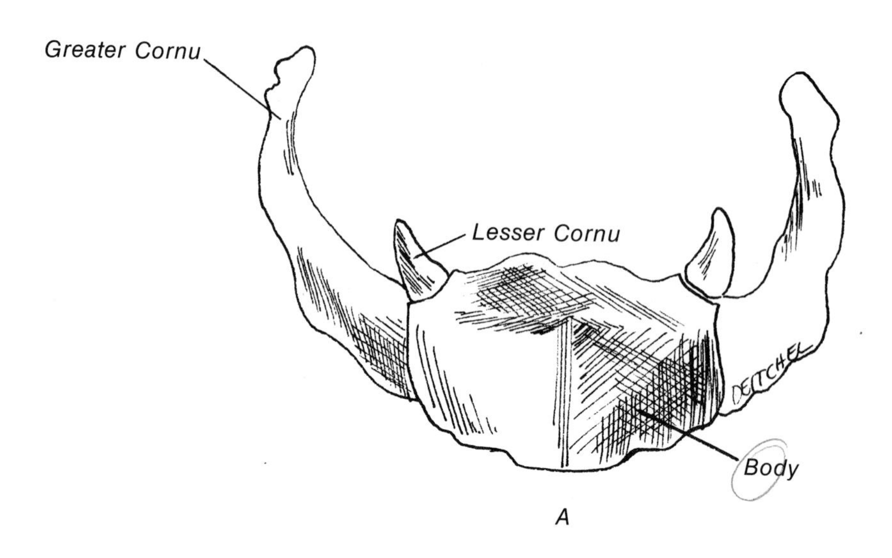

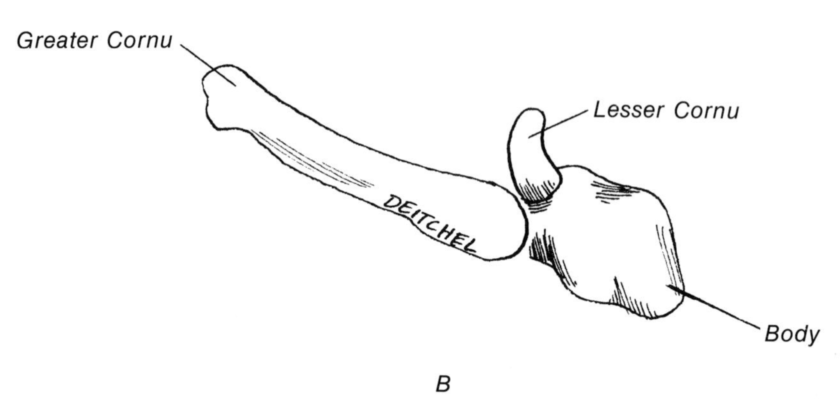

FIG. 1-20. Hyoid bone. *A.* Anterior view. *B.* Lateral view, right side.

lying just below the roots of the premolars. On the superior aspect of the body lies the *alveolar process* which houses the mandibular teeth. The *external oblique line* begins just posterior to the mental foramen and passes posteriorly and superiorly to become the anterior border of the vertical ramus and finally ends at the coronoid tip.

Medially, in the area of the symphysis, the bony *mental* (*genial*) *tubercles* or *genial spines* can be seen. Just posterior and lateral to these, is an oblique ridge of bone extending posteriorly and superiorly to end at the crest of the alveolar ridge near the third molar. This is the *mylohyoid line* or *ridge*. The depression on the medial aspect of the posterior portion of the body, just below the mylohyoid line is the *submandibular fossa.* Above the line and more anteriorly, another depression, the *sublingual fossa,* is seen.

Almost exactly in the center of the medial surface of the vertical mandibular ramus, the *inferior alveolar canal* begins with a wide opening, the *mandibular foramen.* On its anterior surface is a bony process, the *lingula.* The inferior alveolar canal lies within the bone and follows the curvature of the mandible from the inferior alveolar foramen in the ramus to the mental foramen in the body. Here it gives off a short mental canal to the mental foramen and then continues on in the bone to the symphysis region as the *incisive canal.*

Behind the last molar, on the crest of the alveolar process is a small roughened *retromandibular triangle.*

Hyoid Bone (Fig. 1-20)

Contributing to the skeleton of the head and neck and acting as a most important functional structure is the *hyoid bone.* The hyoid bone consists of five parts: an anterior middle part, *the body;* and two pairs of horns, the *greater and lesser cornus.* The shape of the bone is horseshoe like. The two free ends of the horseshoe are the two greater horns which fuse with the body. The two lesser horns project superiorly in a cone-shaped structure with the base attached at the junction of the body of the greater cornu.

The hyoid bone is suspended in the neck and lies superior and anterior to the thyroid cartilage (Adam's apple). It serves as the attachment for many muscles that lie in the anterior portion of the neck and as a point of fixation for the accessory muscles of mastication.

Important Landmarks

Three important landmarks of the skull will often be mentioned in future chapters. As discussions of the blood supply and nerve supply are encountered, the importance of the temporal fossa, the infratemporal fossa, and the pterygopalatine fossa will be obvious.

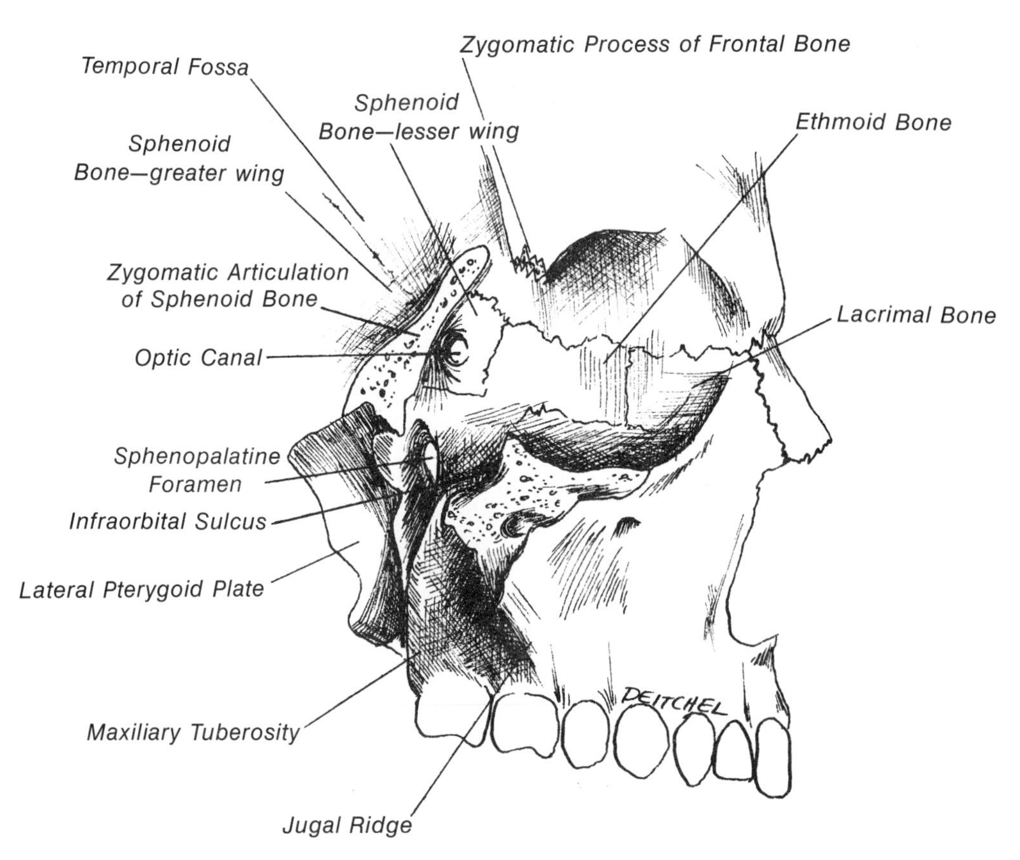

FIG. 1-21. Pterygopalatine fossa (zygoma removed).

The Temporal Fossa (Fig. 1-4A)

The temporal fossa, a flat depression, lies on the lateral surface of the skull. Anteriorly it is bounded by the temporal surface of the frontal process of the zygoma. The inferior temporal line is the superior boundary. It comprises a narrow strip of the parietal bone, the greater part of the temporal squama, the temporal surface of the frontal bone, and the temporal surface of the great wing of the sphenoid bone. This fossa serves as the field of origin for the temporalis muscle.

The Infratemporal Fossa (Fig. 1-4A)

Below and in front of the temporal fossa is the infratemporal fossa. It is limited anteriorly by the maxillary tuberosity. Laterally, the zygomatic arch and vertical mandibular ramus form an incomplete boundary. Medially, the fossa floor is formed by the lateral pterygoid plate.

The medial and lateral pterygoid muscles and portions of the maxillary artery and pterygoid plexus of veins are located in the infratemporal fossa. The mandibular division of the trigeminal nerve also ramifies here. From this fossa, openings to other important areas are noted: the infraorbital fissure leading to the orbit, the foramen ovale for the mandibular nerve, and the pterygopalatine fissure leading to the pterygopalatine fossa.

Pterygopalatine Fossa (Fig. 1-21)

The area of the pterygopalatine fossa is difficult to see, even in a dry skull, because it is covered by the zygoma and a portion of the pterygoid plates. It is a cone-shaped area, deep to the infratemporal fossa. The anterior boundary is the deep portion of the maxillary tuberosity. The pterygoid process of the sphenoid bone limits this space posteriorly, and the greater wing limits the space superiorly.

This cone-shaped area has its base directed superiorly, and the apex leads into the pterygopalatine canal which opens onto the hard palate as the greater and lesser palatine foramina.

Within the pterygopalatine fossa lie the terminal branches of the maxillary artery, the divisions of the maxillary nerve, and the pterygopalatine ganglion.

2

Muscles of the Head and Neck

Muscles are organs of motion. By their contractions, they move the various parts of the body. The energy of their contraction is made mechanically effective by means of tendons, aponeuroses, and fascia, which secure the ends of the muscles and control the direction of their pull. Muscles are usually suspended between two moving parts, such as between two bones, bone and skin, two different areas of skin, or two organs.

Actively, muscles contract. Their relaxation is passive and comes about through lack of stimulation. A muscle is usually supplied by one or more nerves that carry the stimulating impulse and thereby cause it to contract. Muscles can also be stimulated directly or by any electrical activity emanating from any source. Usually, however, the stimulation for muscular contraction originates in the nerve supply to that muscle. Lack of stimulation to the muscle results in relaxation.

When a muscle is suspended between two parts, one of which is fixed and the other movable, the attachment of the muscle on the fixed part is known as the *origin.* The attachment of the muscle to the movable part is referred to as the *insertion.*

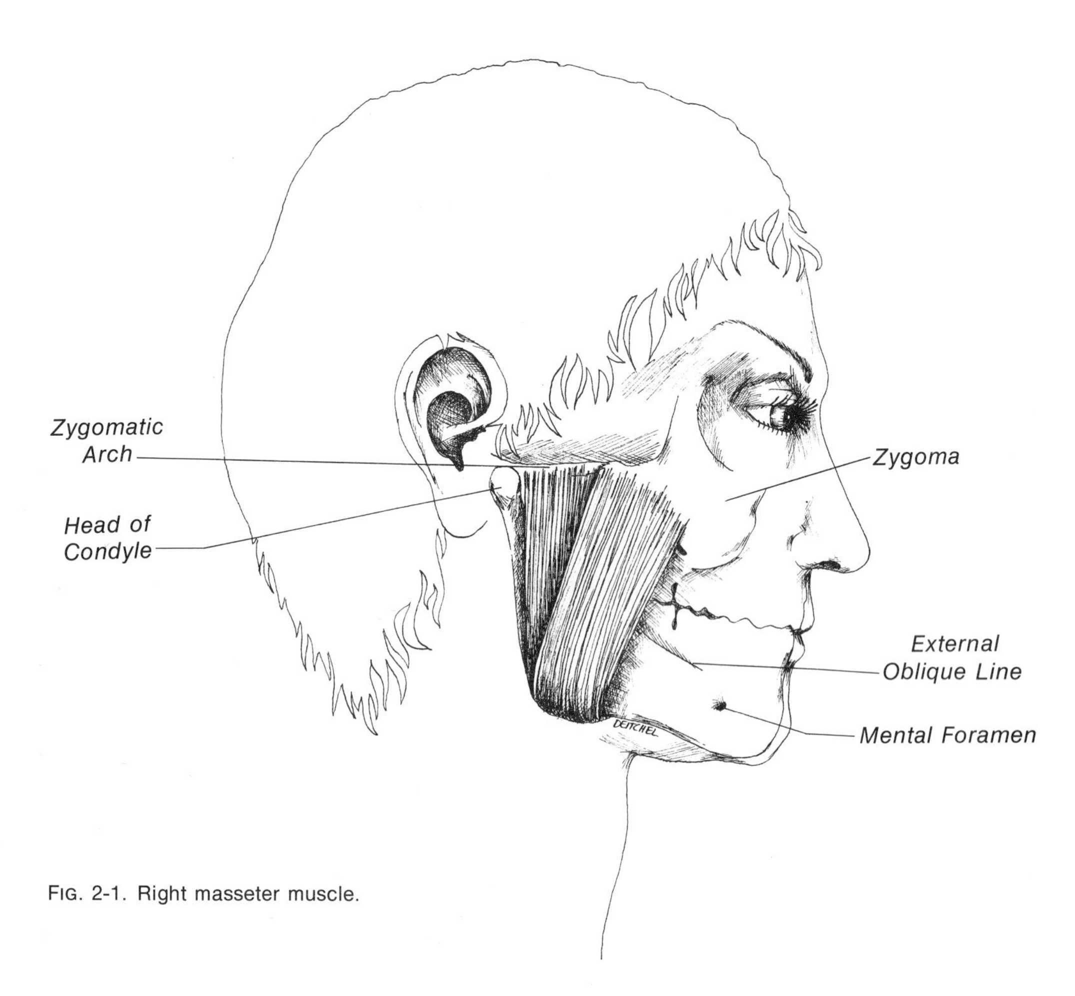

FIG. 2-1. Right masseter muscle.

The Muscles of Mastication

The muscles of mastication originate on the upper two thirds of the skull and insert on the mandible. They are all supplied by the muscular branches of the third division of the trigeminal nerve. Their blood supply is almost entirely by way of the muscular branches of the maxillary artery.

These muscles, like all muscles, work in groups with each other and with other muscles to perform a smooth, balanced, coordinated series of movements of the mandible. We must study these muscles individually, but their cofunction with one another must always be kept in mind.

Masseter Muscle (Fig. 2-1)

The most superficial of the muscles of mastication is the masseter muscle. It is a broad, thick, rectangular plate, which originates on the zygomatic arch and zygoma and passes inferiorly and posteriorly to insert on the lateral surface of the mandibular ramus at the angular region. The more superficial fibers of this muscle arise from the lower border of the zygoma. The more deeply situated fibers arise from the entire length of the zygomatic arch. The superficial fibers originate no further posteriorly than the zygomaticotemporal suture.

The action of this thick, powerful muscle is to elevate the lower jaw. The superficial fibers exert power at a right angle to the occlusal plane, whereas the deep fibers, along with their elevating action, exert a secondary retracting component, particularly when the mandible is in a protruded position. This balanced combination of elevation and retraction (retrusion) is important during the closing movement of the mandible.

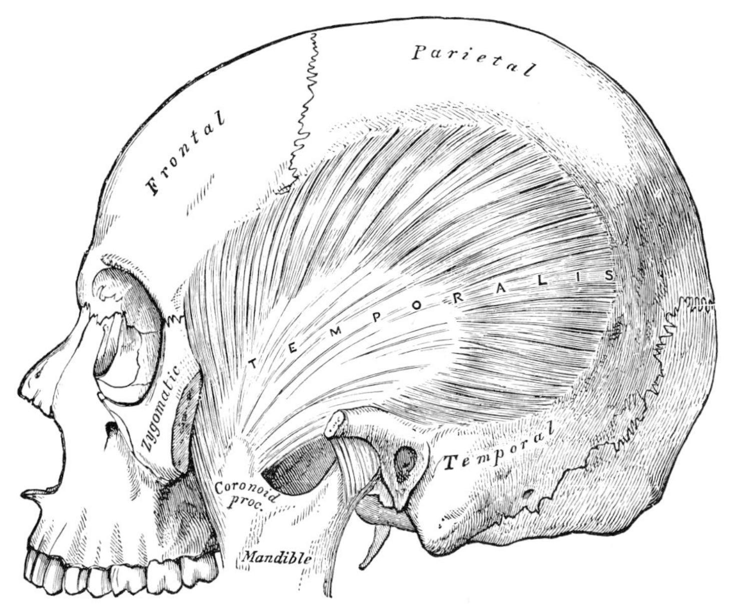

FIG. 2-2. The Temporalis. The zygomatic arch and the Masseter have been removed. (From Gray's Anatomy of the Human Body, 29th ed. C. M. Goss, editor. Philadelphia, Lea & Febiger, 1973.)

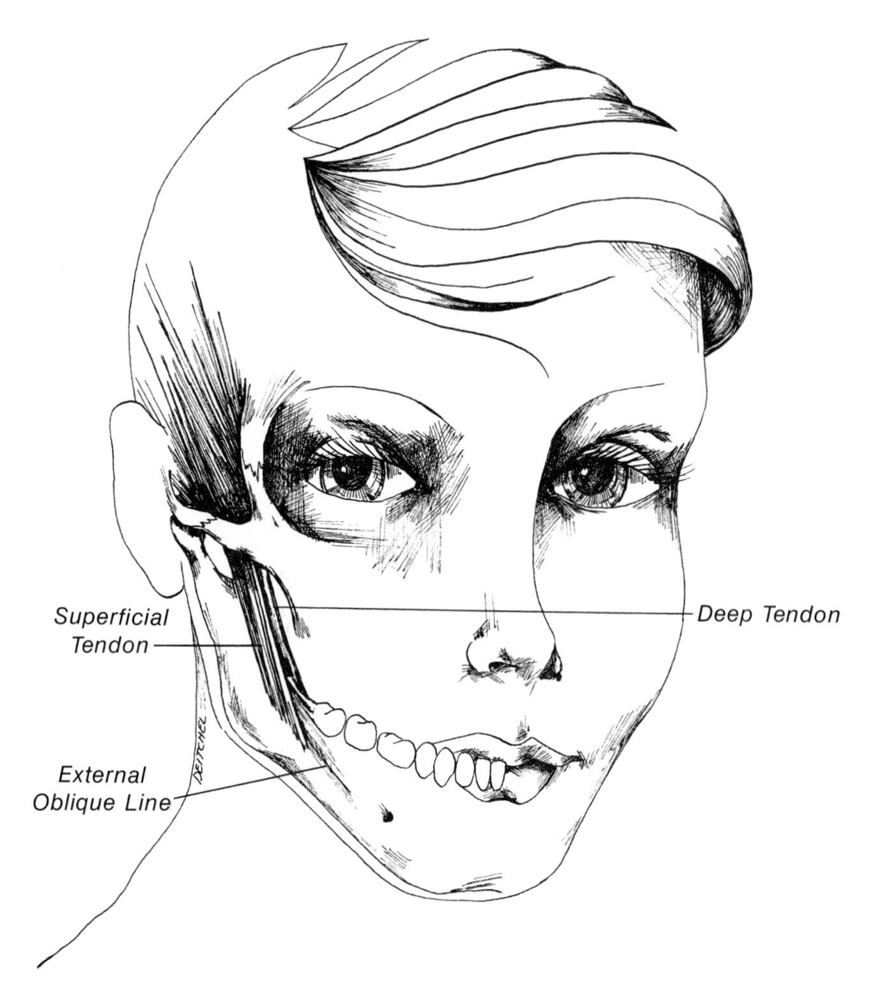

FIG. 2-3. Superficial and deep tendon insertions of the right temporalis muscle.

The Temporal Muscle (Temporalis) (Fig. 2-2)

The Temporalis is a broad, fan-shaped muscle that arises in the temporal fossa, a wide field on the lateral surface of the skull.

The fibers of the Temporalis converge toward the space between the zygomatic arch and the skull. Here, the apex of the coronoid process is located. The anterior fibers are quite vertical, and the posterior fibers become increasingly oblique. The insertion occupies the coronoid process and extends inferiorly along the anterior border of the vertical ramus and external oblique ridge.

Two tendons of insertion are to be considered here (Fig. 2-3). The outer or superficial tendon is attached to the anterior border of the coronoid and mandibular ramus. The deep tendon juts medially and inserts into the region of the lower third molar. The retromolar triangle is free from insertion of the temporal muscle.

The Temporalis is innervated by the temporal branches of the muscular rami of the third division of the trigeminal nerve. Blood is supplied by the middle and deep temporal arteries. The deep temporal artery is a branch of the maxillary artery, and the middle temporal artery is a branch of the superficial temporal artery.

The Temporalis is flatter and broader than the Masseter. The fibers are somewhat longer and therefore the temporal muscle is less powerful than the Masseter, but it provides for rapid movement of the mandible. The posterior fibers have a retracting component to their action, but the Temporalis is still considered primarily as an elevator of the mandible.

Medial (Internal) Pterygoid Muscle (Fig. 2-4)

The medial pterygoid muscle is the deep counterpart of the masseter muscle, but it is weaker. It originates primarily in the pterygoid fossa and the medial surface of the lateral pterygoid plate. Some fibers arise from the palatine bone and maxillary tuberosity.

The medial pterygoid muscle passes inferiorly, posteriorly, and laterally to insert on the deep surface of the mandibular angle. It acts as an elevator of the mandible. The blood supply is by branches of the maxillary artery, and the nerve supply is by the medial pterygoid branch of the third division of the trigeminal nerve.

Lateral Pterygoid Muscle (Fig. 2-5)

The lateral (external) pterygoid muscle has two heads from which it arises. The inferior head, which is the largest, originates from the lateral surface of the lateral pterygoid plate. The infratemporal surface of the great wing of the sphenoid bone serves as the origin of the superior head. The muscle then passes posteriorly to insert on the anterior surface of the condylar neck.

Some fibers insert on the anterior surface of the temporomandibular joint capsule. The nerve and blood supply are from the

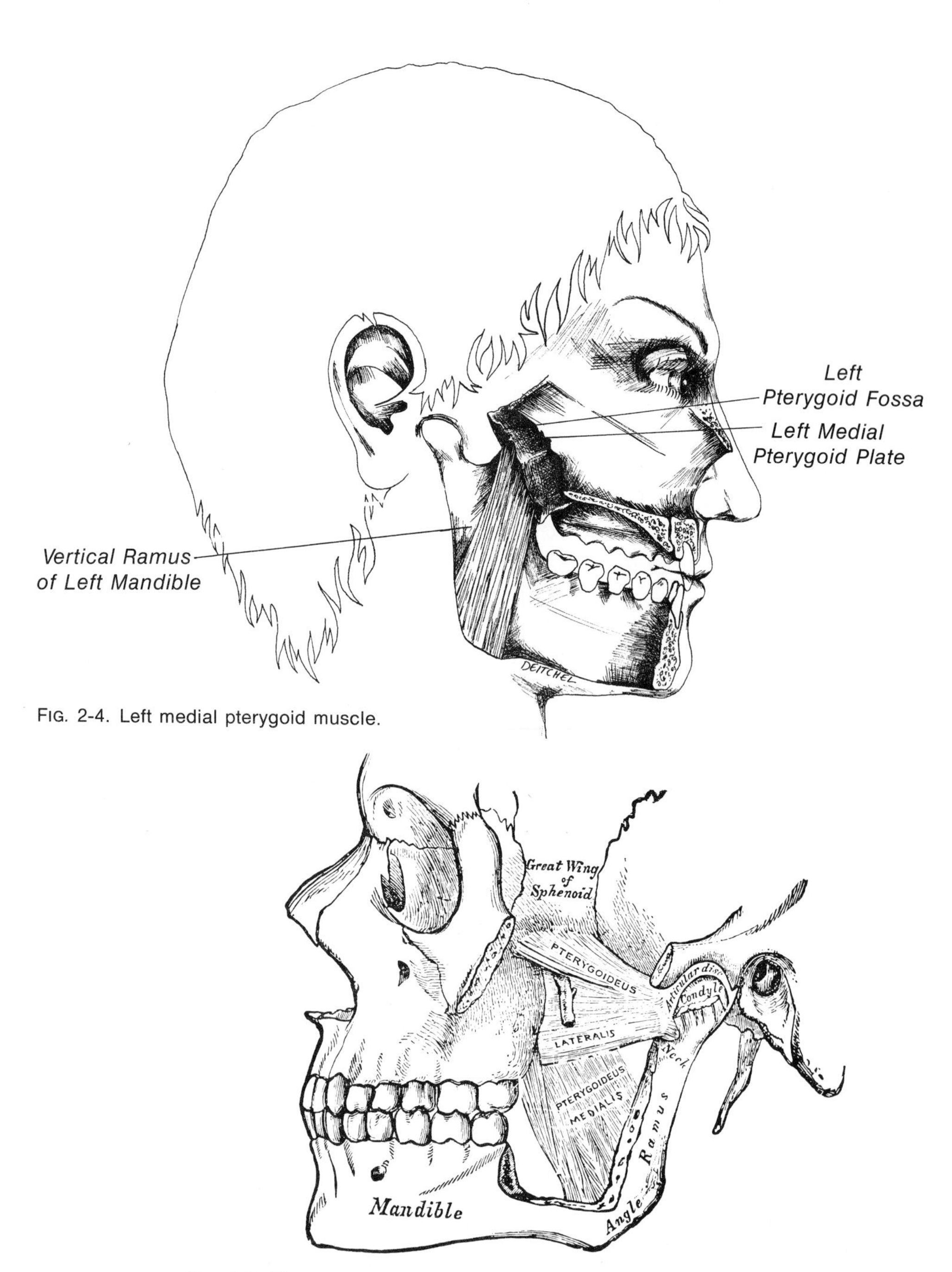

FIG. 2-4. Left medial pterygoid muscle.

FIG. 2-5. The ptyergoid muscles. The zygomatic arch and a portion of the ramus of the mandible have been removed, showing position of the maxillary artery. (From Gray's Anatomy of the Human Body, 29th ed. C. M. Goss, editor. Philadelphia, Lea & Febiger, 1973.)

same source as the other muscles of mastication. (Details are described in Chapters 3 and 4.)

The lateral pterygoid muscle acts to open the jaw. It causes the mandible to shift to the opposite side when it contracts without the aid of its contralateral mate.

Accessory Muscles of Mastication

Primarily, the accessory muscles of mastication involve the muscles of the anterior neck. They can be divided into those muscles that are situated above the hyoid bone, the *suprahyoid muscles,* and those placed below the hyoid bone, the *infrahyoid muscles.*

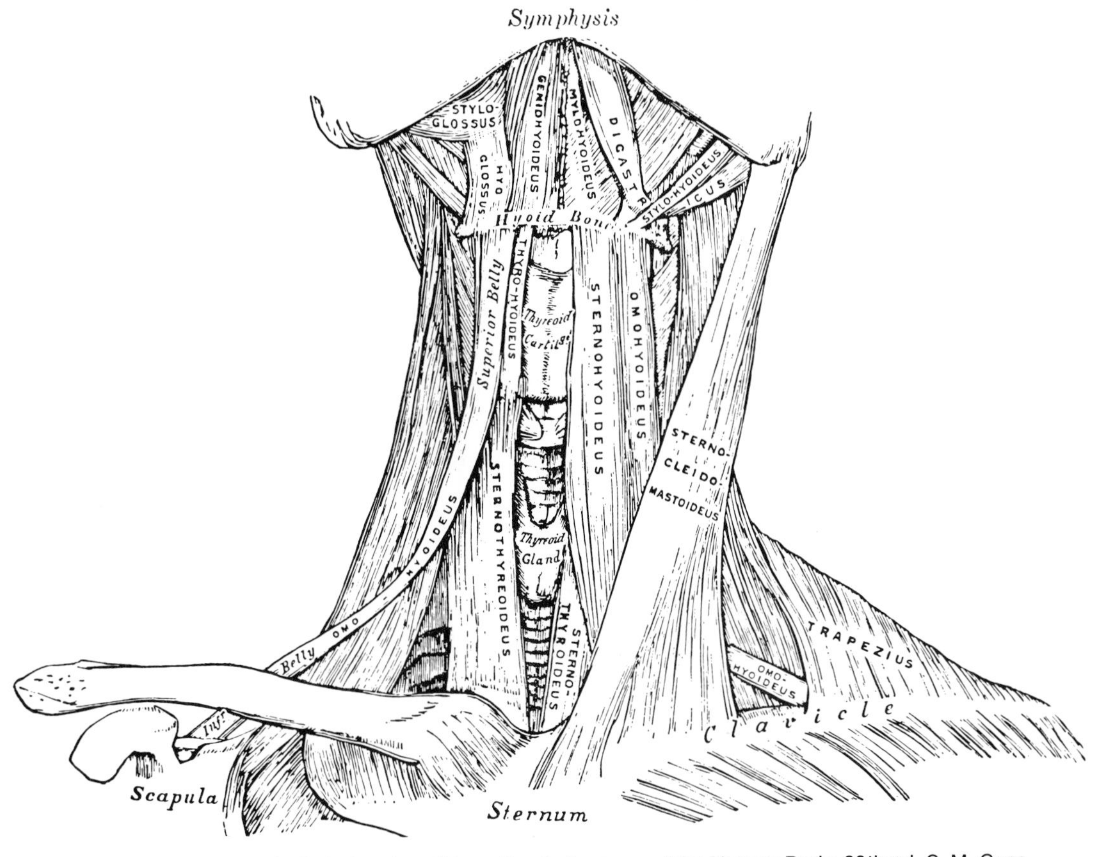

FIG. 2-6. Muscles of the neck. Anterior view. (From Gray's Anatomy of the Human Body, 29th ed. C. M. Goss, editor. Philadelphia, Lea & Febiger, 1973.)

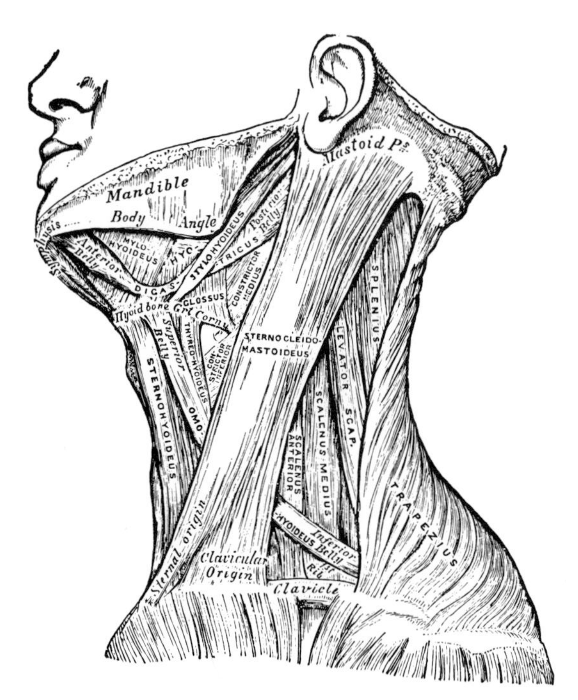

FIG. 2-7. Muscles of the neck. Lateral view. (From Gray's Anatomy of the Human Body, 29th ed. C. M. Goss, editor. Philadelphia, Lea & Febiger, 1973.)

The Infrahyoid Muscles (Fig. 2-6)

The infrahyoid muscles act either to depress the hyoid bone and larynx or fix the hyoid bone so that it cannot be elevated by contraction of the suprahyoids. If the latter occurs, the suprahyoids can act to depress the mandible. The nerve supply to the infrahyoids is via the first, second, and third cervical nerves.

Sternohyoid Muscle (Figs. 2-6; 2-7)

The sternohyoid muscle originates on the sternum (breast bone), near the lateral edge, where the sternum joins the clavicle. It passes superiorly and inserts on the body of the hyoid bone.

Omohyoid Muscle (Figs. 2-6; 2-7)

The omohyoid muscle originates on the scapula (wing bone) and passes anteriorly and superiorly to insert on the body of the hyoid bone, just lateral to the insertion of the sternohyoid. It contains a superior and an inferior belly, both joined by an intermediate tendon.

Sternothyroid Muscle (Fig. 2-6)

The sternothyroid muscle arises medially to the sternohyoid, on the posterior surface of the most superior aspect of the sternum. It passes superiorly to insert on the thyroid cartilage and lies deep to the sternohyoid muscle.

Thyrohyoid Muscle (Fig. 2-6)

The thyrohyoid muscle arises from the thyroid cartilage and inserts on the hyoid bone at the body and a part of the greater cornu.

The Suprahyoid Muscles (Fig. 2-7)

The suprahyoids' function is to either elevate the hyoid bone or depress the mandible. If the mandible is fixed by contraction of the muscles of mastication, the contraction of the suprahyoids will elevate the hyoid bone and the larynx. However, if the hyoid bone is fixed by contraction of the infrahyoids, contraction of the suprahyoids will *open* and/or *depress* the mandible.

Digastric Muscle (Digastricus) (Figs. 2-6; 2-7; 2-8)

The Digastricus is made up of two fusiform, fleshy parts, known as bellies, attached to one another, end to end. The posterior belly originates at the mastoid notch and passes anteriorly and inferiorly to the hyoid bone where its anterior portion ends in a tendon. This tendon slides through a connective tissue pulley, which is attached to the greater horn of the hyoid bone. The anterior belly arises on this *intermediate tendon* and passes superiorly and anteriorly to insert on the medial surface of the symphysis of the mandible, near the inferior border, in a depression known as the *digastric fossa.*

The nerve supply to the posterior belly is via a branch of the seventh cranial nerve. The anterior belly is innervated by the mylohyoid branch of the third division of the fifth cranial nerve.

Geniohyoid Muscle (Figs. 2-6; 2-7)

The geniohyoid muscle arises from the anterior end of the mylohyoid line, near the genial tubercles, and passes posteriorly and

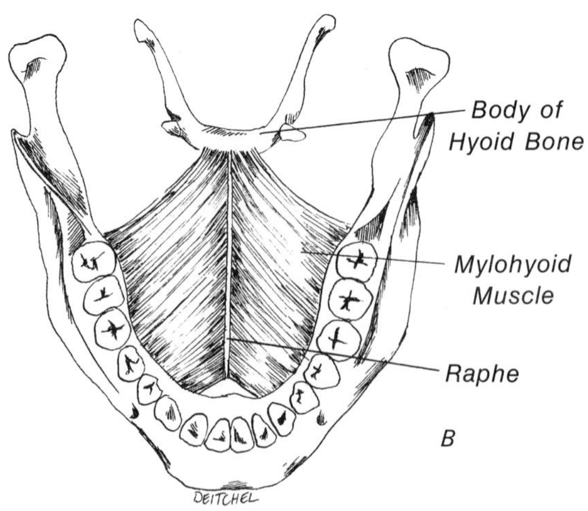

FIG. 2-8. Mylohyoid muscle. A. External surface. B. Internal surface.

inferiorly to insert on the upper half of the body of the hyoid bone. It is supplied by the first and second cervical nerves.

Mylohyoid Muscle (Mylohyoideus) (Figs. 2-7; 2-8)

The mylohyoid muscle forms the floor of the mouth. The right and left muscles pass inferiorly and medially and unite with one another in the midline on a tendinous band, known as the *raphe.*

The origin is the mylohyoid line on the mandible. The posterior fibers insert on the body of the hyoid bone, but most of the fibers of this broad, flat muscle insert on the mylohyoid raphe. The posterior border is free. The mylohyoid muscle is inferior to the bandlike geniohyoid muscle and superior to the digastric muscle. It is innervated by mylohyoid branches of the third division of the trigeminal nerve.

The primary function of this muscle is to elevate the tongue. When the fibers contract, the curvature is flattened; thus the floor of the mouth is elevated and with it, the tongue. Only the posterior fibers of the mylohyoid muscle, running from the mandible to the hyoid bone, influence the position of the hyoid bone or the mandible.

Stylohyoid Muscle (Figs. 2-6; 2-7)

This muscle arises from the styloid process of the temporal bone. It is a thin, round muscle, which passes anteriorly and inferiorly to insert on the hyoid bone at the junction of the body and greater cornu. It is innervated by the facial nerve and acts to elevate and retract the hyoid bone or to fix the hyoid bone in cooperation with the infrahyoid muscles.

Muscles of the Tongue

The tongue is a powerful muscular organ with a fantastic ability to alter its shape, configuration, and position. The tongue is covered by mucous membrane. Other than its covering, the tongue is composed entirely of muscles, nerves, and blood vessels. Blood supply is mainly via the lingual artery. Nerve supply to the muscles of the tongue is by way of the hypoglossal nerve.

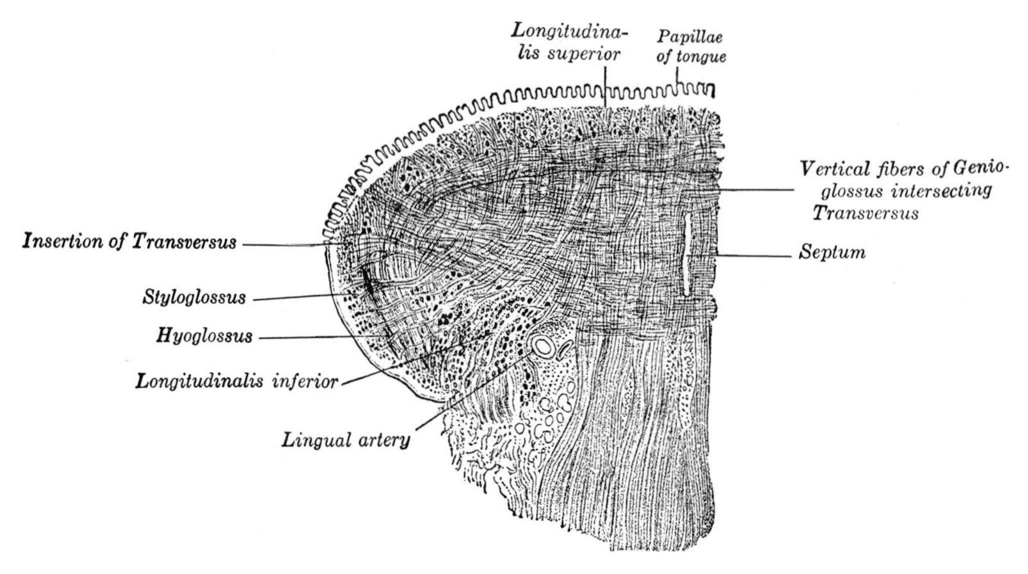

FIG. 2-9. Coronal section of the tongue, showing intrinsic muscles. (Modified from Krause in Gray's Anatomy of the Human Body, 29th ed. C. M. Goss, editor. Philadelphia, Lea & Febiger, 1973.)

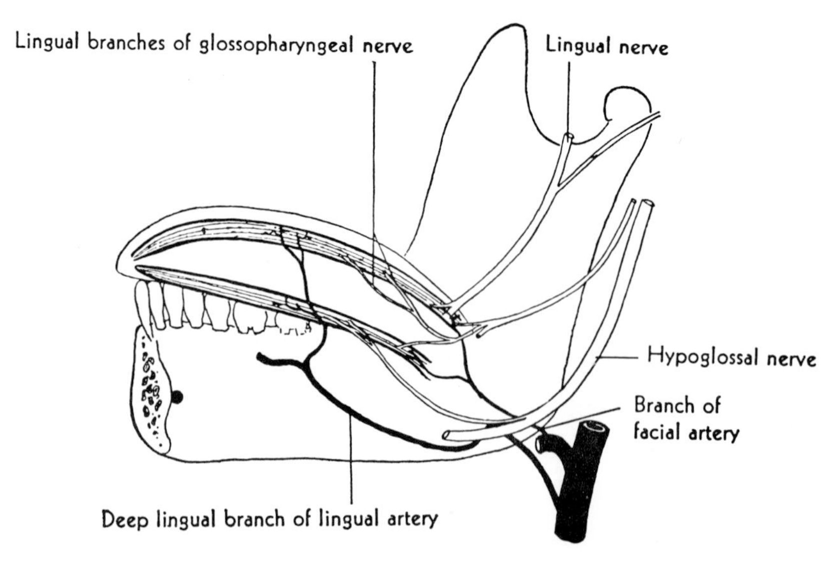

FIG. 2-10. Superior and inferior longitudinal muscles. (From Warfel, J. H.: The Head, Neck, and Trunk, 4th ed. Philadelphia, Lea & Febiger, 1973.)

The muscles of the tongue can be divided into two groups, the *intrinsic* and the *extrinsic* muscles. Those muscles known as the intrinsic lie within the tongue itself, whereas the extrinsic muscles originate at a point distant from the tongue and insert on and interlace with the intrinsic muscles.

The tongue is divided into two halves by a median fibrous septum, which extends throughout its length and is fixed below and behind to the hyoid bone. The intrinsic and extrinsic muscles are thereby also divided into pairs.

Intrinsic Muscles of the Tongue (Fig. 2-9)

The intrinsic muscles of the tongue can be subdivided into longitudinal, transverse, and vertical. The varying shapes that the tongue can assume are quite complex but can be easily predicted once one considers what happens when the various intrinsic fibers contract.

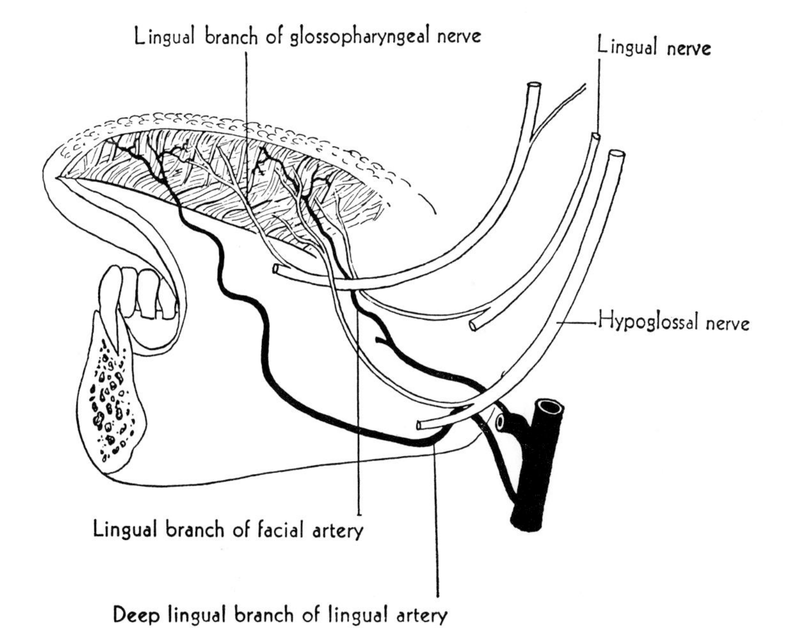

FIG. 2-11. Transverse and vertical fibers. (From Warfel, J. H.: The Head, Neck, and Trunk, 4th ed. Philadelphia, Lea & Febiger, 1973.)

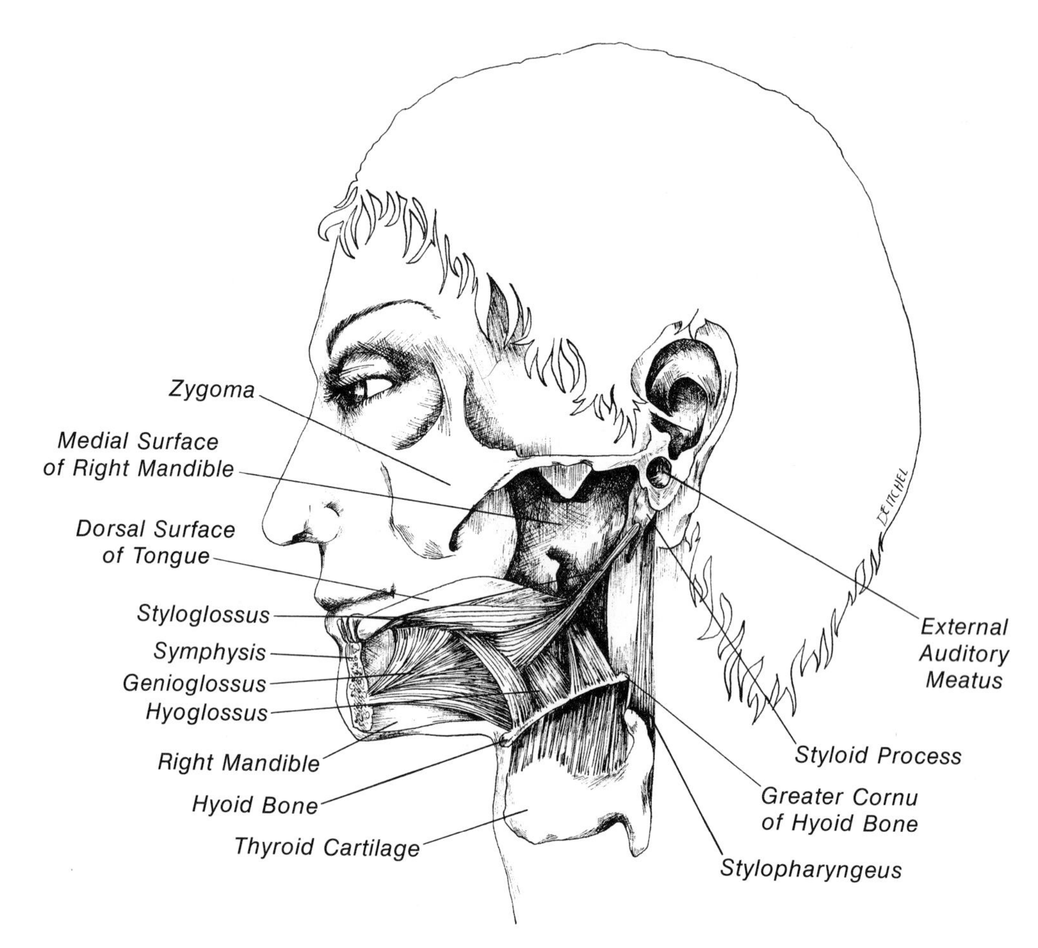

FIG. 2-12. Extrinsic muscles of the tongue.

Superior Longitudinal Muscles (Fig. 2-10)

The superior longitudinal fibers lie directly under the dorsal mucosa. They arise from the posterior submucous fibrous tissue and run forward and obliquely to insert on the edges of the tongue.

Inferior Longitudinal Muscles (Fig. 2-10)

The narrow band of fibers known as the inferior longitudinal muscles is situated on the under surface of the tongue and extends from the root to the apex.

Transverse Fibers (Fig. 2-11)

The transverse fibers originate on the median septum and pass laterally to the lateral borders of the tongue.

Vertical Fibers (Fig. 2-11)

The vertical fibers are found in the forepart of the tongue and extend from the upper to the under surface.

Extrinsic Muscles of the Tongue (Fig. 2-12)

Genioglossus (Fig. 2-12)

The genioglossus muscle is a flat, triangular muscle, with the origin at the superior genial tubercles of the mandible. It fans out and inserts into the tongue and hyoid bone. Only the inferior fibers are inserted on the body of the hyoid bone; the rest insert on the entire undersurface of the tongue from the root to the apex. The posterior fibers protrude the tongue, and the anterior fibers retract the tongue.

Hyoglossus (Fig. 2-12)

The Hyoglossus arises from the hyoid bone, mainly from the greater cornu and body. It passes vertically behind the mylohyoid muscle to insert in the lateral portion of the tongue and mingle with other fibers. It acts to depress the tongue and draw the sides down.

Styloglossus (Fig. 2-12)

The origin of the styloglossus muscle is the styloid process. It passes down and forward and enters the tongue in two heads: one, longitudinal, blends with the inferior longitudinal muscle, and the other overlaps and intermingles with the hyoglossus. The styloglossus draws the tongue upward and backward.

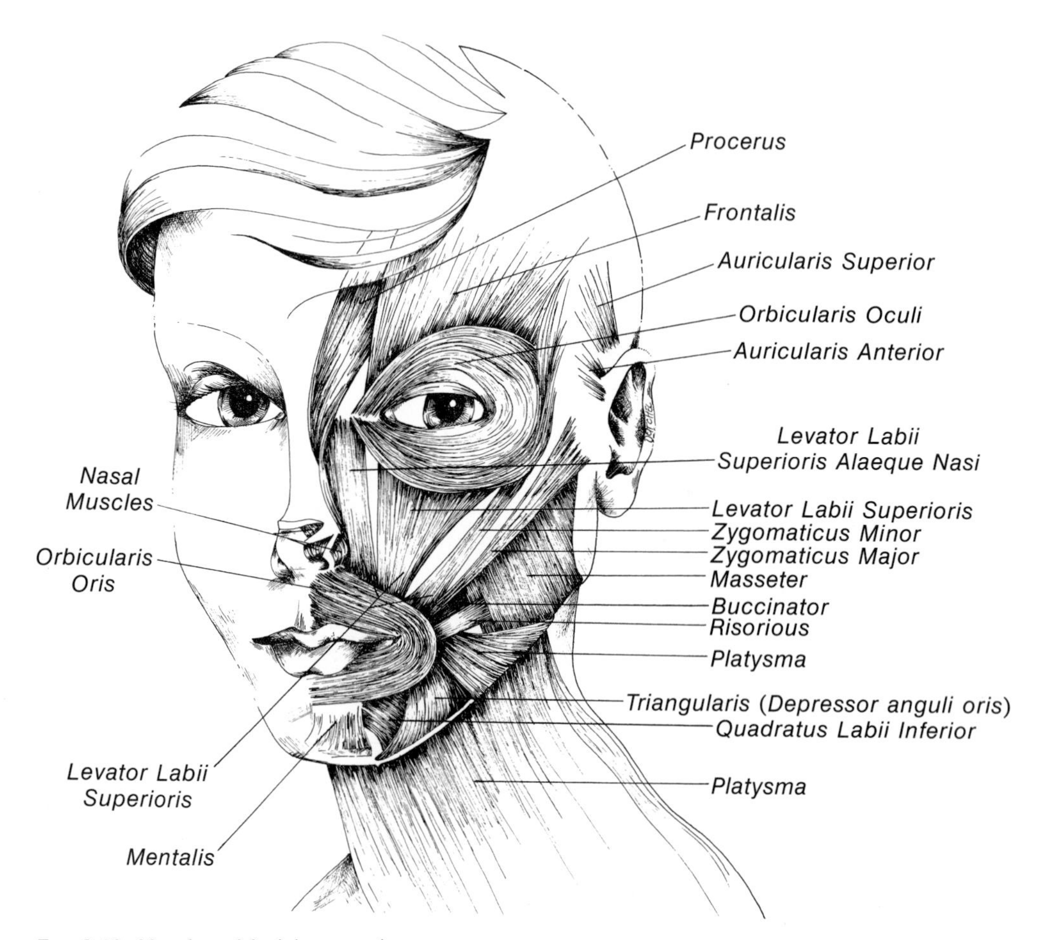

FIG. 2-13. Muscles of facial expression.

The Muscles of Facial Expression (Fig. 2-13)

The muscles of facial expression are very superficial and all of them attach to skin, at least at their insertion. Some attach to skin at both the origin and the insertion. These muscles give character and diversity to the human face. They also perform many other important functions, such as closing of the eyes, moving the cheeks and lips during mastication and speech, and revealing some psychologic activities. All of the muscles of facial expression receive nerve innervation via the seventh cranial nerve. Another interesting characteristic of these muscles is their wide variation in size, shape, and strength.

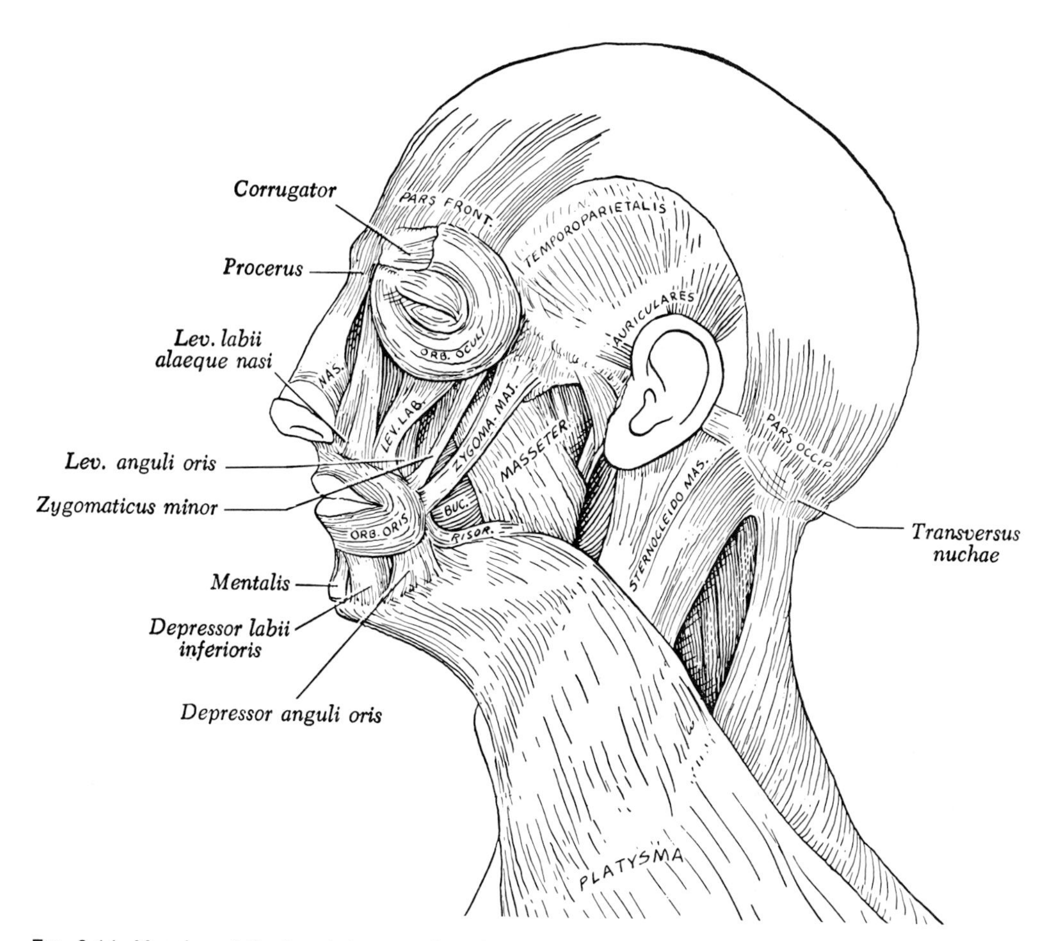

FIG. 2-14. Muscles of the head, face, and neck. (From Gray's Anatomy of the Human Body, 29th ed. C. M. Goss, editor. Philadelphia, Lea & Febiger, 1973.)

Like the muscles of mastication, the muscles of facial expression work synergistically and not independently. The group functions as a well-organized and coordinated team, each member having specified functions, one of which is primary. In addition, these muscles interweave with one another, and it is most difficult to separate the boundaries between the various muscles. The terminal ends of these muscles are interlaced with each other.

The creases of the face and skin lines are indicative of contractions of these muscles. These lines become deeper, sharper, and more permanent with advancing age due to decrease in the elasticity of the skin.

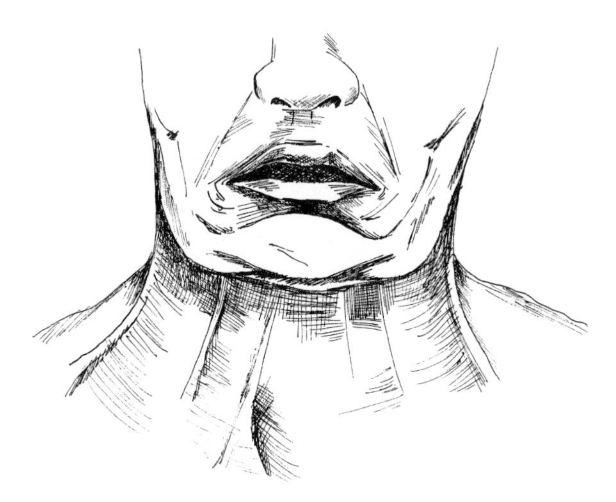

FIG. 2-15. Expression of horror.

Platysma Muscle (Figs. 2-13; 2-14)

The thin, wide, flat muscle plate that covers most of the anterior and lateral portion of the neck is called the Platysma. It lies just deep to the skin. At its upper border it attaches to the lower border of the mandible and contiguous skin. It also interlaces with the *Depressor anguli Oris.* The Platysma arises from the fibrous connective tissue under the skin of the clavicle (collarbone) and shoulder. The fibers ascend anteriorly in an oblique course to their insertions. The nerve supply is by way of the cervical division of the facial nerve.

The action of the Platysma is to raise the skin of the neck, as if to relieve the pressure of a tight collar. Also, it draws the outer part of the lower lip down and back as in an expression of horror (Fig. 2-15).

Muscles of the Scalp (Fig. 2-14)

The four muscles of the scalp are the paired *occipital* and *frontal* muscles. They are related to one another by a common tendon known as the *galea aponeurotica.* The two frontalis muscles, the two occipital muscles, and the galea aponeurotica, as a unit, are known as the *epicranial muscle.*

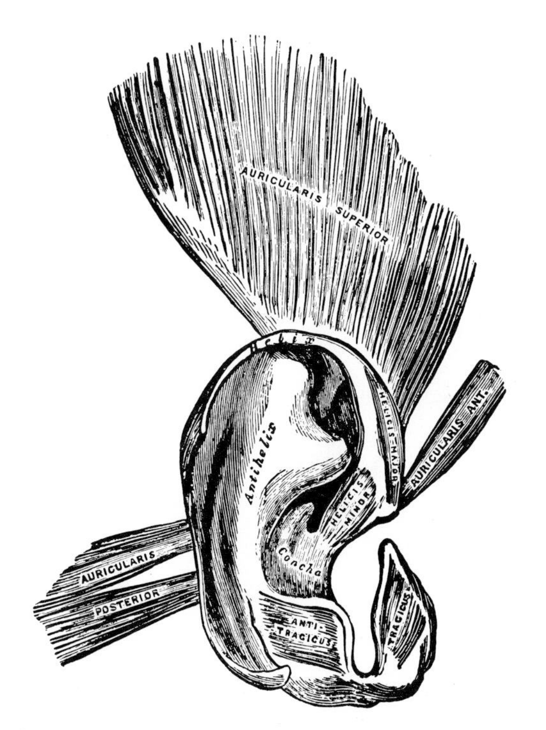

FIG. 2-16. Muscles of the outer ear (auricula). (From Gray's Anatomy of the Human Body, 29th ed. C. M. Goss, editor. Philadelphia, Lea & Febiger, 1973.)

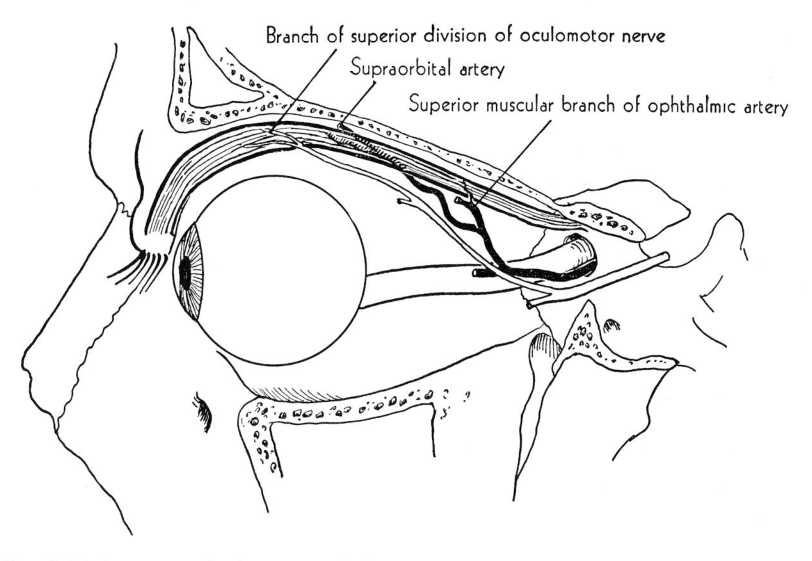

FIG. 2-17. Levator palpebrae superioris. (From Warfel, J. H.: The Head, Neck, and Trunk, 4th ed. Philadelphia, Lea & Febiger, 1973.)

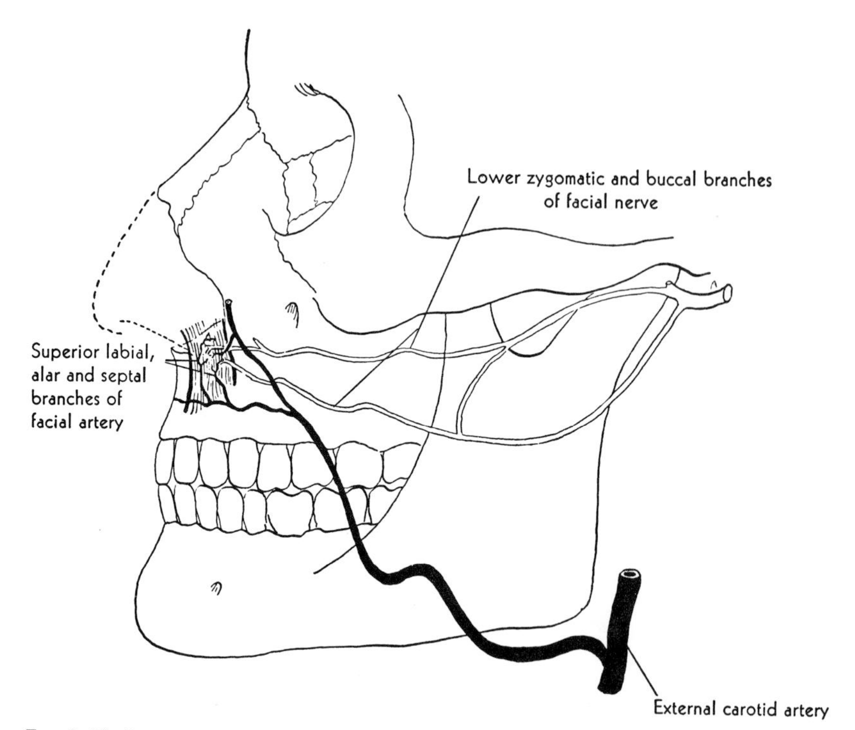

FIG. 2-18. Depressor septi. (From Warfel, J. H.: The Head, Neck, and Trunk, 4th ed. Philadelphia, Lea & Febiger, 1973.)

The Frontalis is attached to the root of the nose and skin of the eyebrow. The Occipitalis originates at the supreme nuchal line from the mastoid process to the midline. The Epicranius covers the calvaria and acts to lift the eyebrow and furrow the skin of the forehead.

The Frontalis is innervated by the temporal branches of the facial nerve; the Occipitalis by the posterior auricular branch.

Muscles of the Outer Ear (Fig. 2-16)

In man, the muscles of the outer ear are vestigial. They serve little or no function. They are the (1) *Musculus auricularis anterior,* (2) *Musculus auricularis superior,* and (3) *Musculus auricularis posterior.*

Only a few people can activate the muscles of the outer ear

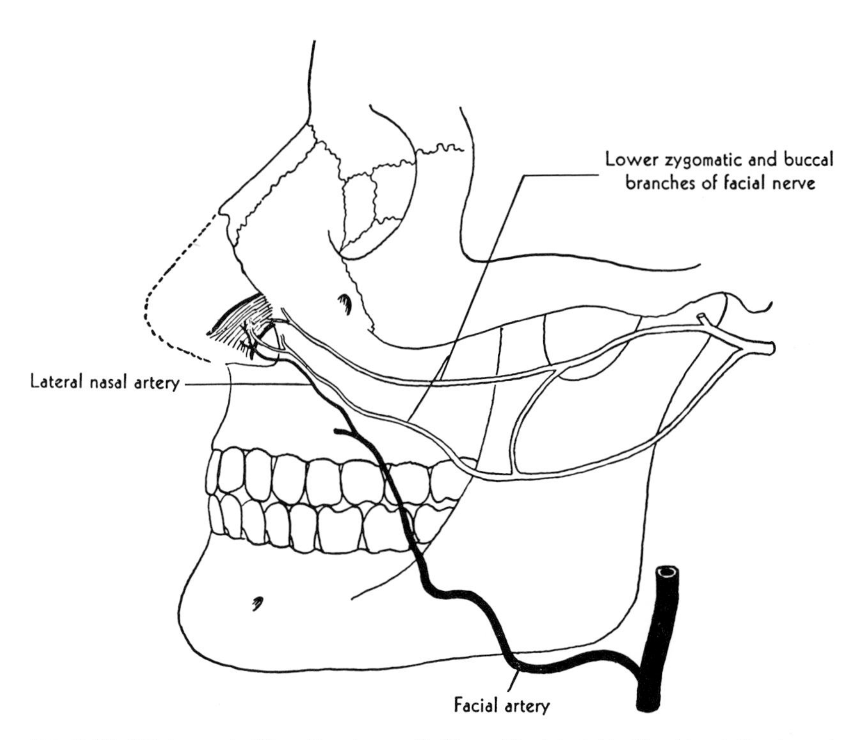

FIG. 2-19. Dilator naris (Nasalis, alar part). (From Warfel, J. H.: The Head, Neck, and Trunk, 4th ed. Philadelphia, Lea & Febiger, 1973.)

voluntarily. Occasionally these muscles contract involuntarily when other muscles of facial expression are activated. The nerve supply is by rudimentary branches of the temporal and posterior auricular divisions of the facial nerve.

Muscles of the Eyelids

Orbicularis Oculi (Fig. 2-13)

The orbicularis oculi muscle encircles the eye. It originates at the inner canthus of the eye from the frontal process of the maxilla and the lacrimal bone. The more superior fibers end in the skin in the lateral region of the eye. The muscle encircles the eye in concentric fibers that act as a sphincter to close the eye.

Levator Palpebrae Superioris (Fig. 2-17)

The Levator palpebrae superioris is considered an extrinsic muscle of the eyeball but for the sake of continuity will be dealt with here. It arises within the orbit above the optic foramen and advances and spreads out to end in the upper lid. When it contracts, it lifts the upper lid. The nerve supply is via the oculomotor nerve.

Muscles of the Nose

The muscles of the nose also are quite rudimentary in man. The principal ones are the Procerus and the Nasalis.

Procerus (Fig. 2-13)

The procerus muscle originates from the nasal bone and passes superiorly to end in the skin of the brow and forehead. It depresses the medial wide part of the eyebrow.

Nasalis (Fig. 2-14)

The Nasalis arises from the alveolar eminence over the lateral incisor and swings around the nose to insert on the superior surface

TABLE 2-1. *Muscles Opening the Lips*

I. Upper lip
  A. Superficial muscles (Quadratus labii superioris*)
    1. Zygomaticus minor
    2. Levator labii superioris
    3. Levator labii superioris alaeque nasi
  B. Deep muscles
    1. Levator anguli oris

II. Lower lip
  A. Superficial muscles
    1. Depressor anguli oris
  B. Deep muscles
    1. Depressor quadratus labii inferioris
    2. Mentalis

*The three superficial muscles of the upper lip are often referred to as one muscle, the Quadratus labii superioris, with three distinct heads.

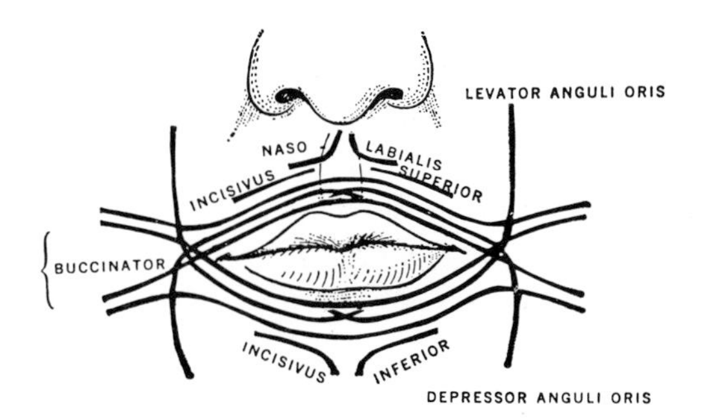

FIG. 2-20. Scheme showing arrangement of fibers of Orbicularis oris. (From Gray's Anatomy of the Human Body, 29th ed. C. M. Goss, editor. Philadelphia, Lea & Febiger, 1973.)

of the bridge, at the tip of the nose and alar cartilages. Also, the *Depressor septi* (Fig. 2-18) and the *Dilator naris* (Fig. 2-19) are members of this group of muscles.

Muscles of the Mouth (Fig. 2-13)

The muscles of the mouth are by far the most important of the muscles of facial expression in man. Although these muscles are numerous, they can be easily understood if divided into groups. One group opens the lips, and one group closes the lips. The muscles closing the lips are the *Orbicularis oris* and the *Incisive* muscles.

The muscles opening the lips are known as the radial muscles. They can be divided into radial muscles of the upper and lower lips, superficial and deep (Table 2-1).

Orbicularis Oris (Figs. 2-13; 2-20)

The Orbicularis oris has no skeletal attachment. Its fibers consist of an upper and lower group, which cross each other at acute angles at the corner of the mouth. The muscle is not anatomically a unit like the Orbicularis oculi. The majority of upper and lower fibers are confined to one side only but interlace at the midline with the fibers of the other side.

The action of the Orbicularis oris is to close the lips. It can also narrow the lips and force them against the teeth, purse the lips, or protrude the lips.

FIG. 2-21. Elevation of the upper lip and nasal wing.

Incisive Muscles

The upper incisive muscle arises from the canine eminence and passes with the Orbicularis to insert in the angle of the mouth. The lower incisive muscle originates on the mandibular canine eminence and passes laterally to insert in the corner of the mouth. They both help to purse and protrude the lips, are very small and blend very closely with the orbicularis muscle.

Zygomaticus Minor (Fig. 2-13)

The origin of the Zygomaticus minor, the weakest of the three heads of the Quadratus labii superioris, is from the body of the zygoma in front of the origin of the Zygomaticus major. It inserts in the skin of the upper lip, lateral to the midline.

Levator Labii Superioris (Fig. 2-13)

The Levator Labii superioris originates from the maxillae just below the orbital rim. The fibers pass down to insert in the skin of the upper lip.

Levator Labii Superioris Alaeque Nasi (Fig. 2-13)

The Levator Labii superioris alaeque nasi elevates the upper lip and nasal wing (Fig. 2-21). It originates from the frontal process of

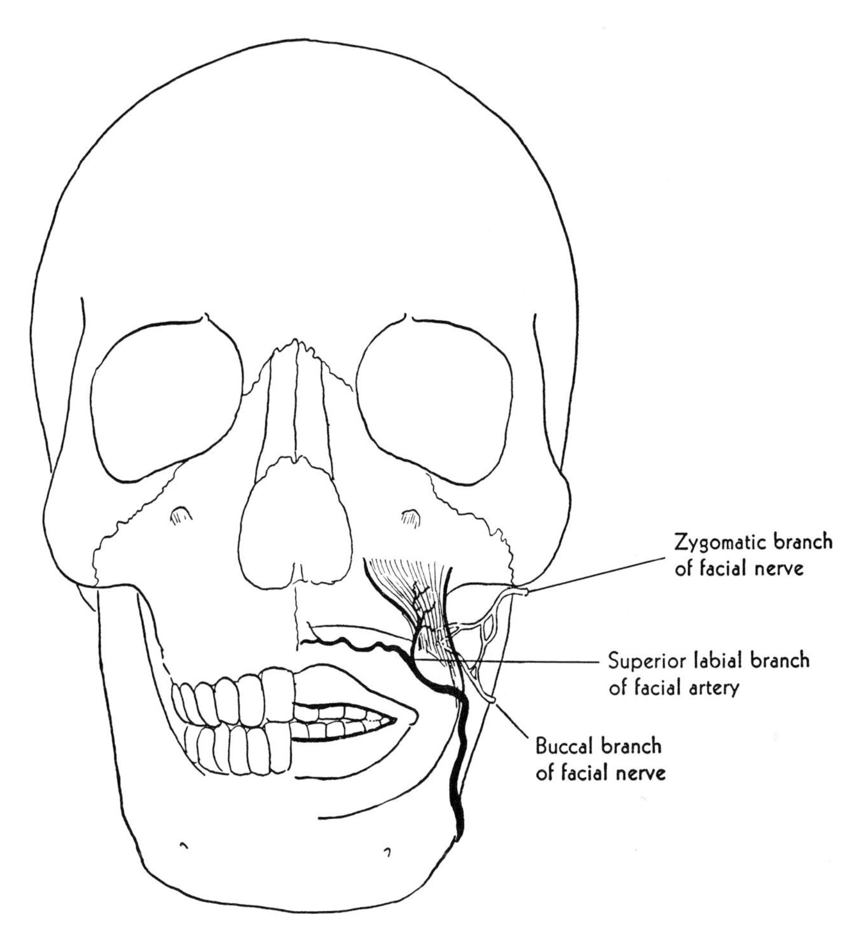

FIG. 2-22. Levator anguli oris. (From Warfel, J. H.: The Head, Neck, and Trunk, 4th ed. Philadelphia, Lea & Febiger, 1973.)

the maxilla and inserts into the skin of the wing of the nose and into the orbicularis oris, near the philtrum.

Zygomaticus Major (Figs. 2-13; 2-14)

The origin of the well-developed zygomaticus major muscle is the temporal process of the zygomatic bone. The fibers pass forward and downward, to insert in the angle of the mouth. It pulls the corner of the mouth up and laterally.

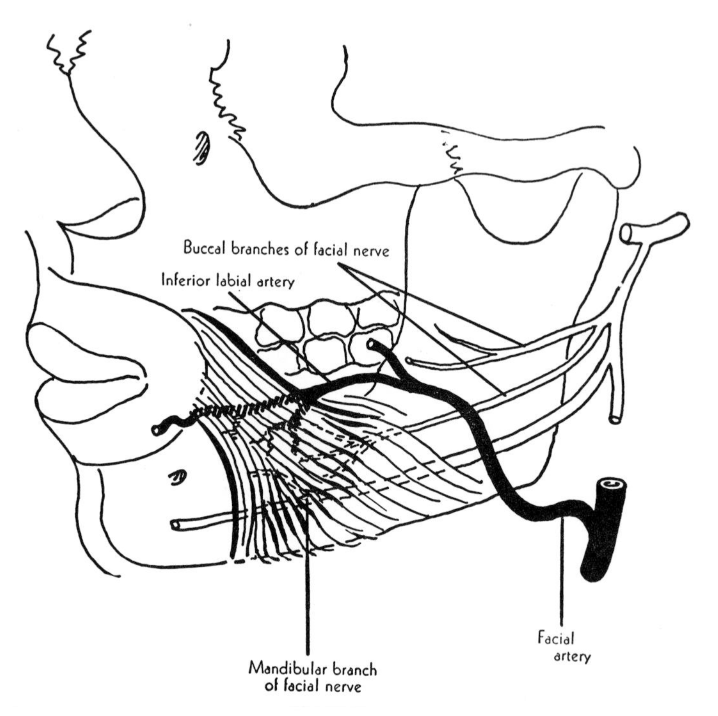

FIG. 2-23. Depressor anguli oris. (From Warfel, J. H.: The Head, Neck, and Trunk, 4th ed. Philadelphia, Lea & Febiger, 1973.)

FIG. 2-24. Corners of mouth turned downward and inward by the Depressor anguli oris.

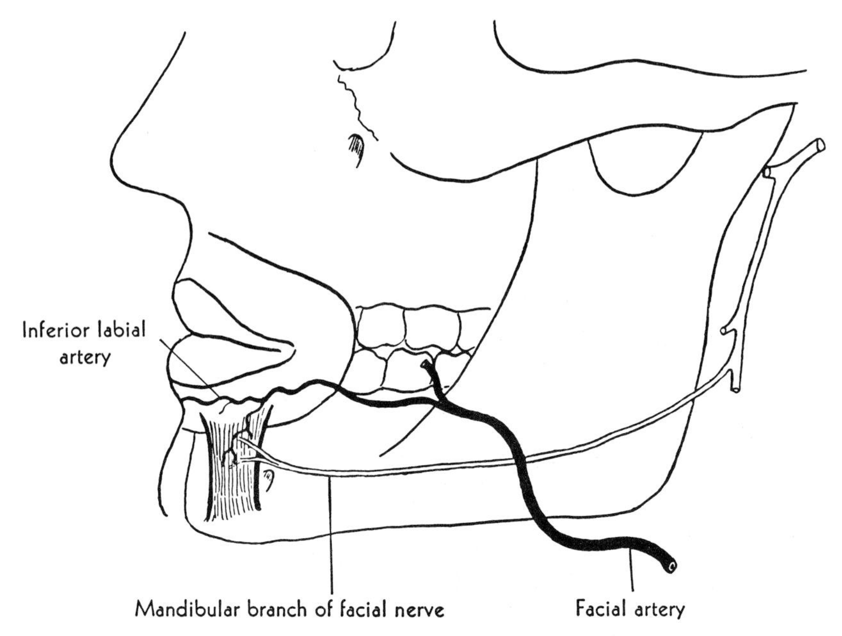

FIG. 2-25. Depressor labii inferioris. (From Warfel, J. H.: The Head, Neck, and Trunk, 4th ed. Philadelphia, Lea & Febiger, 1973.)

Levator Anguli Oris (Fig. 2-22)

The Levator anguli oris, also referred to as the Caninus, is the only muscle in the deep layer of muscles that open the lips. It takes origin from the canine fossa, passes inferiorly and laterally, and inserts in the tendinous node of intertwining muscles at the corner of the mouth. The function of the Caninus is to elevate the angle of the mouth.

Depressor Anguli Oris (Triangular) Muscle (Figs. 2-13; 2-23)

The Depressor anguli oris, or triangular muscle, arises from the area near the insertion of the platysma and inserts into the tendinous node at the angle of the mouth. Its function is to pull the corner of the mouth downward and inward (Fig. 2-24).

Depressor (Quadratus) Labii Inferioris (Figs. 2-13; 2-25)

The Depressor labii inferioris originates near the origin of the triangular muscle. It passes upward to insert in the skin of the lower lip and pulls the lower lip down and laterally.

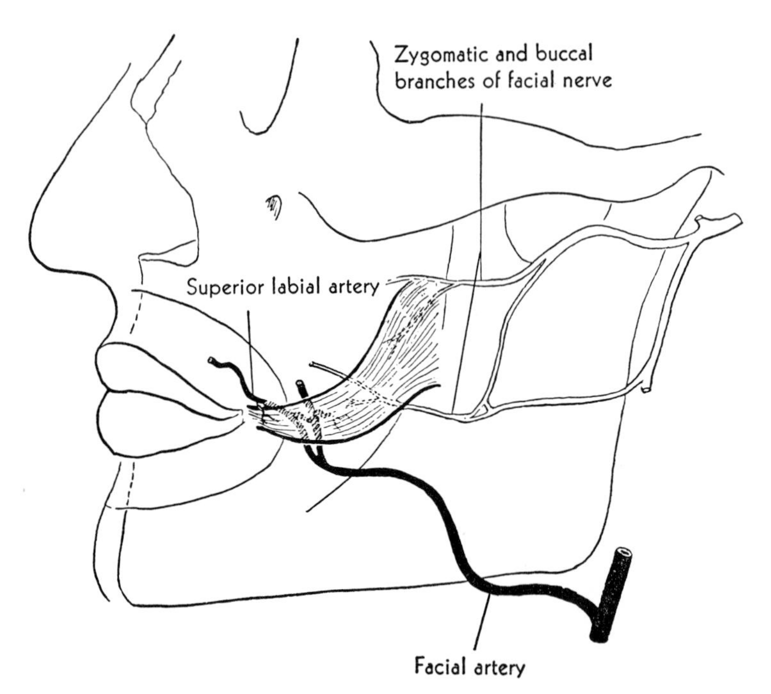

FIG. 2-26. Risorius. (From Warfel, J. H.: The Head, Neck, and Trunk, 4th ed. Philadelphia, Lea & Febiger, 1973.)

Mentalis (Fig. 2-13)

The mentalis muscle takes origin in a somewhat circular area above the mental tuberosity. It passes in a lateral direction toward the skin. Some of the medial fibers converge and cross those of the contralateral mentalis muscle. The insertion is in the skin of the chin, and its action is to elevate the skin of the chin and turn the lip outward. Since the origin extends to a point above the level of the depth of the labial vestibule, when the Mentalis contracts, the vestibule is made shallow. This action often interferes greatly with dental procedures involving the labial surface of the lower anterior teeth.

Risorius (Fig. 2-26)

The Risorius, one of the muscles at the corner of the mouth, originates from the fascia of, and just deep to the anterior border of, the masseter muscle. It passes anteriorly in a horizontal line to

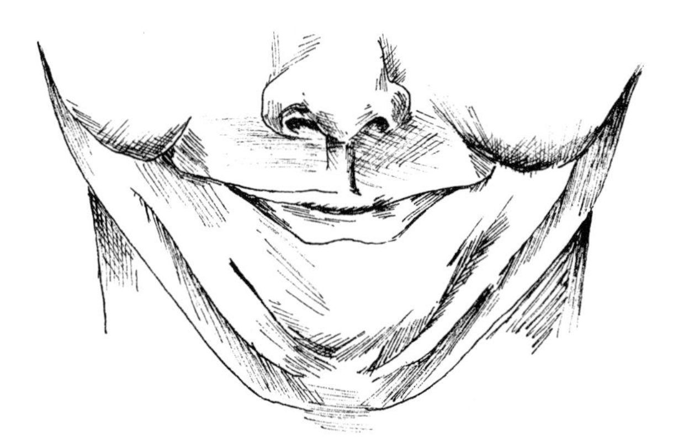

FIG. 2-27. Angle of the mouth pulled laterally as in smiling.

insert under the skin and mucous membrane of the upper lip near the corner of the mouth. The Risorius pulls the angle of the mouth laterally and is often referred to as the "smiling muscle" (Fig. 2-27).

Buccinator (Fig. 2-28)

The buccinator muscle, which also inserts at the corner of the mouth, forms the major portion of the substance of the cheeks. It is quite thin, wide, and flat. Its origin can be traced as a horse-shoe-shaped line starting on the lateral surface of the maxillary alveolar process, opposite the first molar. It passes posteriorly, to the suture between the maxilla and palatine bone. The line crosses the hamular notch to the pterygoid hamulus. It then follows the *pterygomandibular raphe* to a point on the mandible behind the lower third molar. On the mandible, the line then passes laterally over the retromolar fossa, to the external oblique line, which it then follows.

The *pterygomandibular raphe* (Fig. 2-28) is a tendinous band from which not only the Buccinator but also the Superior constrictor of the pharynx take origin. This raphe is attached superiorly to the pterygoid hamulus and inferiorly to the mandible at the posterior end of the mylohyoid line.

The muscle fibers of the Buccinator, pass anteriorly from their origins on the maxillary alveolar process to the corner of the mouth

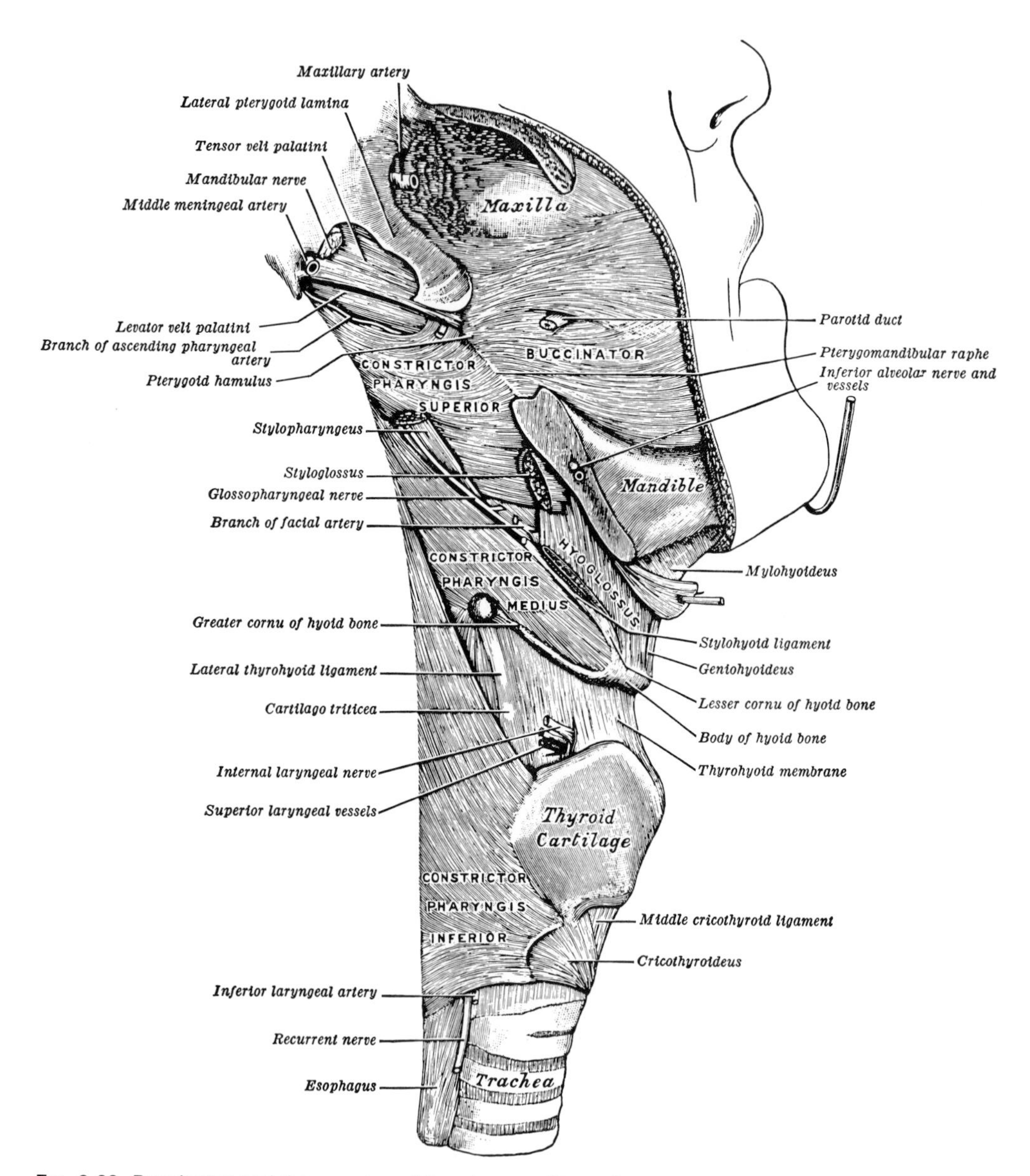

FIG. 2-28. Buccinator and the muscles of the pharynx. (From Gray's Anatomy of the Human Body, 29th ed. C. M. Goss, editor. Philadelphia, Lea & Febiger, 1973.)

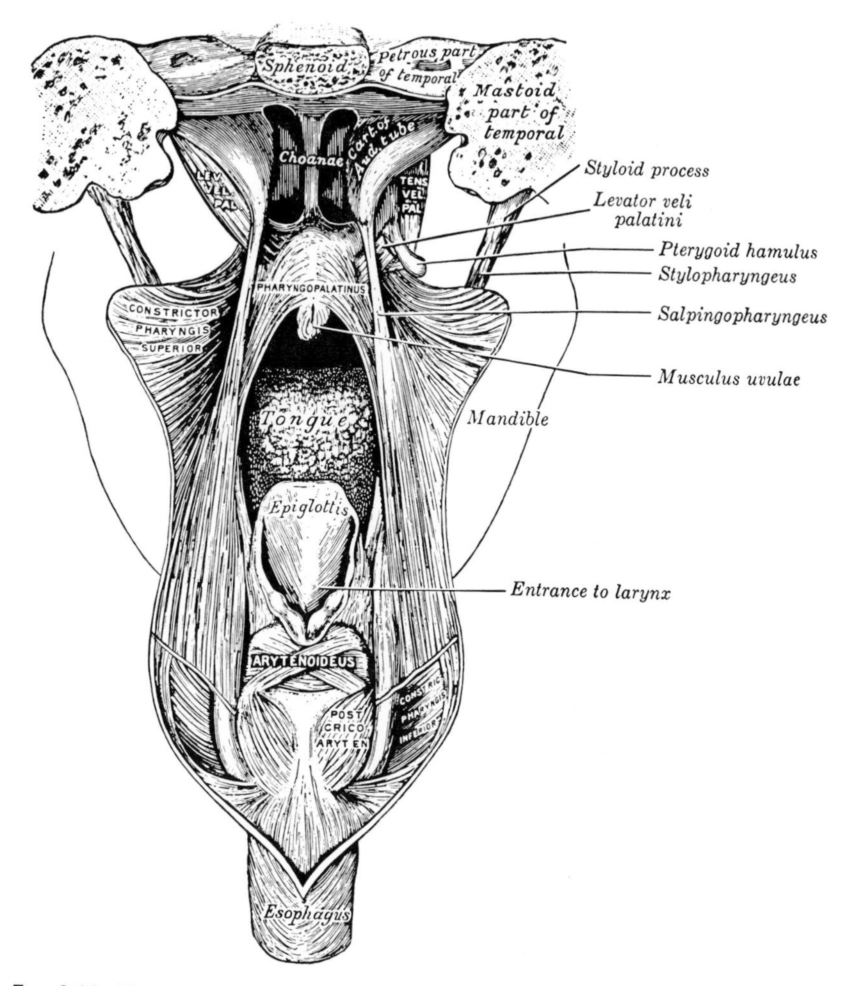

FIG. 2-29. Dissection of the muscles of the palate, posterior aspect. (From Gray's Anatomy of the Human Body, 29th ed. C. M. Goss, editor. Philadelphia, Lea & Febiger, 1973.)

and insert into the mucosa of the cheek near the tendinous node. Some fibers interlace with the neighboring muscles.

The main function of the Buccinator is to keep the cheek in tone. This prevents biting the mobile and adaptive cheek. Some authorities think that the Buccinator aids mastication by working with the tongue to keep the bolus of food on the chewing surfaces of the posterior teeth.

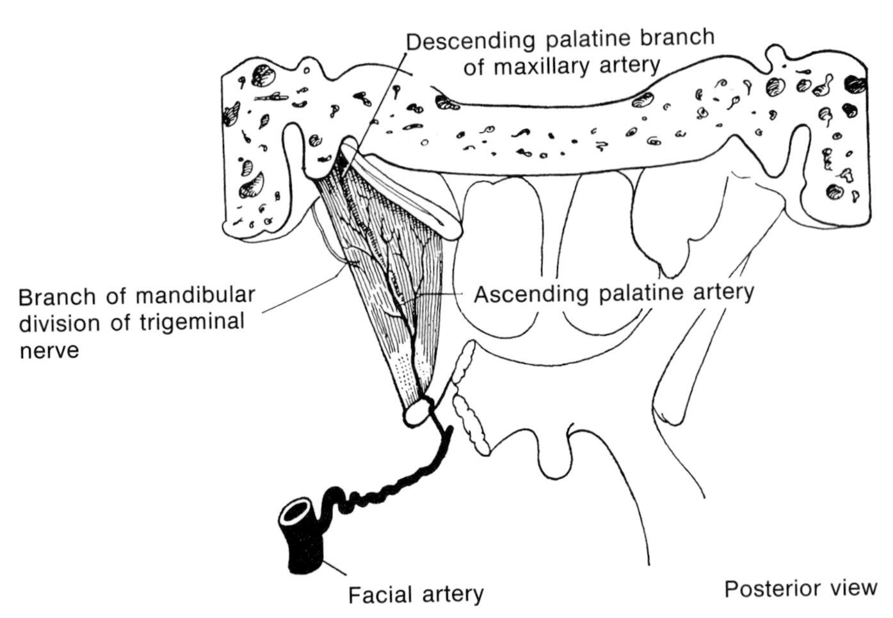

FIG. 2-30. Tensor veli palatini. (From Warfel, J. H.: The Head, Neck, and Trunk, 4th ed. Philadelphia, Lea & Febiger, 1973.)

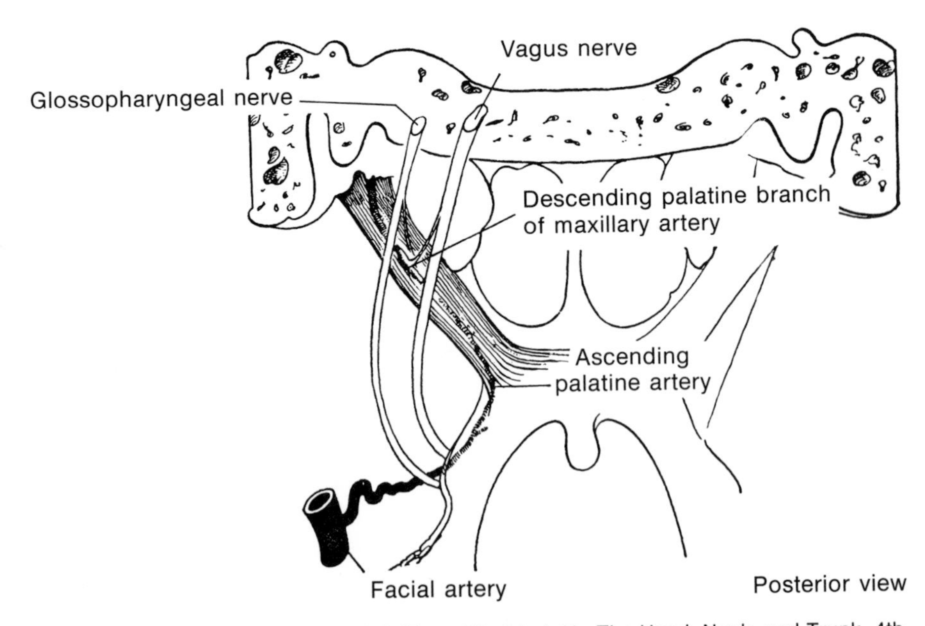

FIG. 2-31. Levator veli palatini. (From Warfel, J. H.: The Head, Neck, and Trunk, 4th ed. Philadelphia, Lea & Febiger, 1973.)

Muscles of the Soft Palate (Fig. 2-29)

The muscles of the soft palate are anatomically and functionally very closely related to those of the pharynx. They work with the muscles of the pharynx during deglutition.

Tensor Veli Palatini (Figs. 2-28; 2-29; 2-30)

The Tensor veli palatini is a little muscle originating from the base of the medial pterygoid plate, the lateral wall of the eustacian (auditory) tube, and the spine of the sphenoid bone. It passes vertically, winds around the pterygoid hamulus, and then passes horizontally into the soft palate. It tenses the soft palate and opens the auditory tube during swallowing. A classic example of this is the act of swallowing to "open the ears" during rapid descent from high altitudes.

The nerve supply is via the third division of the trigeminal nerve.

Levator Veli Palatini (Figs. 2-28; 2-29; 2-31)

The Levator veli palatini arises from the petrous portion of the temporal bone and the auditory tube. It is a round band of muscle, extending inferiorly and medially into the soft palate and joining the muscle fibers on the opposite side. It acts to elevate the vertical posterior part of the soft palate. Along with the tensor, it closes the oral from the nasal pharynx. The nerve supply is via the pharyngeal plexus, which is made up of Cranial nerves IX, X, and XI. Cranial nerve IX provides sensory fibers to the plexus, X provides sensory and motor fibers, and XI provides motor fibers only.

The levator, despite its origin, has no functional effect on the auditory tube.

Uvulae Muscle (Fig. 2-29)

The unpaired uvulae muscle originates on the posterior nasal spine and palatine aponeurosis. This aponeurosis is a thin, firm, fibrous plate that supports and gives strength to the soft palate. The fibers of the uvulae muscle pass posteriorly to the mucous mem-

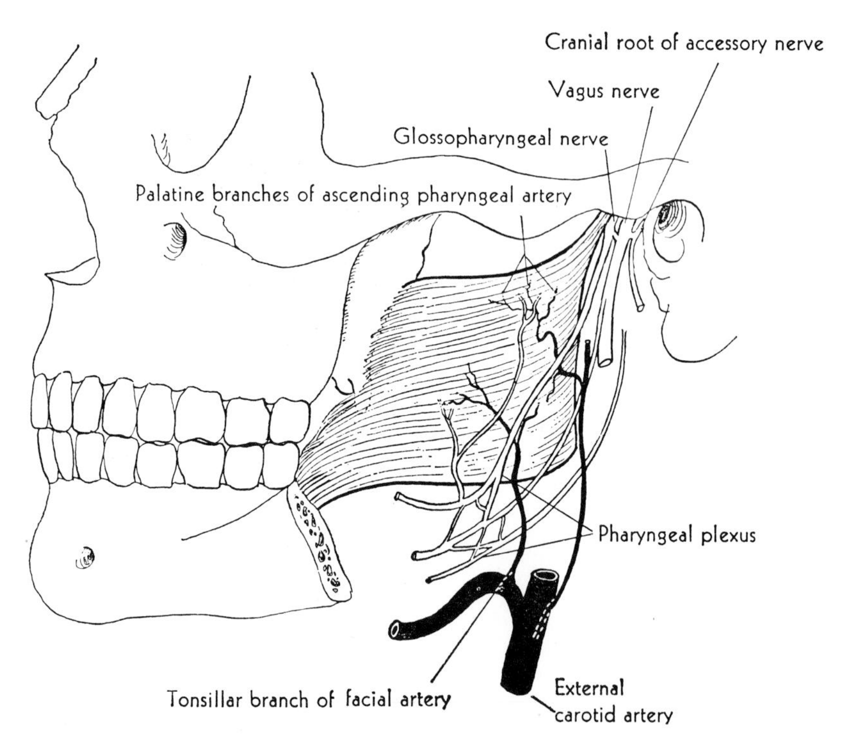

FIG. 2-32. Superior constrictor. (From Warfel, J. H.: The Head, Neck, and Trunk, 4th ed. Philadelphia, Lea & Febiger, 1973.)

brane of the uvula, to shorten it. The nerve supply is the same as for the levator.

Muscles of the Pharynx (Figs. 2-28; 2-29)

The muscles of the pharynx can be divided into the constrictors and the dilators and compare quite readily to the circular muscles of the rest of the digestive tract. They are all innervated by the pharyngeal plexus with one exception, the *Stylopharyngeus,* which is supplied by motor fibers of the glossopharyngeal nerve.

The pharynx is that part of the digestive tract which lies behind the nasal cavities, mouth, and larynx (voice box). It is a muscular conical tube, with its base upward. The pharynx extends from the base of the skull to the level of the sixth cervical vertebra.

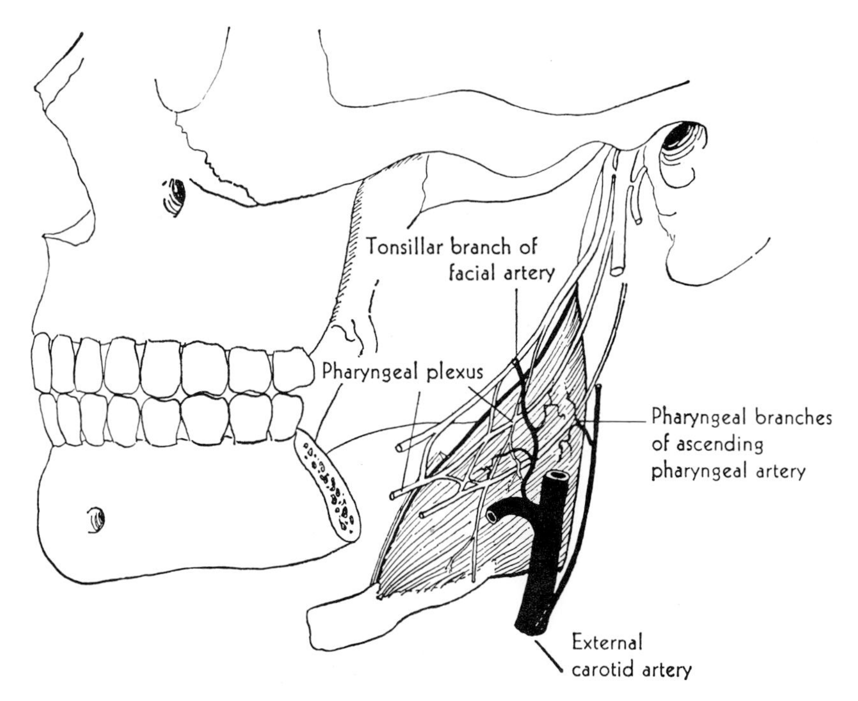

FIG. 2-33. Middle constrictor. (From Warfel, J. H.: The Head, Neck, and Trunk, 4th ed. Philadelphia, Lea & Febiger, 1973.)

Constrictors of the Pharynx

Superior Constrictor (Figs. 2-28; 2-32)

The superior constrictor muscle arises from the lower part of the medial pterygoid plate, the pterygomandibular raphe, and from the alveolar process of the mandible, behind the mylohyoid line. The fibers pass back and superior to insert on a fibrous band called the *median raphe.* This raphe runs from the pharyngeal tubercle of the occipital bone (Fig. 1-5) to the inferior border of the inferior constrictor of the pharynx.

Middle Constrictor (Figs. 2-28; 2-33)

The origin of the middle constrictor, a fan-shaped muscle, is the whole length of the upper border of the greater cornu of the hyoid

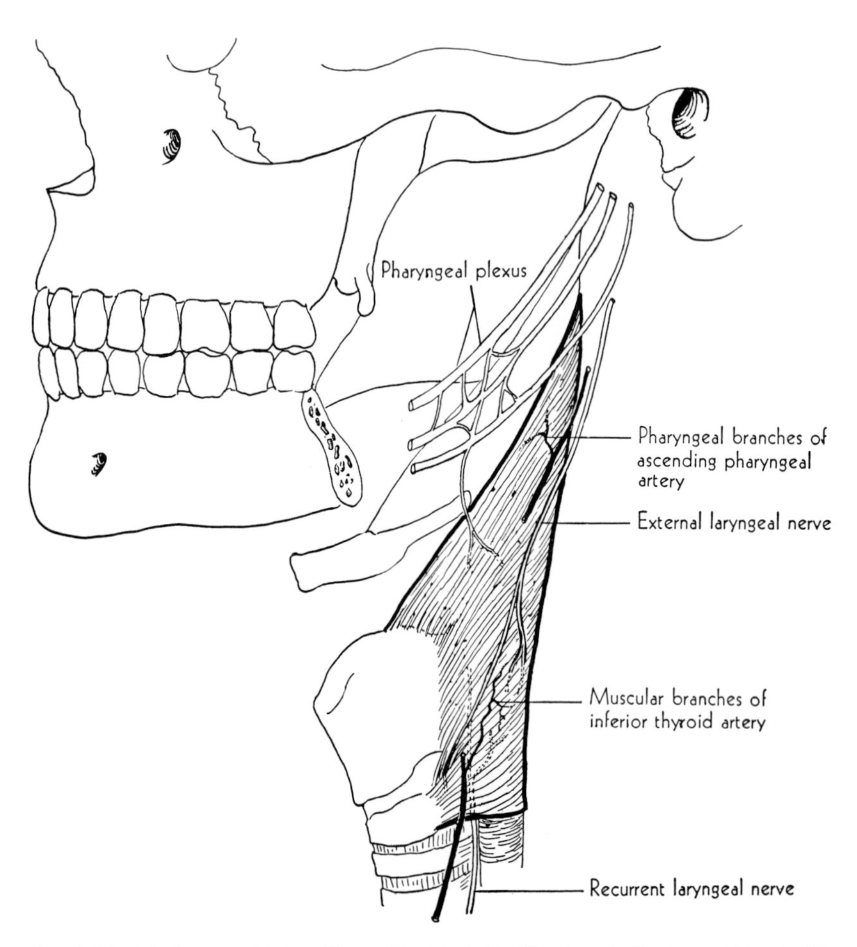

FIG. 2-34. Inferior constrictor. (From Warfel, J. H.: The Head, Neck, and Trunk, 4th ed. Philadelphia, Lea & Febiger, 1973.)

bone, from the lesser cornu and from the *stylohyoid ligament.* The fibers pass posteriorly and superiorly as well as posteriorly and inferiorly. The upper fibers overlap the lower fibers of the superior constrictor, whereas the lower fibers descend *beneath* the upper fibers of the *inferior* constrictor. The middle constrictor inserts posteriorly on the median raphe.

Inferior Constrictor (Figs. 2-28; 2-34)

The Inferior constrictor is the widest and thickest of the three constrictors. It originates from the larynx (cricoid and thyroid carti-

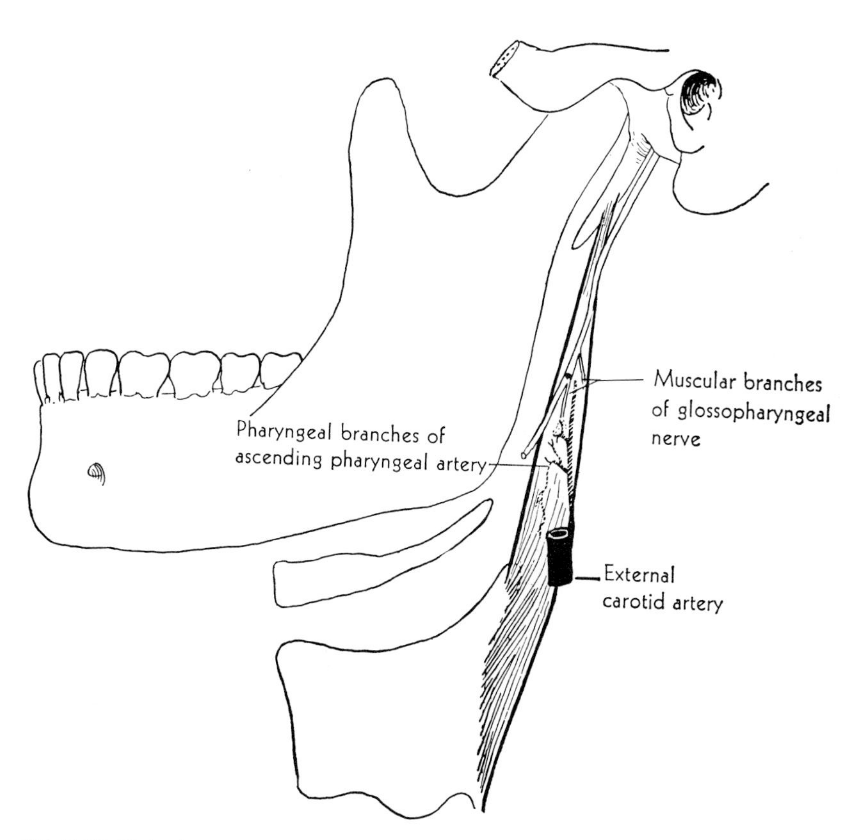

FIG. 2-35. Stylopharyngeus. (From Warfel, J. H.: The Head, Neck, and Trunk, 4th ed. Philadelphia, Lea & Febiger, 1973.)

lages). The fibers fan out to insert on the fibrous median raphe. As can be seen in the illustrations, the most superior fibers of this muscle insert at a point as high as the midportion of the superior constrictor.

Elevators and Dilators of the Pharynx

The Palatoglossus and the Palatopharyngeus are often considered with other muscle groups, but because of their action during deglutition, they are included here.

Palatoglossus

The palatoglossus muscle forms the *anterior tonsillar pillar.* It arises from the anterior surface of the soft palate and inserts into the dorsum and side of the tongue.

Palatopharyngeus (*Pharyngopalatinus*) (Fig. 2-29)

The long fleshy muscle known as the palatopharyngeus originates in the soft palate. It passes laterally and inferiorly and forms the *posterior tonsillar pillar.* It inserts into the posterior border of the thyroid cartilage of the larynx, and into the pharynx.

Salpingopharyngeus (Fig. 2-29)

The Salpingopharyngeus is a thin, vertically oriented muscle, originating near the opening of the auditory tube in the posterior nasal cavity, just above the soft palate. It passes downward and blends with the Palatopharyngeus.

Stylopharyngeus (Figs. 2-12; 2-29; 2-35)

The long, slender, cylindrical stylopharyngeus muscle arises from the styloid process. It passes inferiorly, between the middle and superior constrictor, and spreads out and intermingles with its neighboring muscle fibers in the pharynx.

The actions of these muscles are interesting. As swallowing begins, the pharynx is elevated and dilated. As soon as the bolus of food reaches the pharynx, the elevators and dilators relax and the pharynx descends. Then the constrictors contract harmoniously to propel the bolus into the esophagus and the stomach.

Lateral Cervical Muscles (Fig. 2-7)

An important group of anterolateral muscles of the neck, the accessory muscles of mastication, have been described. Another group of muscles of the neck are the lateral cervical muscles. These large superficial muscles are the *Sternocleidomastoideus* and the *Trapezius.* Although the Trapezius is often considered a muscle of the back and shoulders, it will be included here for completeness.

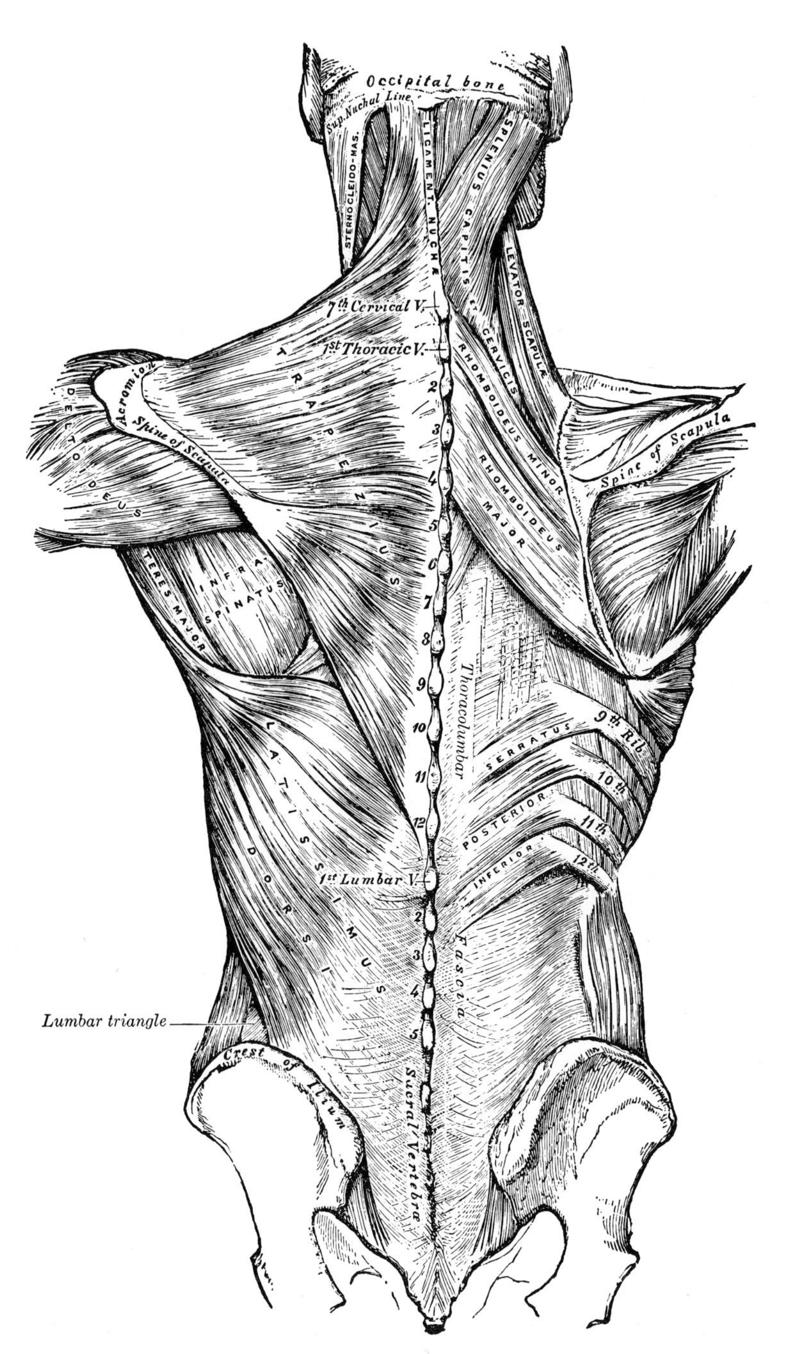

FIG. 2-36. Muscles connecting the upper limb to the vertebral column. (From Gray's Anatomy of the Human Body, 29th ed. C. M. Goss, editor. Philadelphia, Lea & Febiger, 1973.)

The deep muscles of the posterior cervical vertebral region of the neck are complex, intricate, and numerous. They can be studied by consulting an appropriate general anatomy text.

Trapezius (Fig. 2-36)

The Trapezius is a large, flat triangular muscle covering the back part of the neck and the upper trunk. It arises from the external occipital protuberance, the *ligamentum nuchae,* the spinous processes of the seventh cervical vertebra, and from the spines of all the thoracic vertebrae. The *ligamentum nuchae* is a fibrous band, extending from the occipital protuberance to the spinous process of the seventh cervical vertebra.

The fibers of the Trapezius converge as they pass laterally to their insertions on the shoulder girdle. It acts to rotate the scapula (wing bone). It also draws the skull laterally and turns the face away. When both the right and left trapezius muscles contract together, they pull the head backward. The nerve supply is via the eleventh cranial nerve and the third and fourth cervical nerves.

Sternocleidomastoideus (Figs. 2-6; 2-7)

The thick sternocleidomastoideus muscle is not only quite distinguishable by its name but also serves as an important anatomic landmark for functional anatomy and for surgery. It ascends superiorly and posteriorly across the side of the neck from its two-headed origin. One head of origin arises from the uppermost part of the sternum and the other head from the medial third of the clavicle. The two heads blend into one thick muscle that inserts into the mastoid process of the temporal bone and the lateral half of the superior nuchal line of the occipital bone.

When the Sternocleidomastoideus (SCM) contracts, the neck is bent laterally and the head is drawn toward the shoulders, as the head rotates. The chin is thus pointed upward and toward the opposite side. When the right and left SCM contract together, the neck is flexed as the head is drawn forward.

It is innervated by the eleventh cranial nerve and the second cervical nerve.

3

Blood Supply to the Head and Neck

The nutrients and oxygen necessary to sustain life are carried to all the cells of all of the tissues by the arteries. Veins and lymphatics pick up the waste products of cell metabolism and, together with deoxygenated red blood cells, return them to various areas. The blood can then be rejuvenated so that it may return to the tissues for continued cell nutrition. The details of the cell physiology involved are not within the scope of this text. However, a basic knowledge of the general mechanics of systemic and pulmonary circulation is important. The circulation of the head and neck can only be understood if one grasps the concept of the "to-and-fro" system of the closed arteriovenous mechanism found in the human.

Circulatory Concepts

Blood is propelled through the body by the heart, which is a hollow muscular organ containing four compartments or chambers (Fig. 3-1). The heart is situated behind the sternum and somewhat to

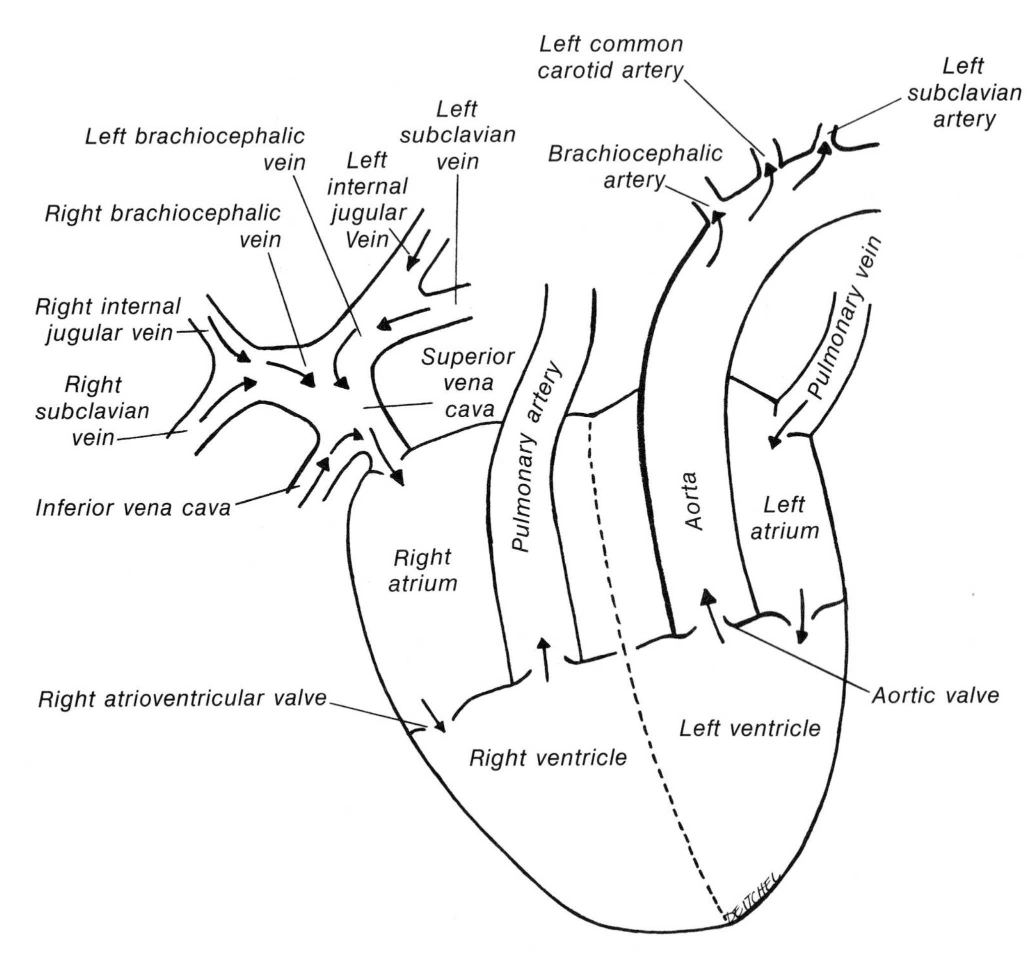

FIG. 3-1. Diagram of blood flow to and from the heart. The arrows denote the flow of blood.

the left (Fig. 3-2). It is about the size of a fist and rests on the diaphragm. The heart is cone-shaped with the apex pointing downward, forward, and to the left.

The four chambers are the right and left *atria* and the right and left *ventricles.* The atria are located superior to their respective ventricles. The right and left ventricles and atria are separated from one another by a fibrous septum which extends from the apex to the base of the heart. Blood returning from the tissues passes into the right atrium. It then flows into the right ventricle through the right *atrioventricular valve.* Blood is then pumped out of the right ventricle into the lungs where it is oxygenated. The oxygenated blood returns to the left atrium of the heart and then passes through the left atrioventricular valve into the left ventricle. From the left ventricle, oxygenated blood is pumped into the body tissues.

It is necessary to understand an important concept here, that is, the cardiopulmonary relationship. Blood going to the lungs from the right ventricle is being sent to the lungs to rid the system of carbon dioxide and to pick up a fresh supply of oxygen (Fig. 3-3). Blood rich in oxygen and other nutrients reaches the *cells* of the lungs, *not* from the right ventricle, but from the *left* ventricle. (See "aorta.")

End products of metabolism (except carbon dioxide) are still in the blood, returning from the lungs to the heart at the left atrium. The

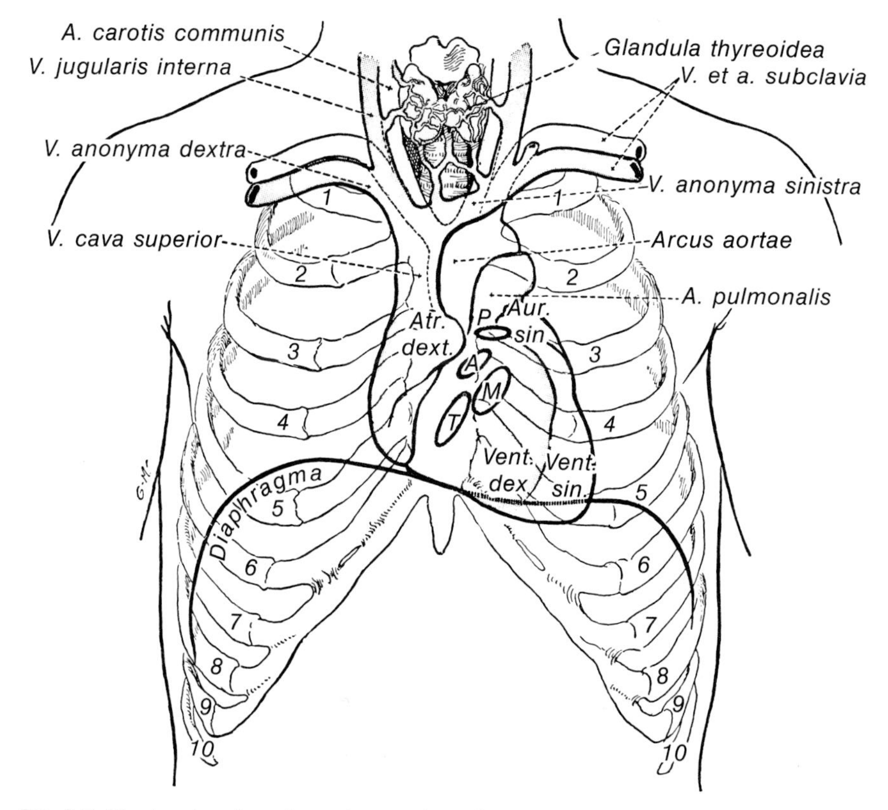

FIG. 3-2. The heart and cardiac valves projected on the anterior chest wall, showing their relation to the ribs, sternum, and diaphragm. (After Eycleshymer and Jones in Gray's Anatomy of the Human Body, 29th ed. C. M. Goss, editor. Philadelphia, Lea & Febiger, 1973.)

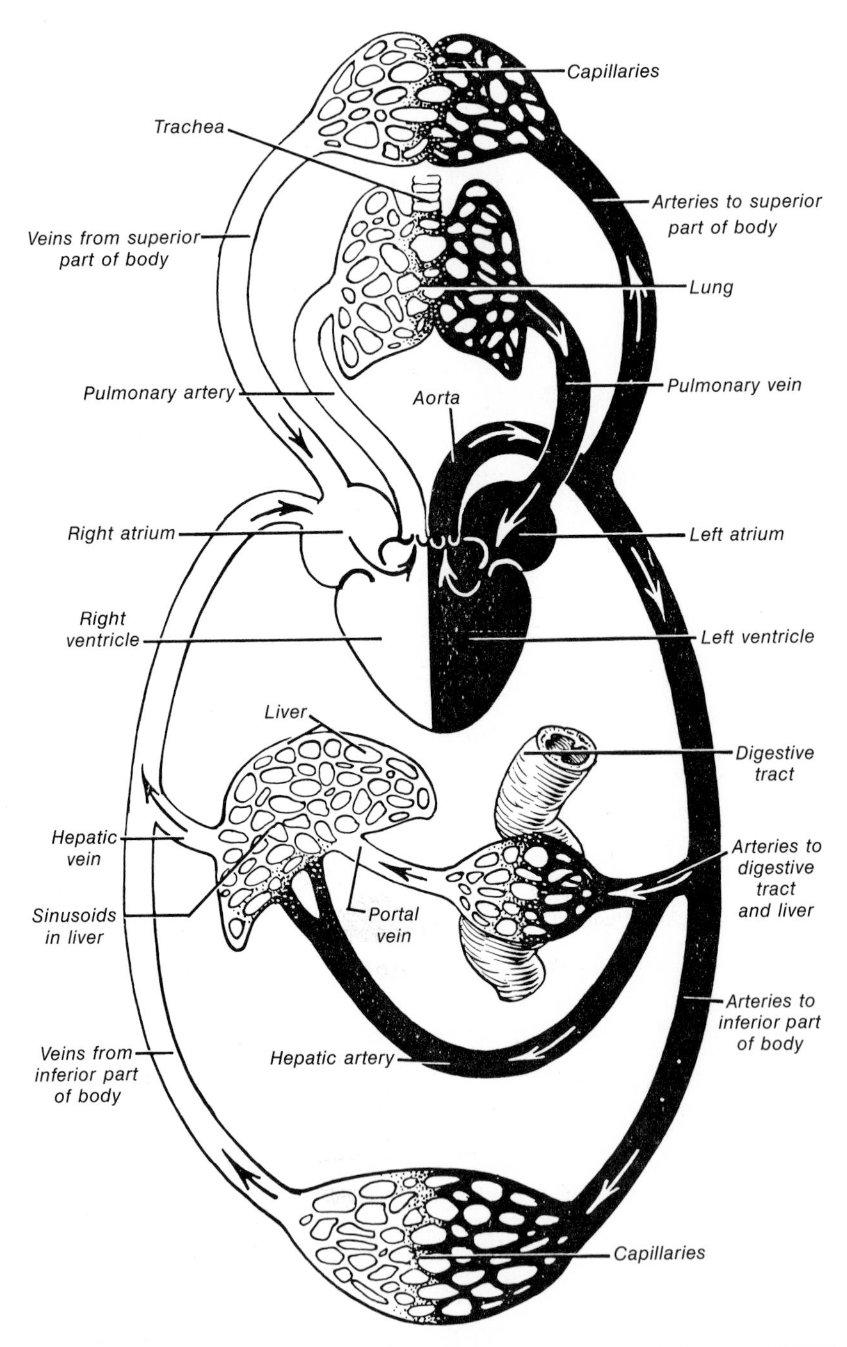

FIG. 3-3. Schematic representation of the double system of circulation. Oxygenated blood is shown in black; nonoxygenated blood in white. The arrows indicate the direction of flow. (From Crouch, J. E.: Human Anatomy, 2nd ed. Philadelphia, Lea & Febiger, 1972.)

blood carrying these end products is then sent to the liver and kidneys from the *left* ventricle. The blood has only been oxygenated and cleared of carbon dioxide in the lungs.

The other end products must be carried in the blood until they reach the kidneys and liver, where the blood is further detoxified. It then returns to the right atrium, via the venous system. One can deduce, therefore, that blood entering the right atrium is a mixture of detoxified blood from the kidneys and liver and blood in *need* of detoxification from other parts of the body. Also, blood returning from the lungs is rich in oxygen but still contains the percentage of other end products of metabolism found in blood returning to the *right* atrium. In addition, *nutrients* enter the blood from the intestine and return to the right atrium, ultimately reaching the cells after leaving the left ventricle.

In summary, blood returning to the right atrium is a mixture of end products of metabolism with more nutrients. It is, however, poor in oxygen. The same blood returns from the lungs, to the left ventricle, via the left atrium, rich in oxygen, having lost carbon dioxide. It is now ready to be sent to the entire body (1) to bring oxygen to the tissues, (2) to bring other nutrients to the tissues, and (3) to be detoxified in the kidneys and liver.

Blood is carried to the various parts of the body via the arterial tree which ultimately ends in arterial capillaries. Blood returns to the heart via the venous system which begins as venous capillaries (Fig. 3-3).

Except for the pulmonary arteries leaving the right atrium, arteries carry oxygenated blood which is bright red. Except for the pulmonary veins, veins carry oxygen-poor blood, which is dark red. Arteries carry blood away from the heart, and veins carry blood toward the heart. For the most part, arteries and veins travel together to and from the various parts of the body. The walls of veins are less muscular than those of arteries, and the pressure therein is also less.

The *arteries* and *veins* are distributed to all parts of the body like branches of a greatly ramified tree. Although the branch of a vessel is smaller than its trunk, the combined cross section of the resulting vessels is greater than that of the trunk. Branches of vessels sometimes open into branches of other vessels of similar size. This

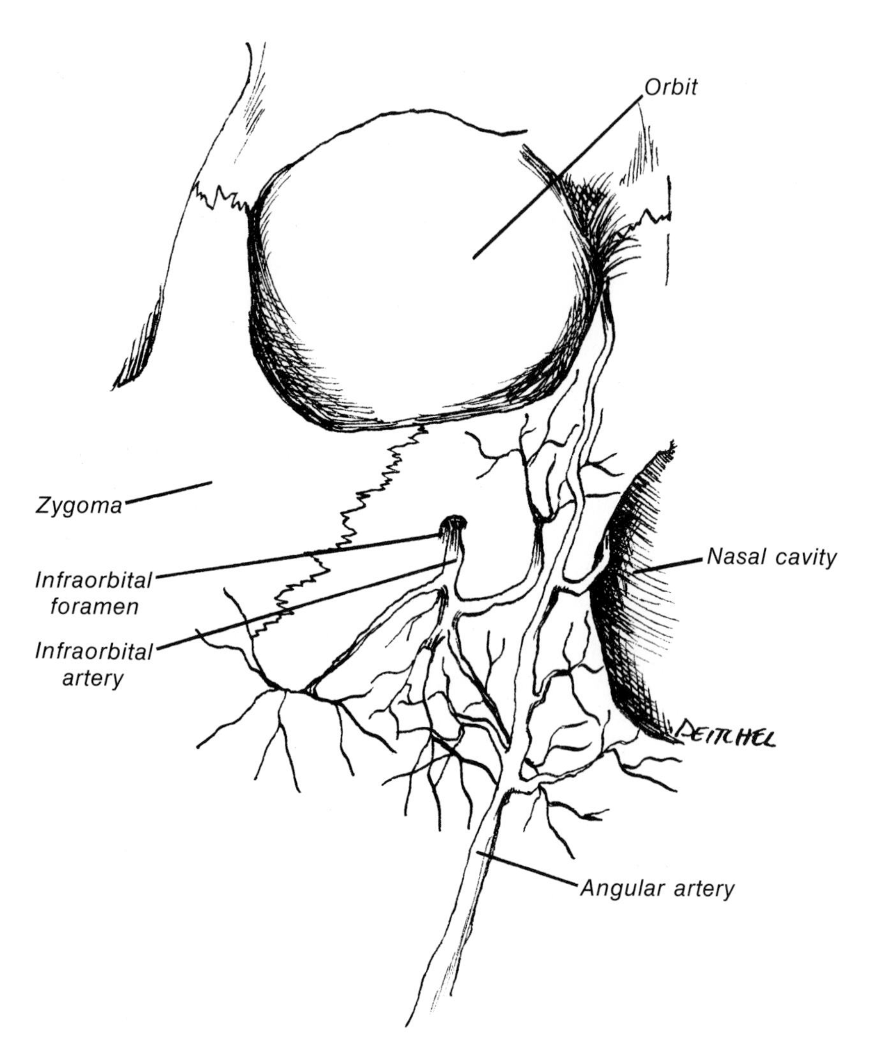

FIG. 3-4. Anastomosis between angular artery and infraorbital artery.

junction of two vessels is known as an *anastomosis* (Fig. 3-4). Anastomoses usually occur in smaller vessels of 0.1 millimeter or less in diameter. When several vessels supply the same area, the area is said to be served by *collateral circulation.*

Arterioles is the term applied to arteries of less than 0.1 millimeter in diameter. They are the smallest arteries. The smallest veins are *venules* (Fig. 3-5).

Capillaries connect the arterioles and venules. They are microscopic in size and their walls are usually only one cell thick (Fig. 3-5).

Multiple small arterioles or venules supplying a highly vascularized area are referred to as venous or arterial *plexuses.* Of importance in the dental profession is the pterygoid plexus, a large network of veins located in the pterygoid plate region of the retromaxillary area between the pterygoid and temporal muscles. Injections of local anesthesia often encounter this plexus, and retrograde dental infection here can be quite hazardous.

Arteries of the Head and Neck

A large artery exists from the left ventricle, carrying all the blood to the systemic circulation. This artery is the *aorta* (Fig. 3-6). Immediately after its exit from the heart, it gives off the right and left *coronary arteries* which supply blood to the musculature of the heart. The aorta then passes upward about 4.0 centimeters, where it curves in a large arch about the level of the first rib and begins to descend in the thorax (Fig. 3-2). Three large vessels arise from the aortic arch: (1) the *brachiocephalic artery,* (2) the *left common carotid artery,* and (3) the *left subclavian artery.* The brachiocephalic artery branches into the right *subclavian* artery and the right *common carotid* artery. Both the subclavian arteries are destined for the arms and the thorax, but they also give rise to a few arteries to the neck. The *vertebral arteries,* which are branches of the subclavian

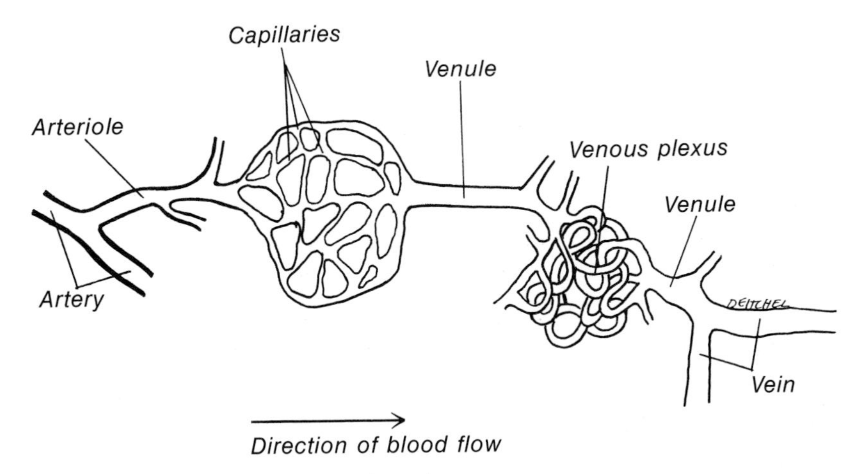

FIG. 3-5. Arterioles, capillaries, and venules.

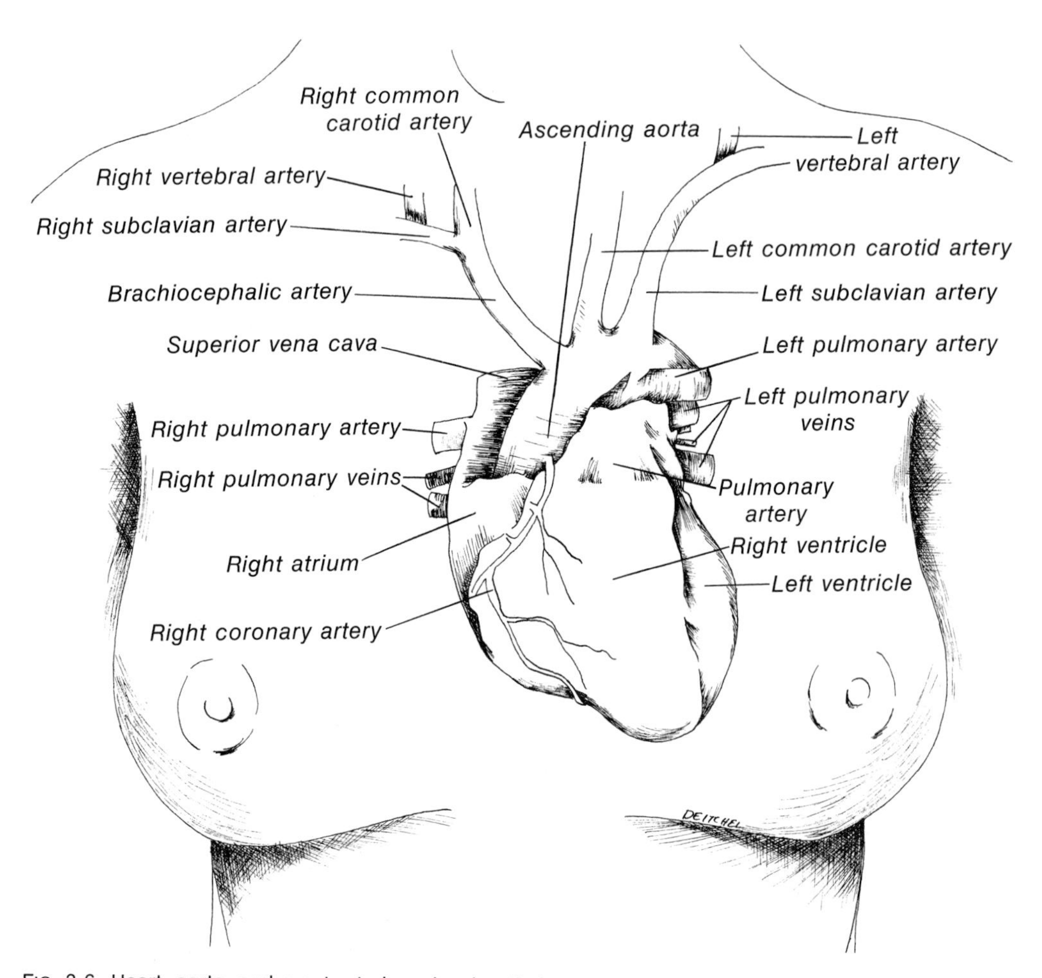

FIG. 3-6. Heart, aorta, and great arteries, showing their positions in the chest.

arteries, ascend in the posterior part of the neck through the foramina in the transverse processes of the upper six cervical vertebrae (Fig. 3-7). They enter the skull through the foramen magnum and contribute to the blood supply of the brain (see also circle of Willis).

The Common Carotid Arteries

The common carotid arteries supply almost all the blood to the head and neck. These arteries ascend on the lateral aspect of the

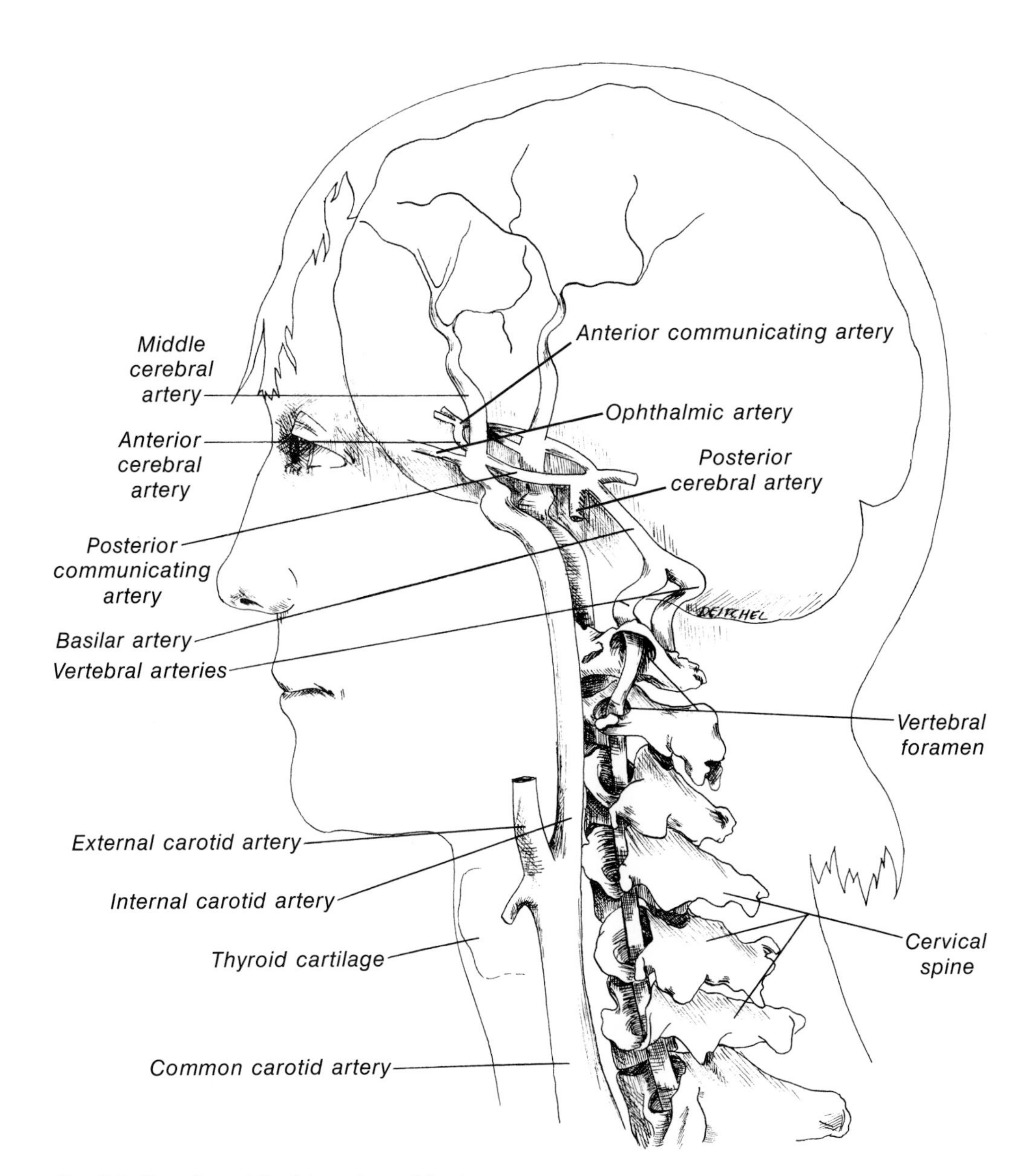

FIG. 3-7. Branches of the internal carotid artery.

neck, deep to the sternocleidomastoid muscle. As the common carotid artery reaches the level of the thyroid cartilage, it divides (bifurcates) into the *internal* and *external carotid arteries.*

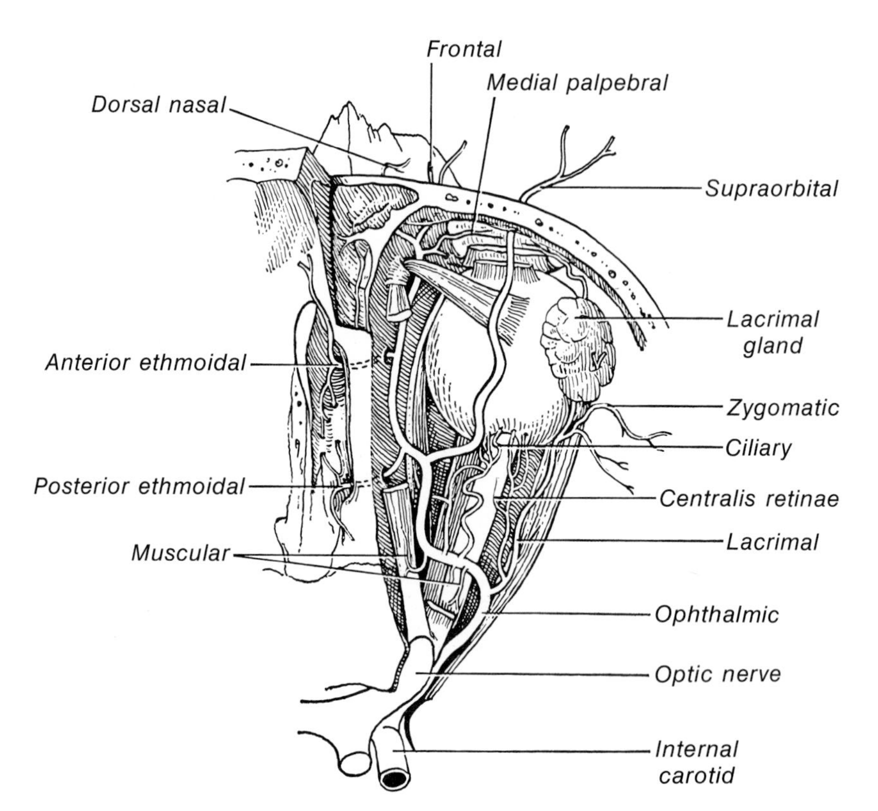

FIG. 3-8. The ophthalmic artery and its branches. (From Gray's Anatomy of the Human Body, 29th ed. C. M. Goss, editor. Philadelphia, Lea & Febiger, 1973.)

Internal Carotid Artery (Fig. 3-7)

The internal carotid artery has no branches in the neck. When it enters the brain case at the carotid canal, located in the petrous portion of the temporal bone, the internal carotid gives off several branches, eleven in number. These branches supply various vital structures, including the brain and the eye.

OPHTHALMIC ARTERY (Figs. 3-7; 3-8). This vessel supplies the muscles and tissues associated with the eye. Along with the optic nerve, it enters the orbit through the optic foramen. It supplies the muscles and globe of the eye by way of its *ocular group* of branches. Also, by way of its *orbital branches,* it supplies the orbit and surrounding parts.

The *supraorbital artery* is a branch of the ophthalmic artery in the orbit that exits from the orbit at the supraorbital notch and supplies the associated area of the forehead and scalp. The ophthalmic artery also gives off the *ethmoidal arteries* to the ethmoid air cells, the lateral wall of the nose, and the nasal septum. The ophthalmic artery terminates as a *frontal artery,* exiting from the orbit at the medial angle of the supraorbital ridge, and the *dorsal nasal artery,* which supplies the root and dorsum of the nose. The lacrimal gland is supplied by the *lacrimal artery,* another intraorbital branch of the ophthalmic artery.

CEREBRAL BRANCHES. *The cerebral branches* terminate the internal carotid artery after the ophthalmic artery is given off. Two branches, the *anterior cerebral* and *posterior communicating arteries,* supply the *circle of Willis* (Fig. 3-9). The *middle cerebral artery* continues up along the side of the brain.

THE CIRCLE OF WILLIS (Figs. 3-7; 3-9). The circle of Willis, an arterial anastomosis of great importance, lies at the base of the brain and is composed of branches of the internal carotid and vertebral arteries.

The right and left vertebral arteries join one another at the base of the brain, posteriorly. The resultant artery from this junction is the *basilar artery.* The basilar artery passes forward and divides into two *posterior cerebral arteries.* The posterior communicating artery of the internal carotid artery, passing posteriorly, enters the posterior cerebral artery. The *anterior cerebral arteries* of the internal carotid pass anteriorly and anastomose with one another by way of a *short anterior communicating artery.* Thus an elaborate, anastomotic, well-designed network of arterial supply is available to the brain.

The External Carotid Artery (Fig. 3-10)

The external carotid is the branch of the common carotid artery that supplies the face, jaws, and scalp. It passes superiorly, but is more superficial than the internal carotid. It parallels the internal carotid artery somewhat and enters the parotid gland with its terminal branch, the *superficial temporal artery.* It gives off this branch about the level of the condylar neck. Here it lies deep to the condyle

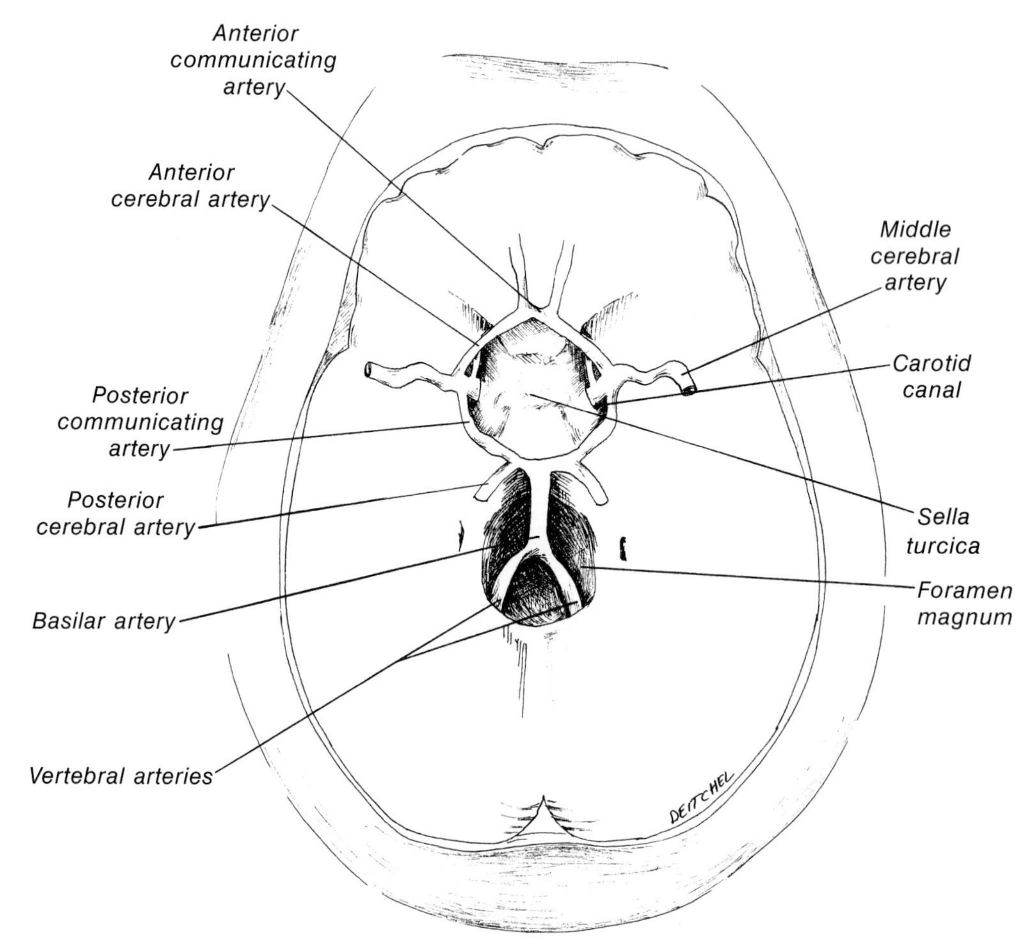

FIG. 3-9. The arterial circle of Willis (from above).

and turns medially and anteriorly and is known as the *maxillary artery.*

The branches of the external carotid begin to arise near the bifurcation of the common carotid. They are eight in number.

ASCENDING PHARYNGEAL ARTERY (Fig. 3-10). This vessel usually arises immediately above the bifurcation of the common carotid. It ascends along the lateral pharyngeal wall to the skull. The blood supply to the pharynx and adjacent muscles is provided, in part, by this vessel. It is small and anastomoses with pharyngeal branches of other arteries.

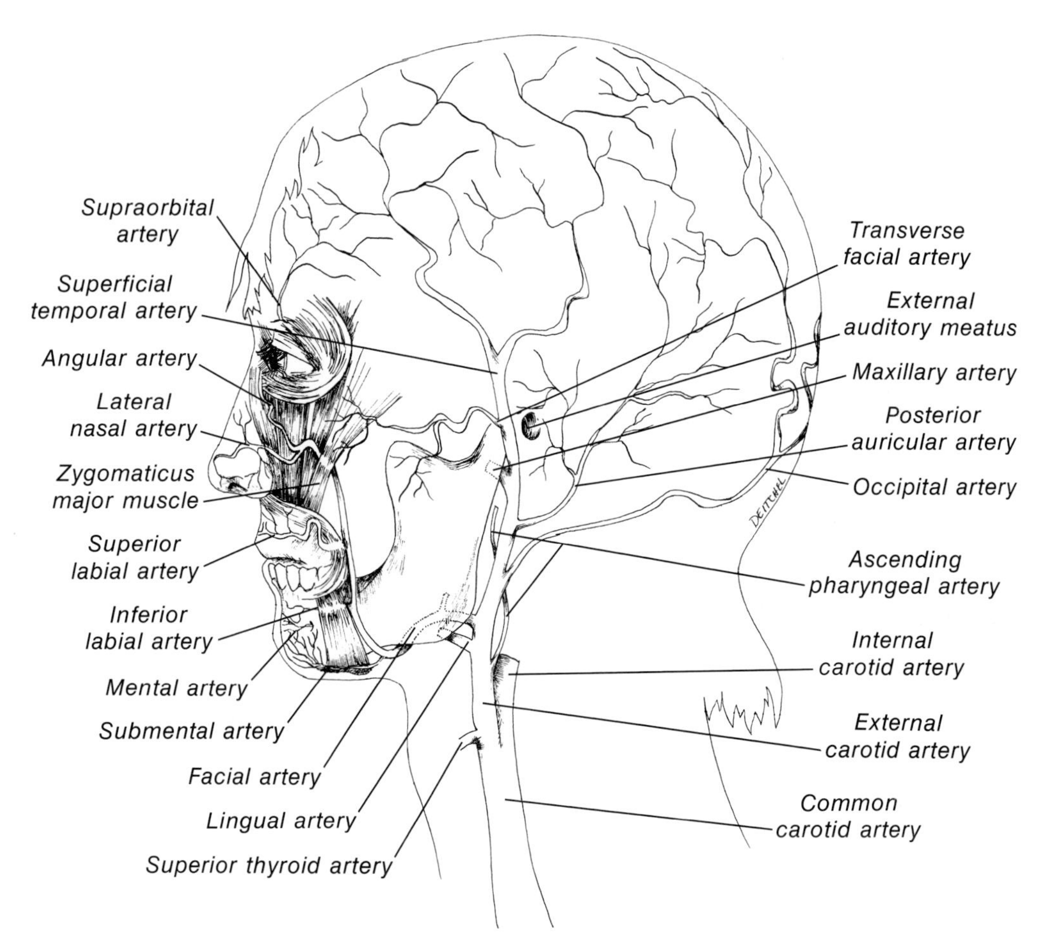

FIG. 3-10. Branches of the external carotid artery.

SUPERIOR THYROID ARTERY (Fig. 3-10). The bifurcation of the common carotid is also the location for the origin of the superior thyroid artery. It curves anteriorly and downward to supply the thyroid gland. The mucous membrane and muscles of the larynx also receive blood from the superior thyroid artery. It dispenses some blood via a small branch to the hyoid region.

LINGUAL ARTERY (Fig. 3-10). Destined for the tip of the tongue via a tortuous course, the lingual artery arises from the external carotid artery at the level of the hyoid bone. Often it arises in common with the facial artery via the *lingual-facial trunk.* From its origin, the lingual artery courses anteriorly to the posterior border of the hyoglossus muscle. It passes deep to this muscle and courses upward to the mouth, where it enters the base of the tongue. It terminates at the tip of the tongue. (See also Fig. 3-12.)

The lingual artery gives off the *sublingual artery* to the floor of the mouth, prior to entering the substance of the tongue. The sublingual artery supplies the sublingual gland, mucosa of the floor of the mouth, the mylohyoid muscle, and the lingual gingiva. It anastomoses with the *submental branch* of the facial artery. After giving off the sublingual artery, the lingual artery, now in the body of the tongue, is termed the *deep lingual artery.*

THE FACIAL ARTERY (EXTERNAL MAXILLARY ARTERY) (Fig. 3-10). The facial artery supplies, for the most part, the superficial structures of the face. It rises from the external carotid artery just above the origin of the lingual artery. It may arise from the *lingual-facial trunk.* Passing forward and upward, it passes deep to the digastric muscle and enters the submandibular triangle. It then enters the substance of the submandibular salivary gland or it may pass deep to the gland. After it reaches the superior border of the gland, it arches upward toward the floor of the mouth and then turns downward and laterally to pass below the inferior border of the mandible. The artery then turns sharply upward on the *lateral* surface of the mandible and crosses in front of the anterior border of the masseter muscle. It is then directed toward the angle of the mouth, deep to the muscles of facial expression, but lateral to the buccinator. After passing under the zygomaticus major muscle, it usually lies superficial to the infraorbital muscles. At the corner of the mouth, it turns upward along the lateral border of the nose to the inner corner (canthus) of the eye. Here it anastomoses with the terminal branches of the ophthalmic branch of the internal carotid artery.

The two most important branches of the facial artery under the jaw are the *ascending palatine* and the *submental* arteries. The ascending palatine comes off close to the origin of the facial artery and supplies the soft palate, pharynx, and tonsils with the ascending pharyngeal artery. The submental artery originates from the facial artery before that artery turns onto the face. It supplies the submandibular region and anastomoses with the sublingual artery.

The facial artery gives off the *inferior labial* and *superior labial arteries* at the corners of the mouth. The inferior labial artery anastomoses with the *mental artery* to supply the chin and lower lip. The superior labial artery anastomoses with the terminal nasal branches of the ophthalmic artery and with the infraorbital arteries. The facial artery then courses superiorly, giving branches to the side of the nose and cheek to end at the medial canthus of the eye as the *angular artery.*

OCCIPITAL ARTERY (Fig. 3-10). The occipital artery arises close to the origin of the facial artery. It, however, runs upward and backward toward the occipital area of the scalp. After it crosses the mastoid process in the occipital groove of the temporal bone, it becomes more superficial. It supplies the scalp.

POSTERIOR AURICULAR ARTERY (Fig. 3-10). The posterior auricular artery originates from the external carotid artery at a level just opposite the lobe of the ear. It then passes laterally and posteriorly to supply the outer ear and adjacent scalp behind the ear. The superficial temporal artery and occipital artery both give branches that anastomose with the posterior auricular artery.

SUPERFICIAL TEMPORAL ARTERY (Fig. 3-10). The superficial temporal artery is anatomically, but not embryologically, the continuation of the external carotid artery. It ascends vertically in front of the ear to the temporal region of the scalp. It passes through the substance of the parotid gland in the retromandibular fossa, releasing the *transverse facial artery,* which passes horizontally and ends near the lateral canthus of the eye. Also, it sends branches to the outer ear and a *middle temporal* branch to the temporalis muscle. In the scalp, the branches of the superficial temporal artery anastomose with branches of the occipital artery, posterior auricular artery, frontal artery, supraorbital artery, and across the midline with arteries of the opposite side.

MAXILLARY (INTERNAL MAXILLARY) ARTERY (Fig. 3-11). This complex and most interesting vessel supplies the deep tissues of the face. It originates at the condylar neck and passes deep. The course is horizontal and anterior as it heads for the pterygopalatine fossa. It is close to the medial surface of the condylar neck when it first originates, and as it passes deeply, it runs between the condylar neck and the *sphenomandibular ligament.* There, it lies either lateral or medial to the external pterygoid muscle. If it lies lateral, it must reach the pterygopalatine fossa by turning medially between the two heads of origin of the external pterygoid muscle.

It is easy to understand the divisions of the maxillary artery if one classifies it into sections and organizes its distribution. The artery is divided into four sections: mandibular, muscular, maxillary, and sphenopalatine.

Mandibular Section. The *middle meningeal artery* passes superiorly into the spinous foramen and into the cranial cavity. It supplies the dura mater and adjacent bones.

The *inferior alveolar artery* (Fig. 3-11) passes vertically downward and enters the inferior alveolar foramen of the mandible. It supplies the bone of the mandible and the teeth. The mental artery is given off within the inferior alveolar canal and exits from the mental foramen to supply the soft tissues of the chin and to anastomose with the inferior labial artery.

The inferior alveolar artery continues on as the *incisive artery* inside the mandible and passes within the bone to the midline, where it anastomoses with the incisive artery of the opposite side. The small arteries that branch from the inferior alveolar artery in the canal supply the teeth (*dental arteries*) and bone, periodontium, and gums (*alveolar branches*).

Muscular Section (Fig. 3-11). This part supplies the muscles of mastication and the buccinator muscle. The Temporalis receives the *posterior* and *anterior deep temporal* arteries. The *masseteric artery* passes laterally through the condylar notch to supply the masseter muscle. A variable number of small *pterygoid branches* are released to the pterygoid muscles. The last branch of this section is the *buccal artery,* which crosses the anterior border of the ascending mandibular ramus at the level of the occlusal plane. It is sometimes encountered during surgery of this area and can be quite troublesome.

Maxillary Section. In the maxillary section the *posterior superior alveolar artery* crosses the maxillary tuberosity and here gives off *posterior superior arteries,* which enter the posterior superior alveolar foramina and supply the maxillary posterior teeth and bone.

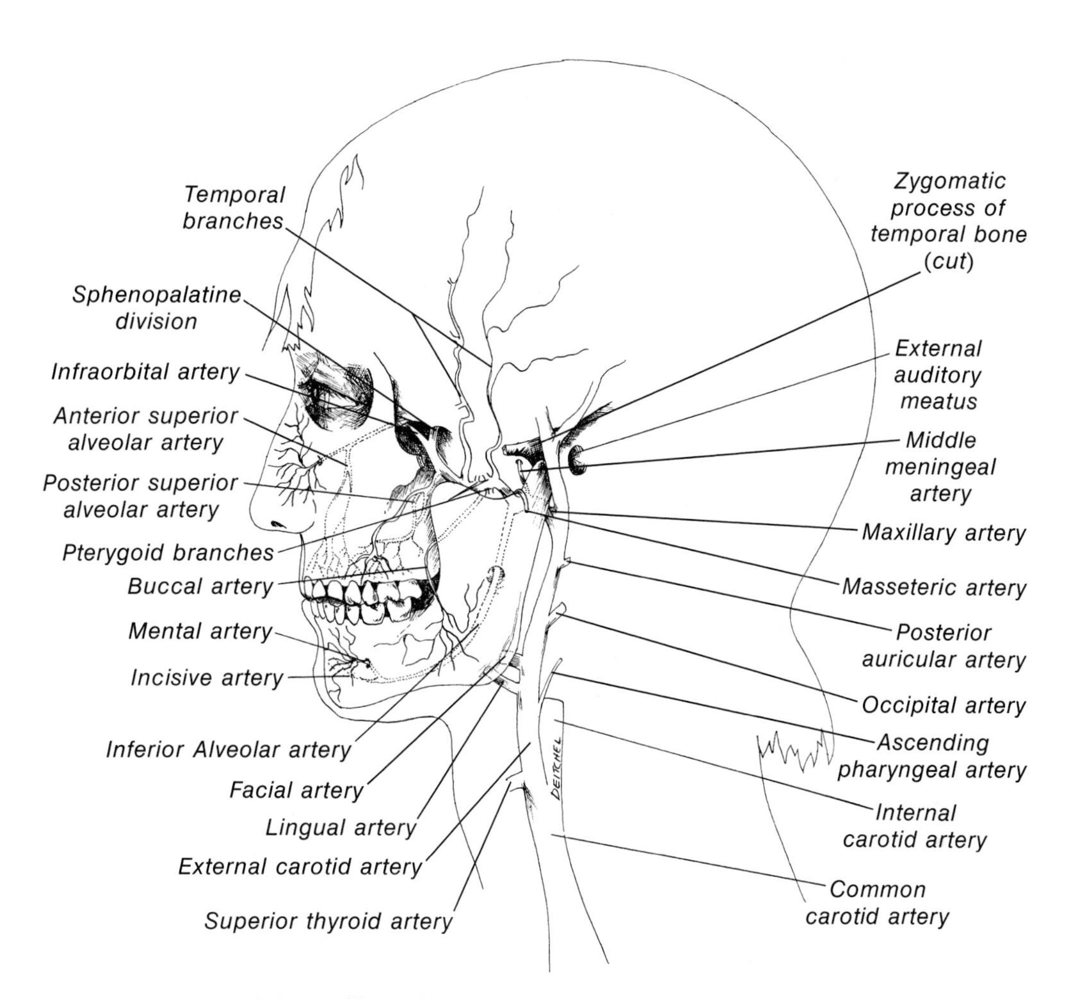

FIG. 3-11. Branches of the maxillary artery.

It continues on as the *gingival artery* and supplies the alveolar process and gingiva.

The *infraorbital artery* is next released by the maxillary artery. This vessel enters the orbit through the inferior orbital fissure and lies in the infraorbital sulcus and then the infraorbital canal. In the orbit, it supplies contiguous structures and releases the *anterior superior alveolar artery* to the anterior teeth, bone, and gingiva. The infraorbital foramen is its exit. It thus supplies the anterior part of the cheek and upper lip, anastomosing with the superior labial and angular arteries.

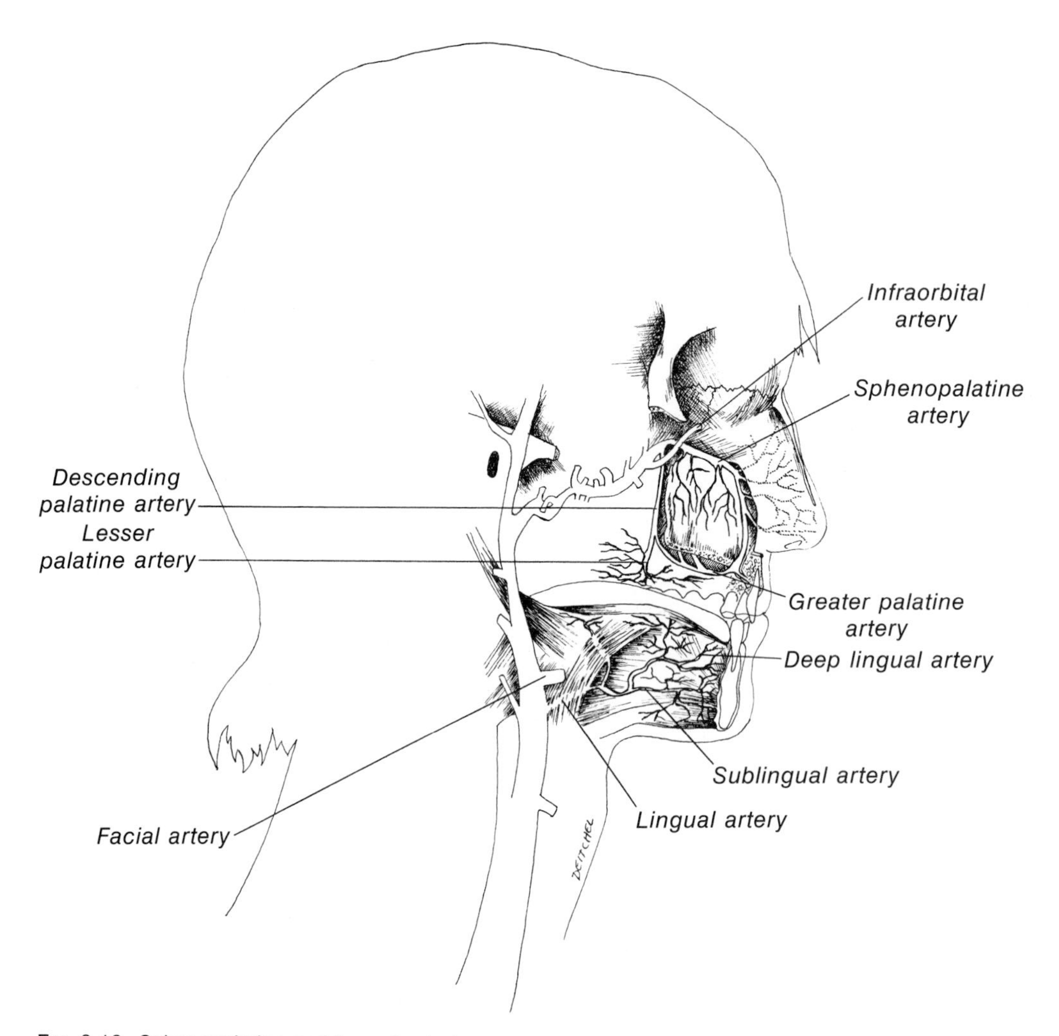

FIG. 3-12. Sphenopalatine and lingual arteries.

Sphenopalatine Section (Fig. 3-12). This section is short and it is from here that the terminal branches occur. The ramifications occur in the pterygopalatine fossa.

The *descending palatine artery* (Fig. 3-12) arises in the pterygopalatine fossa and passes inferiorly to enter the oral cavity through the greater (major) palatine foramen. In the pterygopalatine canal, it may give off some small *nasal branches* to the lateral wall of the nasal cavity. Also it gives off *lesser palatine branches* which exit from the lesser palatine foramina and supply the soft palate and tonsil.

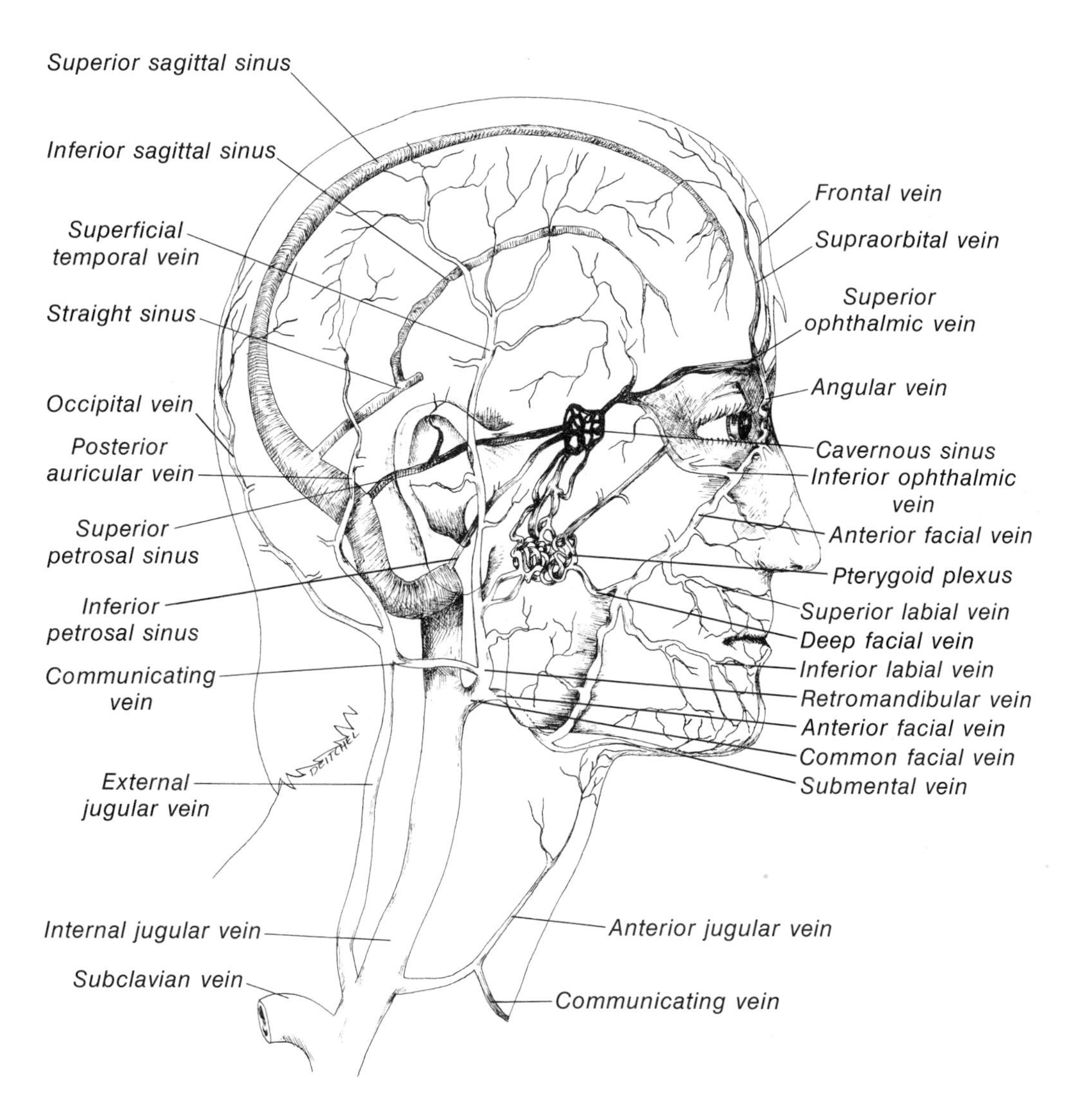

FIG. 3-13. Veins of the head and neck.

Once it exits from the greater palatine foramen, the descending palatine artery is known as the *anterior (greater) palatine artery,* which turns forward in the substance of the palatal mucosa and passes anteriorly to the nasopalatine foramen. In its course it gives branches to the associated bone, glands, and mucosa. Once it reaches the nasopalatine foramen, it turns upward and passes into the nose and anastomoses with the sphenopalatine artery on the septum.

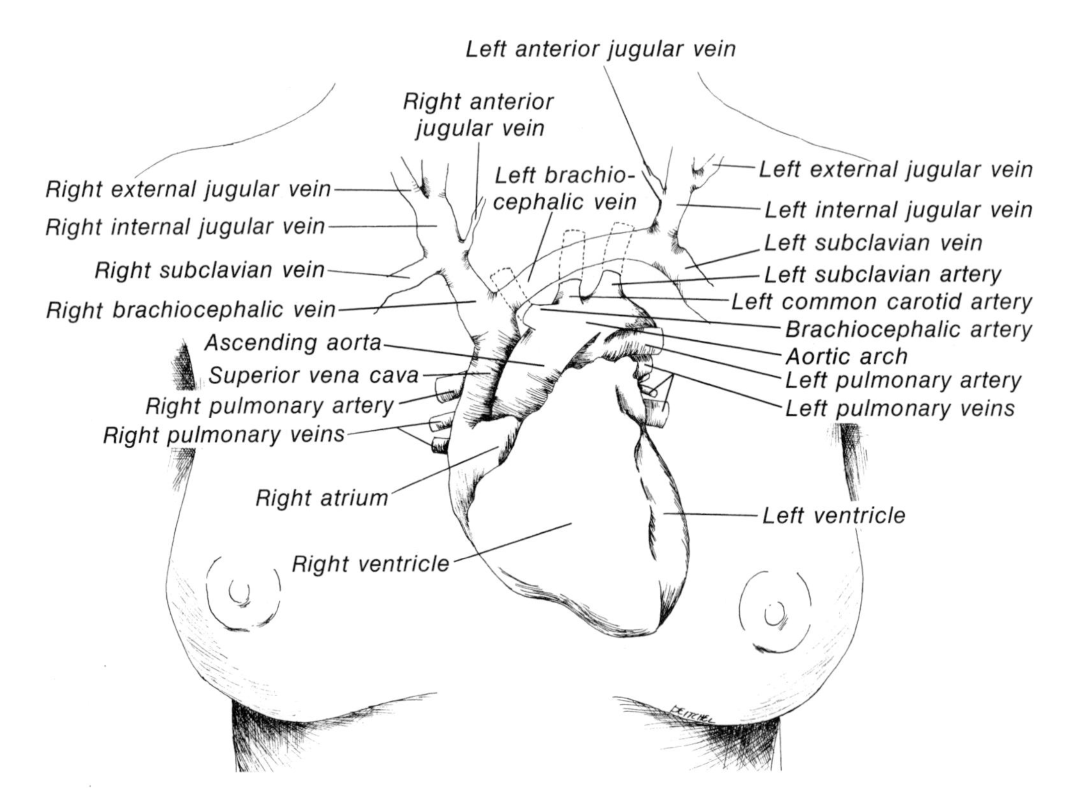

FIG. 3-14. Heart, aorta, and great veins.

The *sphenopalatine artery* (Fig. 3-12) also arises in the pterygopalatine fossa. It enters the nasal cavity through the sphenopalatine foramen. Here it divides into branches supplying the lateral wall of the nasal cavity and septum. On the septum it anastomoses with the septal branch of the *anterior palatine artery.*

Veins of the Head and Neck (Fig. 3-13)

The venous blood of the head and neck is drained almost entirely by the internal jugular vein. Veins usually accompany arteries and carry the same or similar names. The venous network is more variable than the arterial tree, but a definite pattern still exists. Deep veins are united with superficial veins by several anastomoses. These multiple anastomoses present a potential danger by increasing the ease of the spread of infection. In addition, because the veins of the face have few, if any, valves, backflow of blood can easily occur. Thus, bacteria can readily flow from superficial to deep veins. Prior to the advent of antibiotics, brain infection, secondary to facial or dental infection, was not uncommon.

Once blood reaches the internal jugular vein, it drains inferiorly to the *brachiocephalic vein* (Fig. 3-14). The brachiocephalic vein is formed by the confluence of the internal jugular vein and the *subclavian vein,* which drains the upper extremity. The right and left brachiocephalic veins then join to form the *superior vena cava.* The *inferior vena cava,* draining the lower portion of the body, joins the superior vena cava in an area known as the *confluence* of the *cava,* located at the right atrium (Fig. 3-15).

Venous Sinuses (Figs. 3-13; 3-16; 3-17)

The sinuses of the dura mater in the brain empty their contents into the internal jugular vein which commences at the jugular foramen. Blood from the eye and the brain is collected in the dural sinuses.

Superior Sagittal Sinus

The superior sagittal sinus begins in the area of the cribriform plate of the ethmoid bone and passes posteriorly in the midline along the inner plate of the frontal, parietal, and occipital bones. It drains some of the veins of the brain as it runs posteriorly.

Inferior Sagittal Sinus

The inferior sagittal sinus.is enclosed in the lower free border of a vertical fold of dura which separates the two halves of the cerebrum.

Straight Sinus

The straight sinus joins the inferior sagittal sinus and the superior sagittal sinus. It lies in a horizontal fold of dura, separating the cerebrum from the cerebellum.

Transverse Sinus

The transverse sinus begins where the straight sinus and the superior sagittal sinuses join. It runs horizontally across the inner table of the occipital and temporal bones to become an S-shaped sinus, the *sigmoid sinus,* and then into the internal jugular vein at the jugular foramen.

Cavernous Sinuses (Figs. 3-13; 3-16; 3-17)

The cavernous sinuses, which lie on either side of the sella turcica, are pools of venous blood, subdivided internally by thin strands and trabeculae of connective tissue. The right and left cavernous sinuses communicate with one another via anterior and posterior chambers known as *intercavernous sinuses.* The cavernous sinuses drain associated portions of the brain. Also, the *ophthalmic vein* draining the eye empties posteriorly into the cavernous sinus. Clinically, it is of great importance to know that the internal carotid artery, the first two divisions of the trigeminal nerve the abducent nerve, the oculomotor nerve and the trochlear nerve all pass through the cavernous sinus. Retrograde infection into this sinus can lead to a grave clinical problem.

Petrosal Sinuses

The blood in the trabeculated cavernous sinus is drained posteriorly into the *superior and inferior petrosal sinuses.* The superior petrosal sinus ends in the sigmoid sinus. The inferior petrosal sinus empties more inferiorly, directly into the jugular vein near the foramen.

Internal Jugular Vein

The blood from the brain empties into the internal jugular vein which commences at the jugular foramen. The vein descends in the neck to the brachiocephalic vein.

Contributing veins include the inferior petrosal sinus, communications from the pharynx and tongue, common facial vein, veins of the larynx and the thyroid gland, the external jugular vein, and the anterior jugular veins.

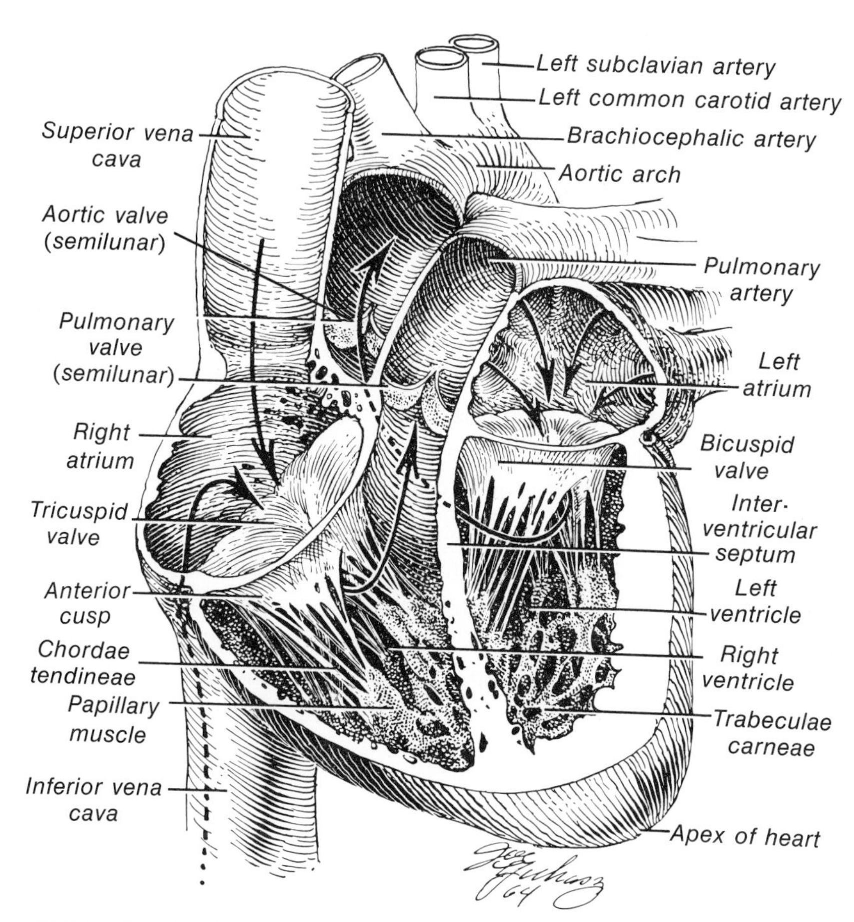

FIG. 3-15. Anterior view of opened heart to show its chambers and valves. The arrows indicate the direction of flow of blood through the heart to and from its major vessels. (From Crouch, J. E.: Functional Human Anatomy, 2nd ed. Philadelphia, Lea & Febiger, 1972.)

Common Facial Vein (Fig. 3-13)

The common facial vein is a short, thick vessel made up of the *anterior facial vein* and the *retromandibular vein.* It enters the internal jugular vein at the level of the hyoid bone.

Anterior Facial Vein (Fig. 3-13)

The anterior facial vein follows a course similar to that of the facial artery. It originates from the veins of the forehead and nose.

The *frontal vein, supraorbital vein,* and veins from the lids and nose all contribute to the first part of the facial vein, the *angular vein.* The angular vein then becomes the facial vein as it drains inferiorly. A wide anastomosis takes place between the angular vein and the *superior ophthalmic vein.* As the facial vein descends, it picks up communicating branches from the associated area. An anastomotic branch from the *pterygoid plexus* opens into the facial vein at the level of the upper lip. This communicating vein is the *deep facial vein.* Once the facial vein has picked up the *superior and inferior labial veins,* it reaches the inferior border of the mandible where it turns posteriorly and deep. Here it receives the *submental* and *palatine veins* and enters the common facial vein.

Retromandibular Vein (Fig. 3-13)

The areas supplied by the maxillary artery and the superficial temporal artery are drained by the retromandibular vein. This vessel is sometimes called the *posterior facial vein.* It is formed by the union of the *superficial temporal* vein with the deep veins of the maxilla. It emerges from the substance of the parotid salivary gland and courses vertically down to the common facial vein.

Lingual Veins

The lingual veins, three or four in number, accompany the lingual arteries. They drain the tongue and floor of the mouth and empty into the anterior facial vein, common facial vein, or retromandibular vein.

Pterygoid Plexus (Fig. 3-13)

The venous network known as the pterygoid plexus lies between the temporal and pterygoid muscles. It drains the muscles of mastication, the nasal cavity, the temporomandibular joint, the external ear, and a small portion of the dura. It communicates with the facial vein by way of the deep facial vein. Also, it drains the maxillae and the palate and communicates with the cavernous sinus. It empties posteroinferiorly, joining the superficial temporal vein to form the retromandibular vein.

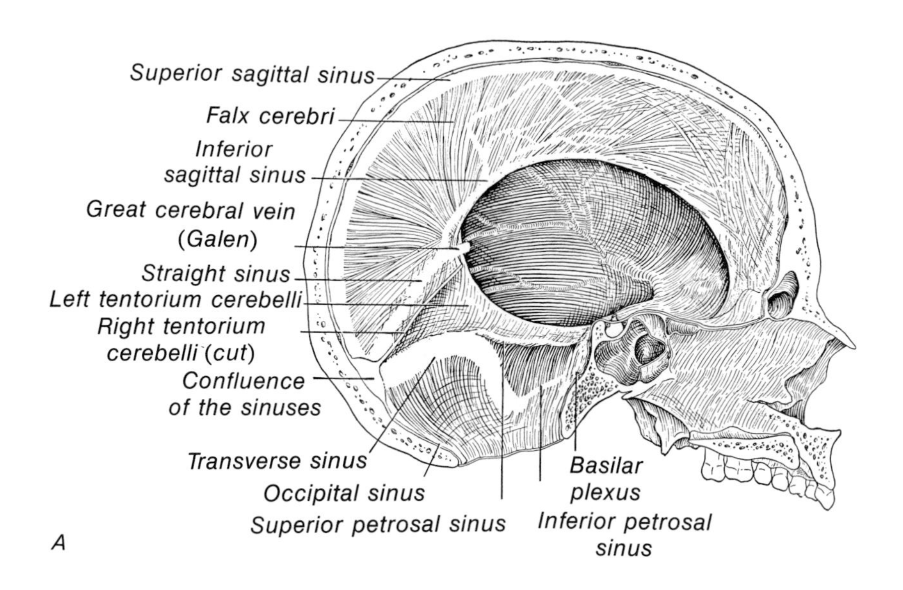

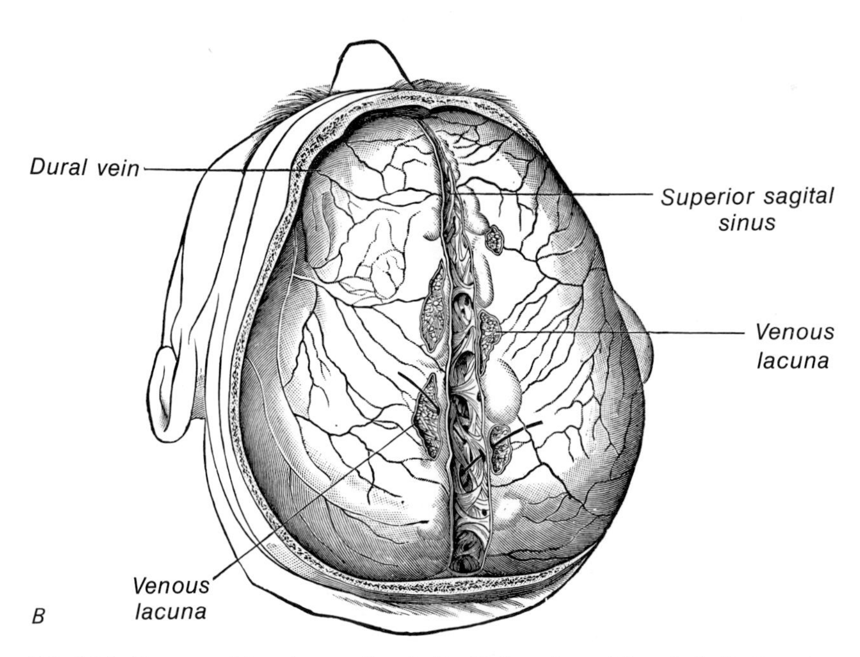

FIG. 3-16. Sinuses of the dura mater. A. Sagittal section of the skull. B. Superior sagittal sinus laid open after removal of the skull cap. Probes are passed from two venous lacunae into the sinus. (After Poirier and Charpy in Gray's Anatomy of the Human Body, 29th ed. C. M. Goss, editor. Philadelphia, Lea & Febiger, 1973.)

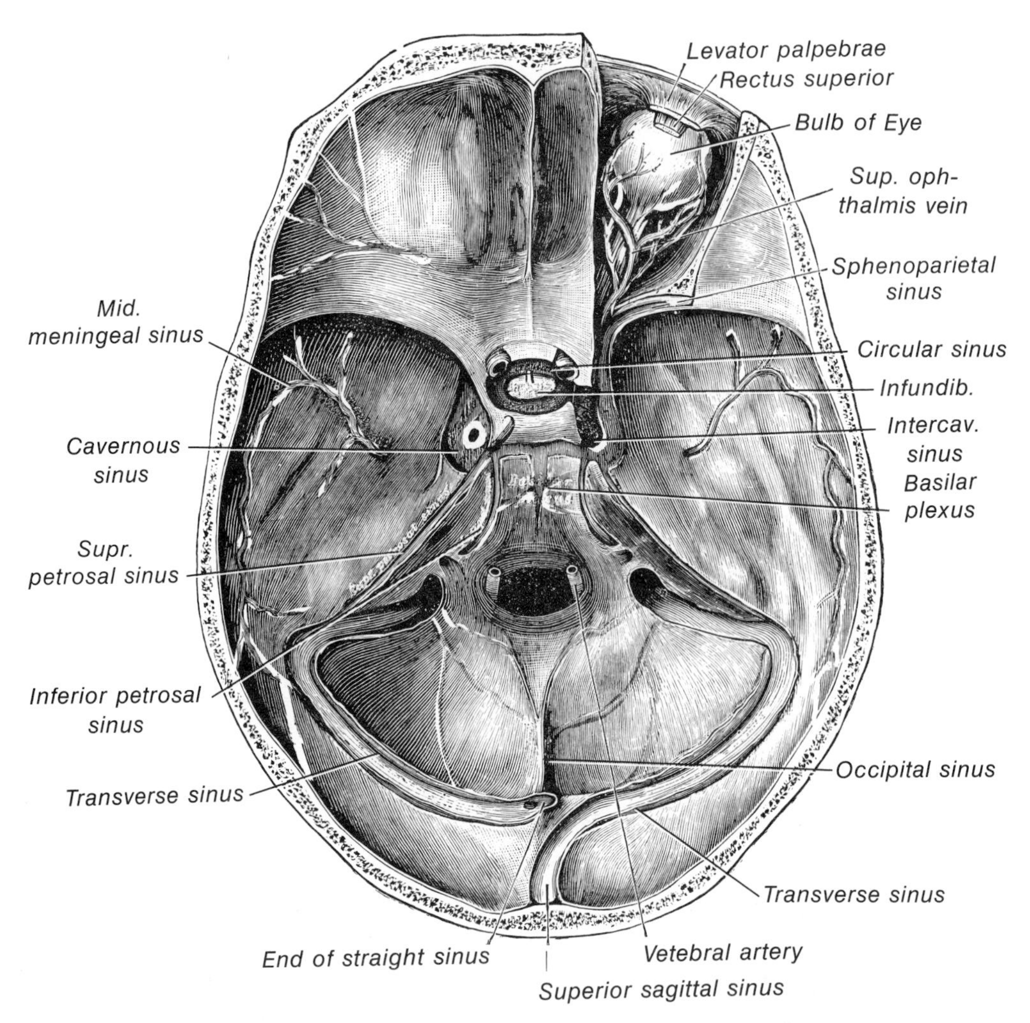

FIG. 3-17. The sinuses at the base of the skull. (From Gray's Anatomy of the Human Body, 29th ed. C. M. Goss, editor. Philadelphia, Lea & Febiger, 1973.)

External Jugular Vein (Fig. 3-13)

The external jugular vein is formed by the junction of the *posterior auricular* vein with the *occipital vein.* It passes across the Sternocleidomastoideus, superficially, and then passes deep to enter the internal jugular vein low in the neck. It often anastomoses with the common facial vein or retromandibular vein.

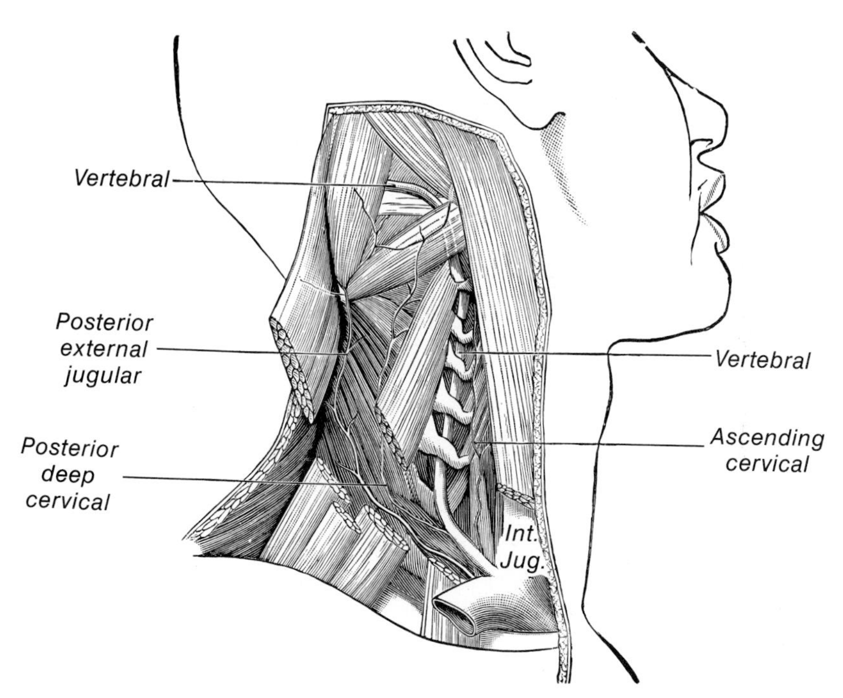

FIG. 3-18. The vertebral vein. (After Poirier and Charpy in Gray's Anatomy of the Human Body, 29th ed. C. M. Goss, editor. Philadelphia, Lea & Febiger, 1973.)

Anterior Jugular Vein (Fig. 3-13)

Often absent, the anterior jugular vein may be a single midline vein or two midline veins. It empties into the internal jugular vein near the junction of the internal jugular vein with the subclavian vein. The suprasternal *space of Burns,* above the superior border of the sternum, is where profuse anastomoses of the two anterior jugular veins often take place.

Variations in Venous Drainage

As noted earlier, many variations in the venous drainage are common. Some of the most common variations, as listed by Sicher, are:

1. The common facial vein may not exist. The retromandibular vein empties into the external jugular vein and the facial vein opens into the internal jugular vein.

2. The common facial vein opens into the external jugular vein.
3. The facial vein empties into the anterior jugular vein, and the retromandibular vein empties into the external jugular vein or the internal jugular vein.

Vertebral Vein (Fig. 3-18)

The vertebral vessel arises in the suboccipital region. It descends in the vertebral foramina and empties into the back of the ipsilateral brachiocephalic vein.

4

Nerve Supply to the Head and Neck

The nervous system is akin to a large network of electrical and telephone wires carrying messages to and from a control center. This system can emit and receive thousands of bits of information, integrate them, and determine the response to be made by the body.

The Peripheral and the Central Nervous Systems (Fig. 4-1)

The *central nervous system* is the control center or "home base" for the entire mechanism. It is divided into two parts: the *brain* and the *spinal cord.* The brain is housed in the cranium, and the spinal cord is housed in the vertebral column. These two parts are divided only for the purposes of description. Functionally and anatomically they are connected and are coordinated.

The *peripheral nervous system* refers to the cranial and spinal nerves, which are made up of several neurons, the cell bodies of which are located in ganglia. There are 31 pairs of spinal nerves and

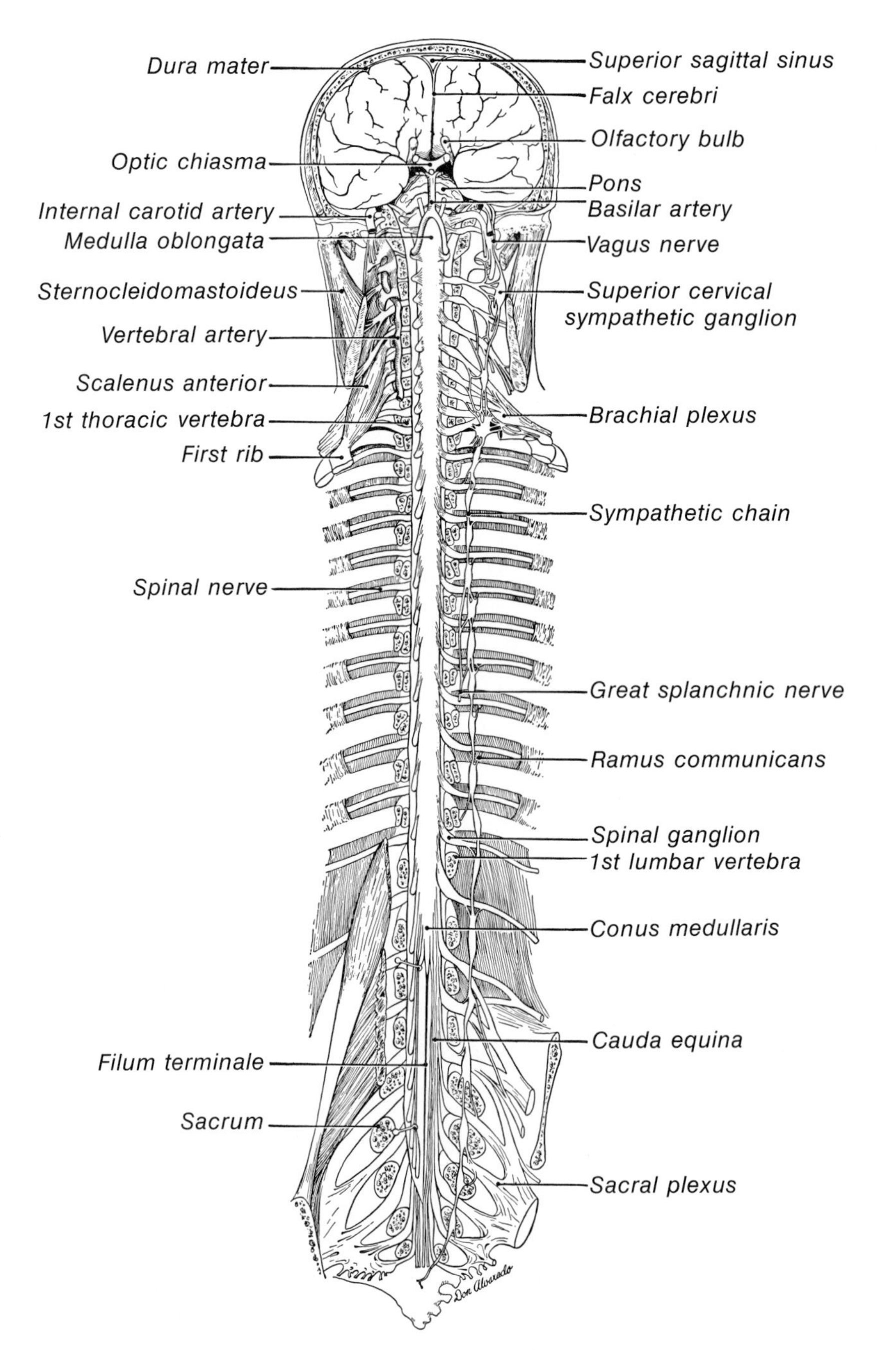

FIG. 4-1. Dissection to show the ventral or anterior aspect of the brain and spinal cord in situ. (Redrawn from Hirschfeld and Leveille in Gray's Anatomy of the Human Body, 29th ed. C. M. Goss, editor. Philadelphia, Lea & Febiger, 1973.)

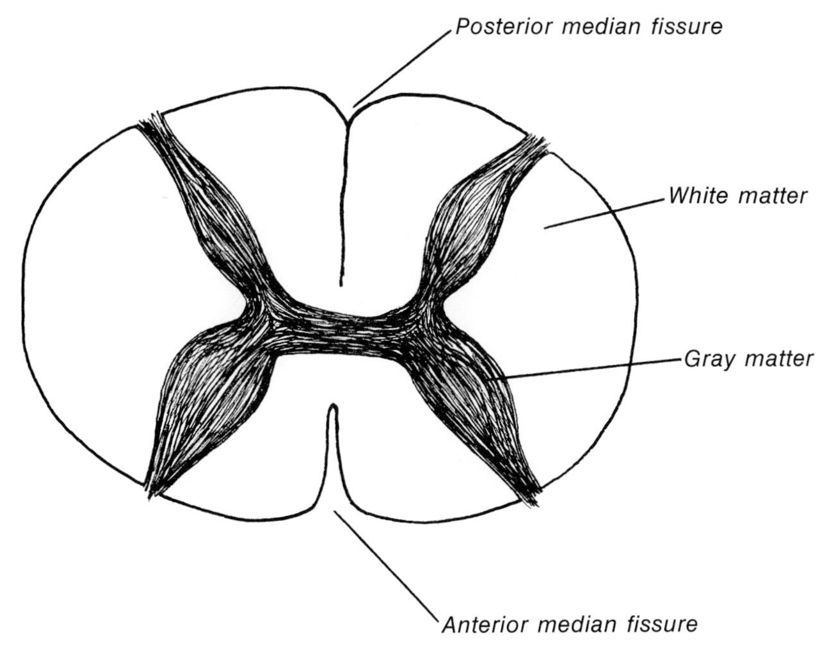

FIG. 4-2. Cross section of the spinal cord.

12 pairs of cranial nerves connecting the central nervous system to all parts of the body.

The Spinal Cord

When viewed in cross section, the spinal cord is seen to be divided into symmetrical halves by the *anterior median fissure* and the *posterior median fissure* (Fig. 4-2). Also one notes in the cross section that the substance of the cord appears to be comprised of two types of tissue. The internal, H-shaped area is the *gray matter* and contains nerve cell bodies. The outside area is the *white matter* and contains the nerve cell processes.

Nerve cells contain a cell body and several processes. Those processes conducting impulses toward the cell body are called *dendrites.* They may be very short or they may be quite long. Those processes carrying impulses away from the cell body are known as *axons.* These processes also vary greatly in length. A nerve cell body with its dendrites and axons is known as a *neuron* (Fig. 4-3).

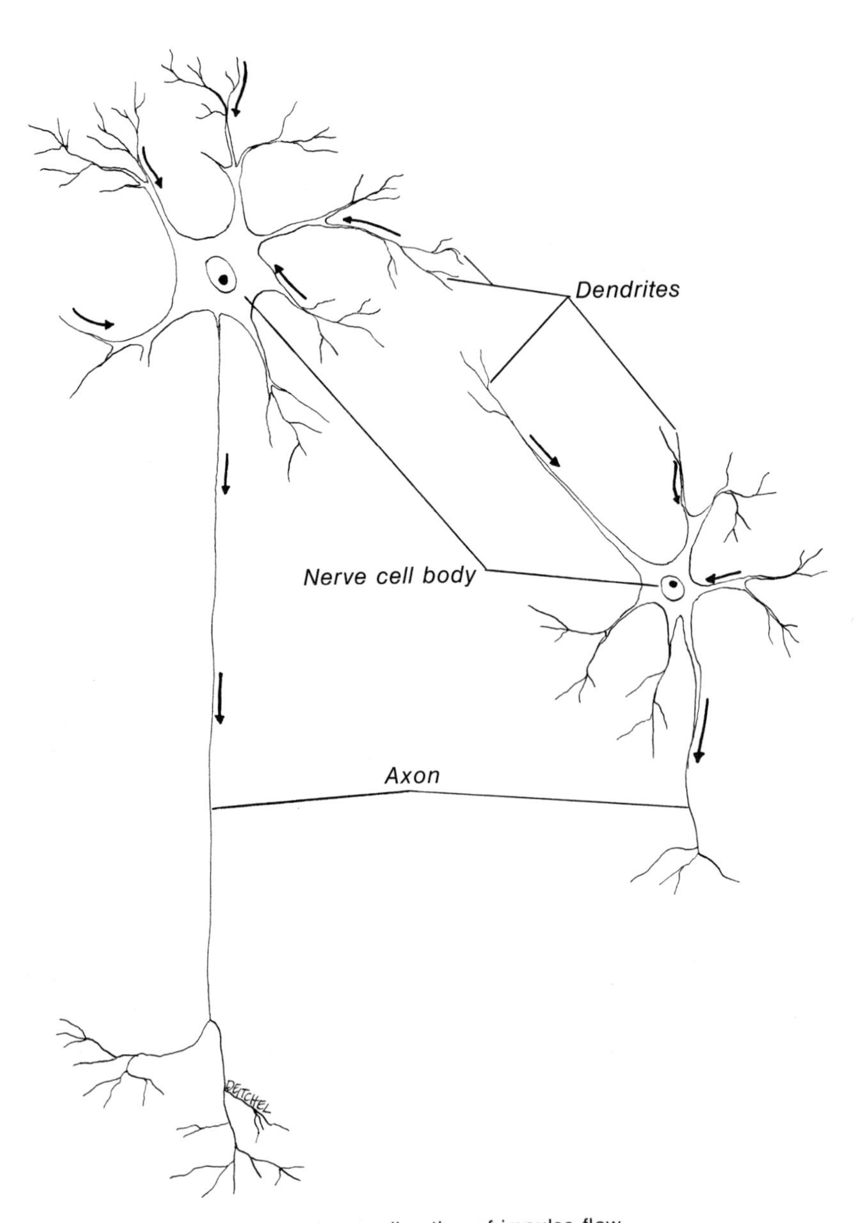

FIG. 4-3. Neurons. Arrows denote direction of impulse flow.

Emanating from the posterior aspect of each side of the cord is a branch called the *posterior* or *dorsal root* (Fig. 4-4). It contains a swelling, the *dorsal root ganglion* (spinal ganglion). The dorsal root contains only axons. From the anterior aspect of the cord is another branch: the *ventral root.* It does not contain a ganglion and is

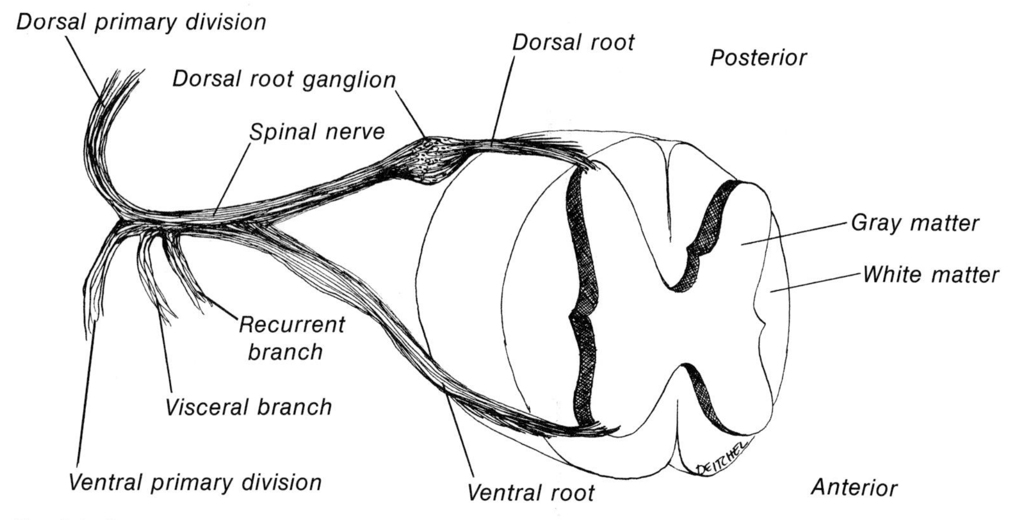

FIG. 4-4. Cross section of spinal cord with roots and ganglion.

composed only of dendrites. A *ganglion* is a collection of nerve cell bodies. The two roots join on the other side of the dorsal root ganglion to form a *spinal nerve.* The spinal nerve then redivides into a *dorsal* (posterior) *primary division* and *ventral* (anterior) *primary division.* Also two more small nerves leave this area: a *recurrent branch* to the vertebral canal and meninges and a *visceral branch* to the *sympathetic chain.* The fibers in the ventral root are *efferent fibers.* That is, they are carrying impulses *away* from the cord. Since the dorsal root is carrying impulses toward the cord, the fibers are referred to as *afferent.* In contrast, the dorsal (posterior) primary division of the spinal nerve carries impulses to and from the posterior aspect of the body and the ventral (anterior) primary division carries impulses to and from the anterior portion of the body. Thus *both* primary divisions of the spinal nerve are carrying afferent and efferent fibers (Fig. 4-5).

Nerve impulses are transmitted from one area to another along neurons. A *synapse* refers to the region of contact between processes of two neurons. It is at this region that an impulse is transmitted from one neuron to another (Fig. 4-6). There may be one or more synapses between the peripheral nervous system and the central nervous system.

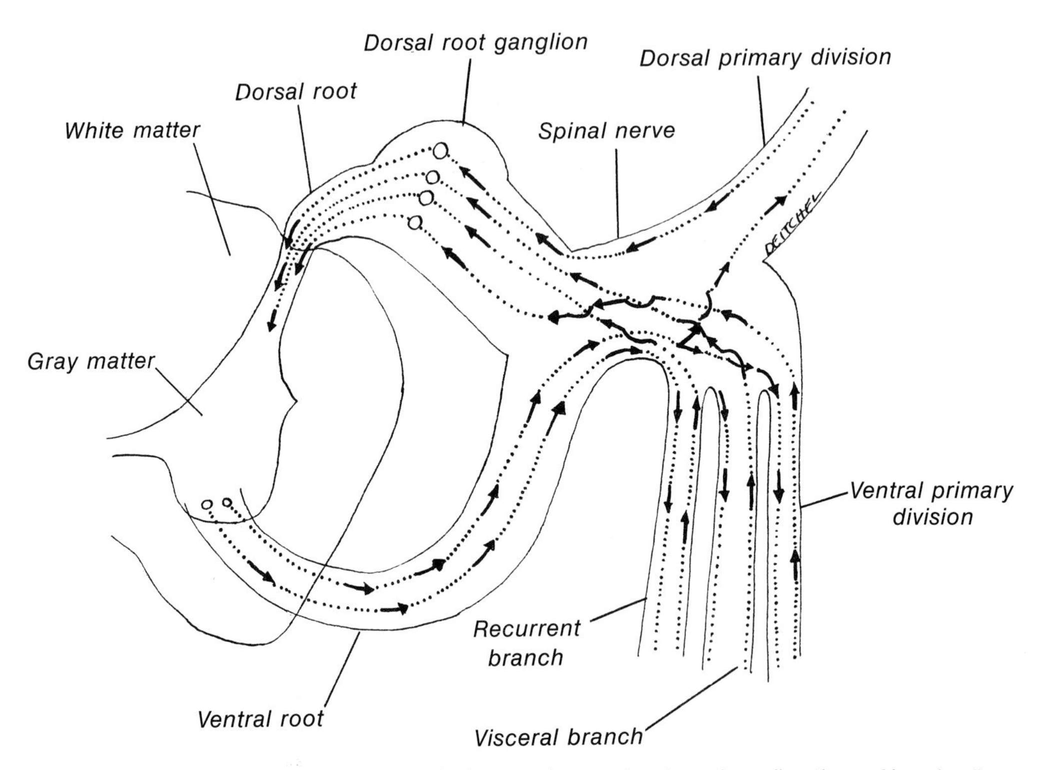

FIG. 4-5. Schematic diagram of a typical spinal nerve. Arrows denote various directions of impulse flow.

The Brain

The brain is a bulbous enlargement of one end of the spinal cord. It also consists of gray and white substance. However, for the most part, the gray matter is the *outer* layer and is known as the *cortex.* The reason for this reversal of the gray and white matter lies in the complex neuroanatomy of the brain.

The two largest areas of the brain are the right and left *cerebral hemispheres.* Various convolutions and sulci are seen traversing the cortex of these hemispheres (Fig. 4-7). They have specific names and functions. The convolutions are known as *gyri* (singular, *gyrus*). There are two major grooves on the lateral surface of the brain, the *lateral fissure* and the *central sulcus.* The cortex is estimated to contain 15 billion nerve cells. Behind and below the cerebral hemispheres is the bilobed *cerebellum,* which is responsible for muscle

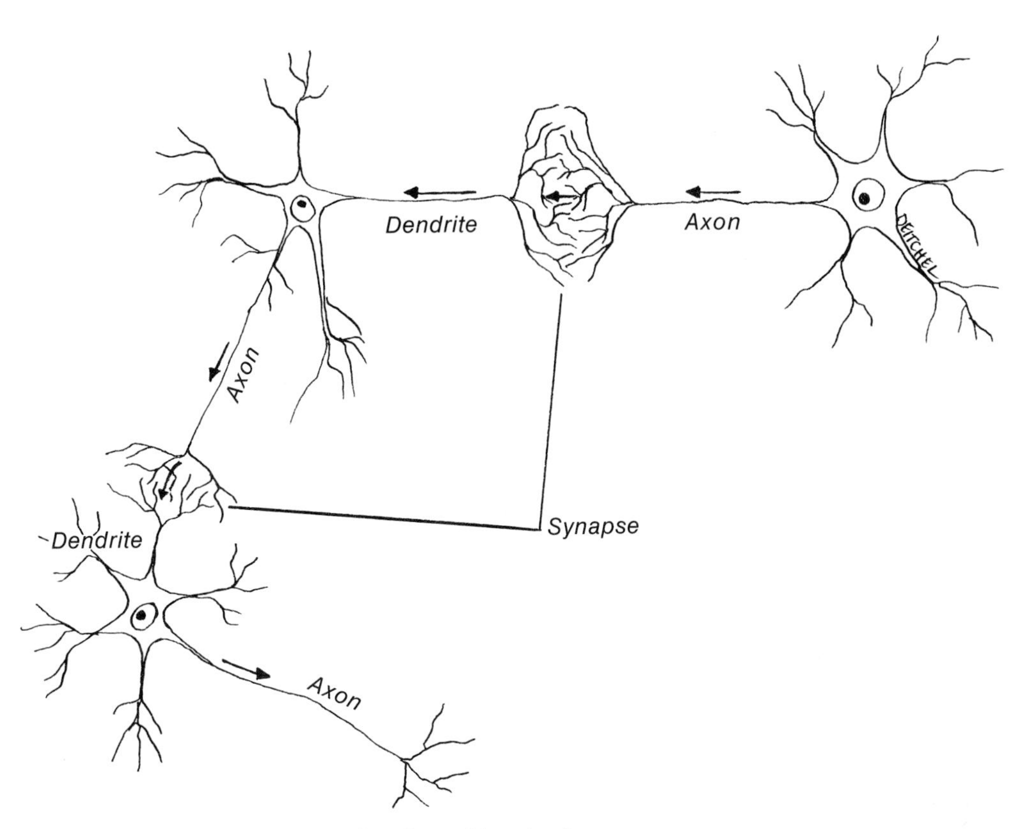

FIG. 4-6. Synapse. Arrows denote directions of impulse flow.

coordination throughout the body. It coordinates the activity of muscles and muscle groups so that physical and visceral activity is performed smoothly and with accuracy. Thus the cerebrum activates muscles and muscle groups; the cerebellum coordinates the muscles' activity.

Cerebrospinal Fluid

The brain and spinal cord are suspended in a clear liquid, which occupies a space known as the *subarachnoid space.* This space lies between the *pia mater* and the *arachnoid.* The pia mater is a delicate film of tissue surrounding the brain and spinal cord and intimately adhering to the outer surface (Fig. 4-8). It dips into the sulci and fissures. The arachnoid is a delicate membrane lying under the *dura*

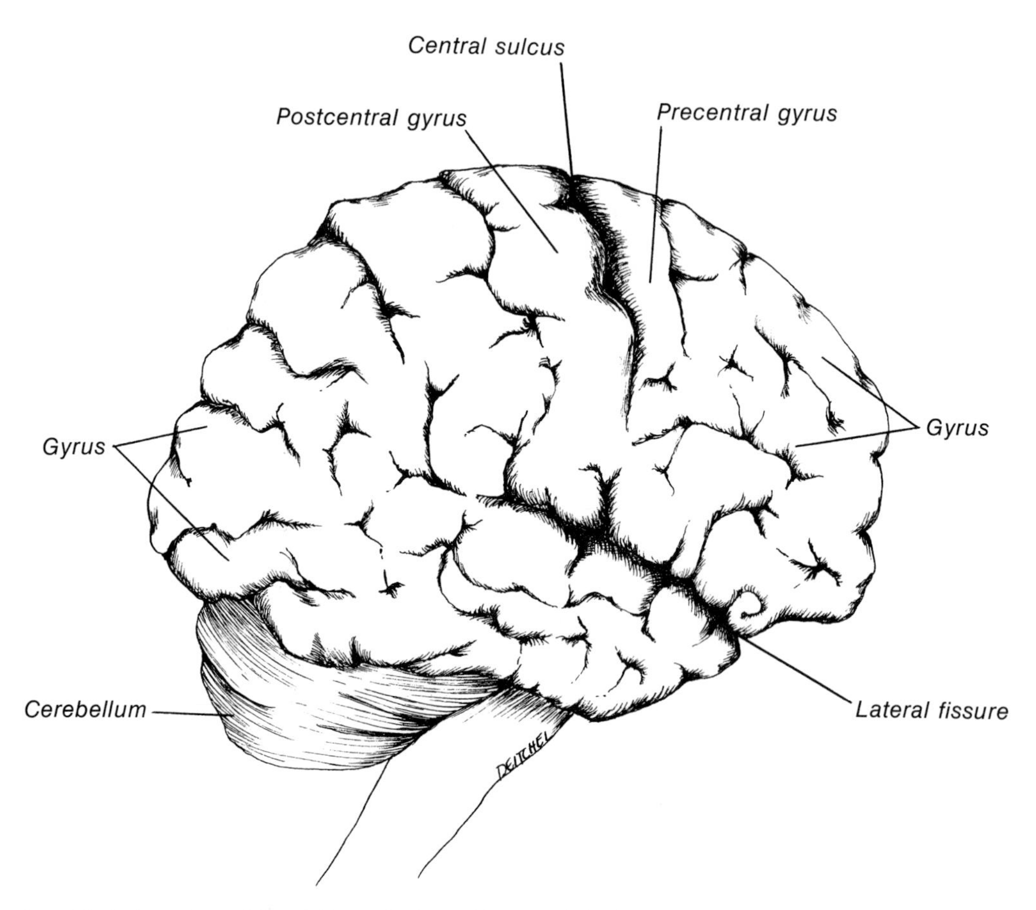

FIG. 4-7. Lateral view of the right cerebral hemisphere of the brain.

mater. The dura mater is a loose sheath around the spinal cord and brain. It is located just beneath the internal surface of the skull and the spinal canal. It is thicker and stronger than the arachnoid and the pia. These various coverings of the brain and spinal cord are collectively referred to as the *meninges.*

There is also a space or cavity within the central core of each cerebral hemisphere. This space is the *lateral ventricle.* In the central portion of the junction of the two cerebral hemispheres is a third space known as the *third ventricle,* which communicates with the lateral ventricles (Figs. 4-9; 4-10). The fourth ventricle is a widened area that lies under the cerebellum. It communicates anteriorly with the third ventricle and posteriorly and inferiorly with the *central canal*

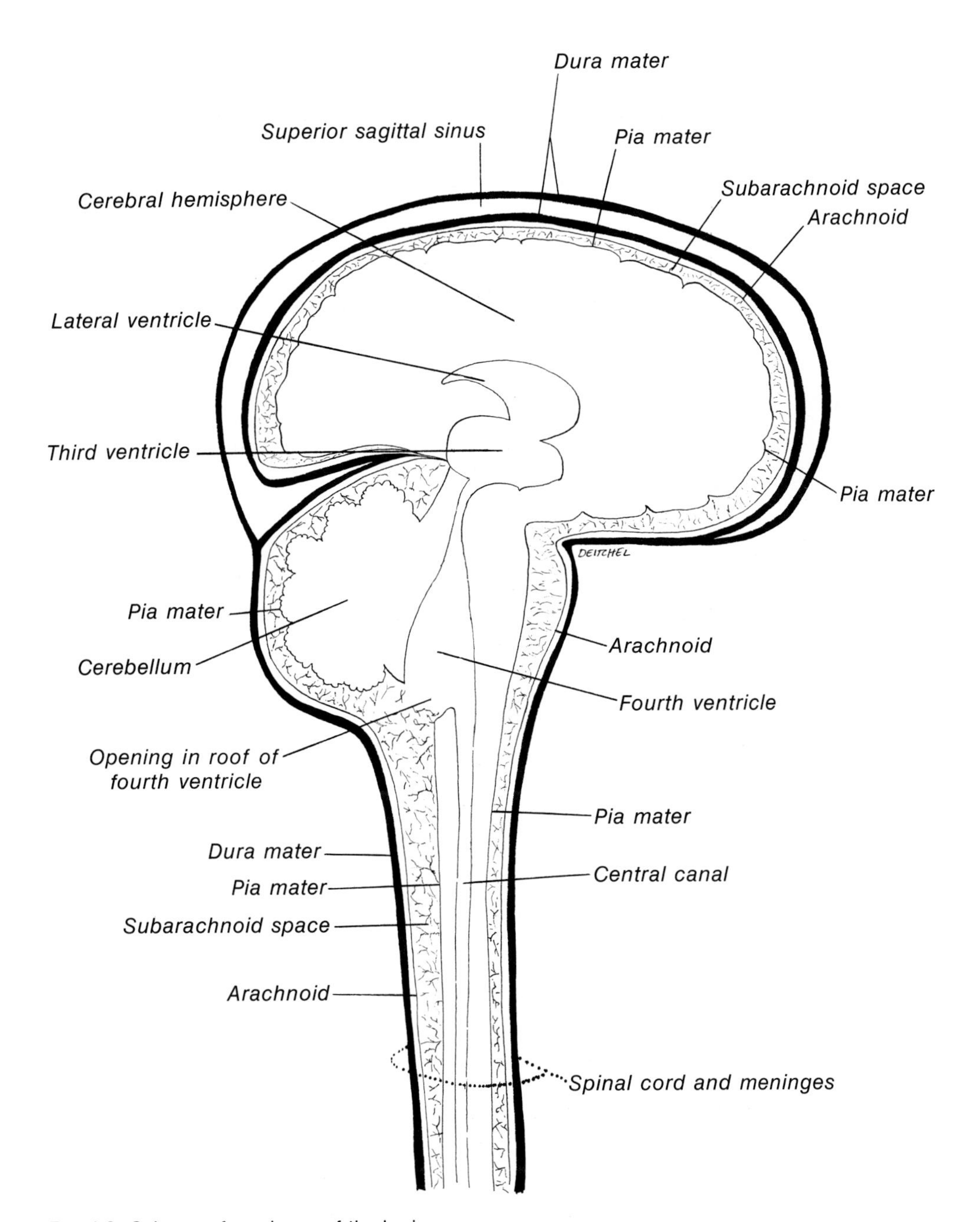

FIG. 4-8. Scheme of meninges of the brain.

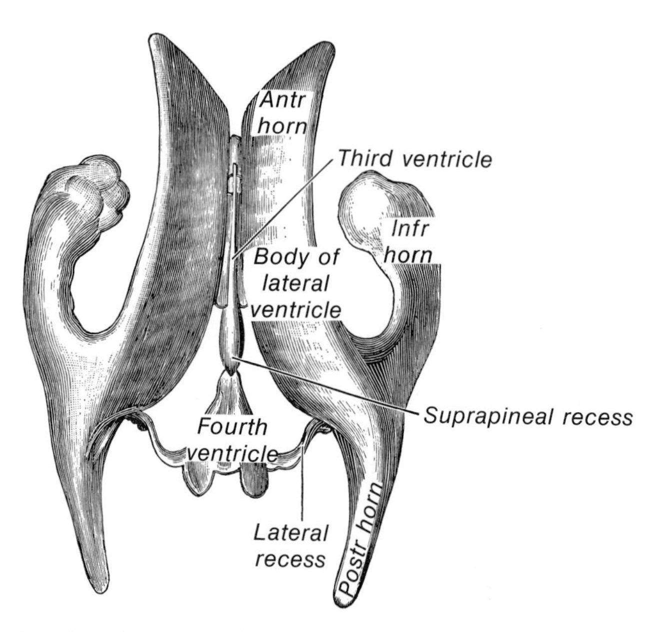

FIG. 4-9. Drawing of a cast of the ventricular cavities, viewed from above. (After Retzius in Gray's Anatomy of the Human Body, 29th ed. C. M. Goss, editor. Philadelphia, Lea & Febiger, 1973.)

of the spinal cord (Fig. 4-8). The central canal and the cerebral ventricles contain the same cerebrospinal fluid as that in the subarachnoid space.

Thus, the cerebrospinal fluid occupies the subarachnoid space, the central canal of the spinal cord, and the ventricles of the brain. It is found within *and around* the central nervous system. The total volume of this clear, watery fluid is from 90 to 150 cc in adults. It is an excellent indicator of various types of systemic and neurologic disease and is often examined to augment diagnosis of patients. A sample may be withdrawn by aspiration of a small amount via a needle introduced into the subarachnoid space of the spinal cord.

Most cerebrospinal fluid is formed in the *choroid plexuses,* tufts of capillaries projecting into the ventricles in various areas. Plasma-like fluid leaves the capillary tufts and passes into the cerebral ventricles. The fluid drains from the ventricles and central canal of the cord by way of an opening in the roof of the fourth ventricle. This opening communicates with the subarachnoid space of the cerebellum. It then diffuses out of the subarachnoid space, over the brain, and is collected by the superior sagittal sinus. From there it is returned to the blood stream (Fig. 4-8).

Cerebrospinal fluid is thought to serve three main functions. Since it communicates with venous drainage, it acts to remove metabolic waste products. It is formed from arterial capillaries and therefore acts in a nutritive capacity. Thirdly, it is an excellent shock absorber.

The Autonomic Nervous System

In contrast to those body functions over which we have conscious control, there are many body functions over which we have no control. We are able to control arm and leg movements but cannot control heart rate or pupil constriction. The autonomic nervous system supplies efferent nerves for the "automatic" activities of glands, the muscle of the cardiovascular system, and the intrinsic muscles of the eye. The mechanism is purely efferent and is

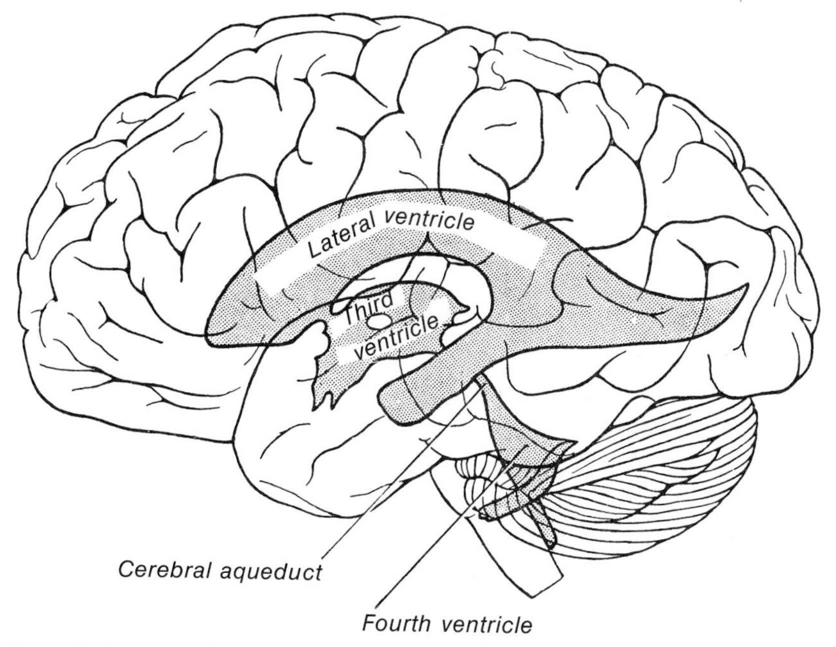

FIG. 4-10. Scheme showing relations of the ventricles to the surface of the brain. (From Gray's Anatomy of the Human Body, 29th ed. C. M. Goss, editor. Philadelphia, Lea & Febiger, 1973.)

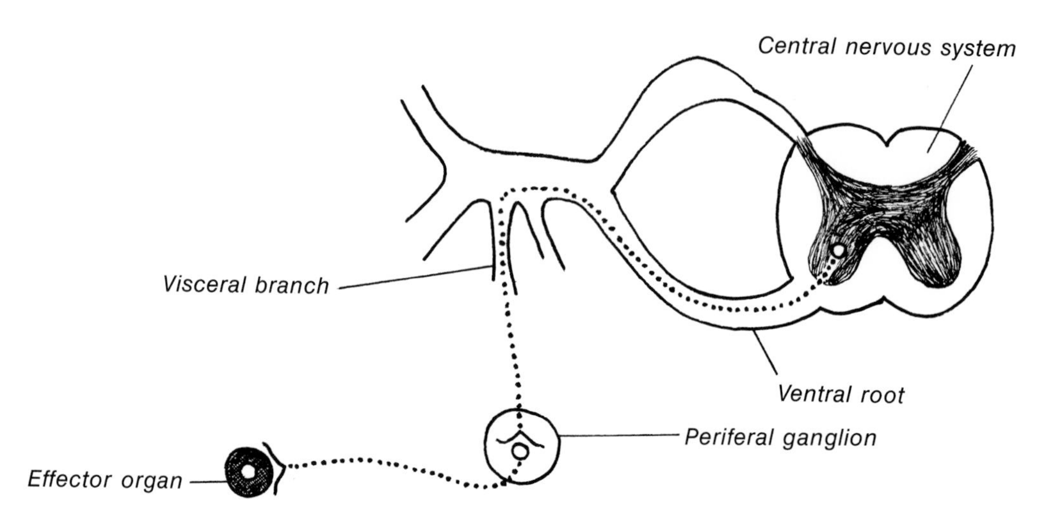

FIG. 4-11. Diagram of the two-neuron pathway.

based on a two-neuron pathway (Fig. 4-11). The first neuron originates in the central nervous system. It leaves via a spinal or cranial nerve and synapses in a ganglion distant from the central nervous system. The second neuron then goes to the effector organ.

There are two divisions of the autonomic nervous system: the *sympathetic nervous system* and the *parasympathetic nervous system.* They oppose and compliment one another.

The Sympathetic Nervous System

The sympathetic nervous system is also called the *thoracolumbar* portion, since the first order neuron exits with the spinal nerves in the thoracic or lumbar regions of the spinal column. The axons of these neurons then go to the *sympathetic trunks* and synapse with the second order neuron. The sympathetic trunks are long ganglionated chains extending along either side of the vertebral column (Figs. 4-12; 4-13). The second order neurons then go to the effector organs. Some organs of the gut are supplied by a three-neuron pathway. The second order neuron goes to a small ganglion near the effector organ where a third order neuron arises and innervates the effector organ. This nervous system is designed to increase the body's response to emergency situations; therefore, it dilates blood

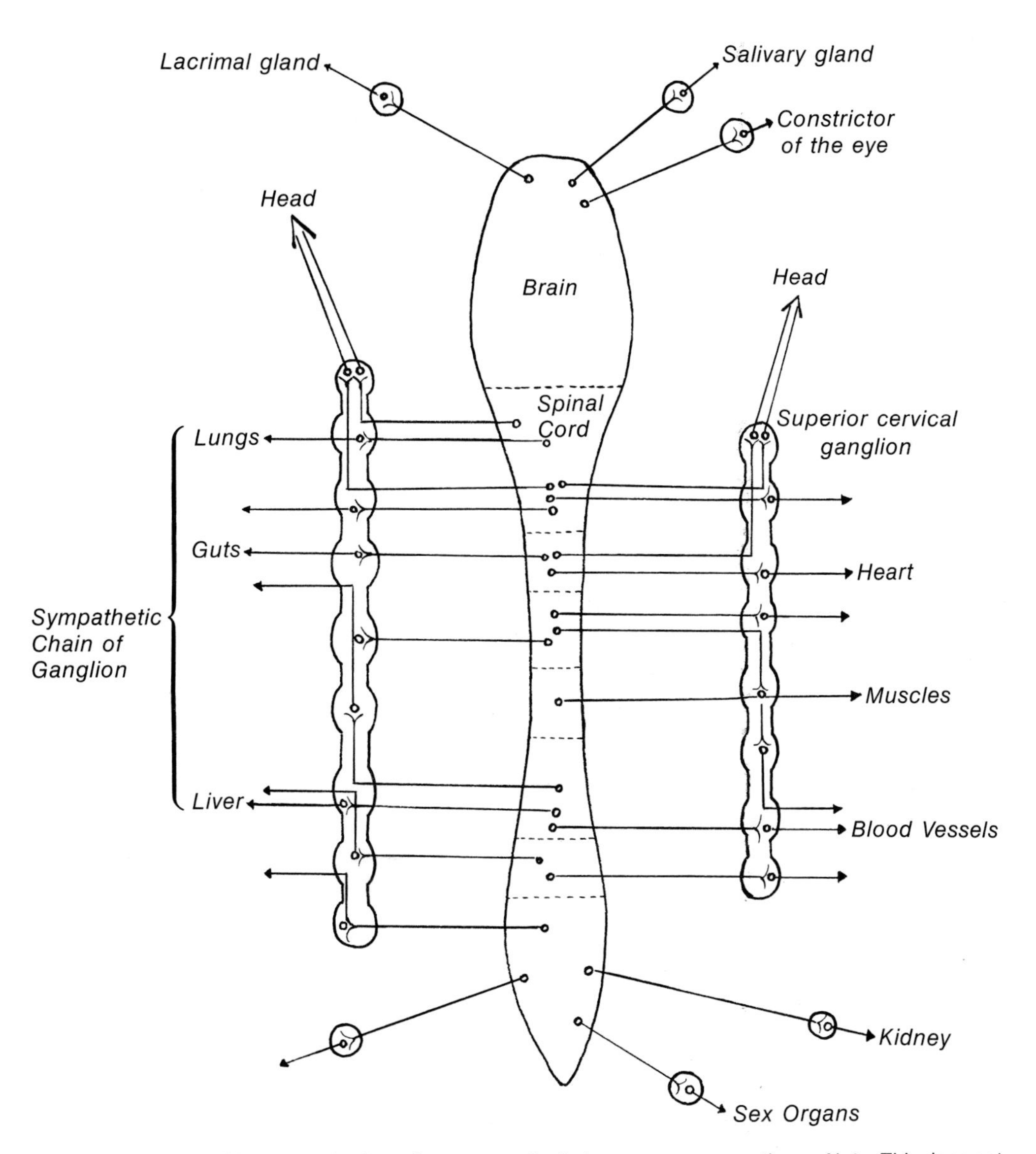

FIG. 4-12. Diagram of the sympathetic and parasympathetic two-way neuron pathway. Note: This does not accurately represent the anatomic features of size and texture.

vessels in muscles, increasing blood supply to the muscles and thus permitting that organ to function more effectively. The sympathetic nervous system also dilates the pupil of the eye and causes the heart to beat more rapidly and with more force.

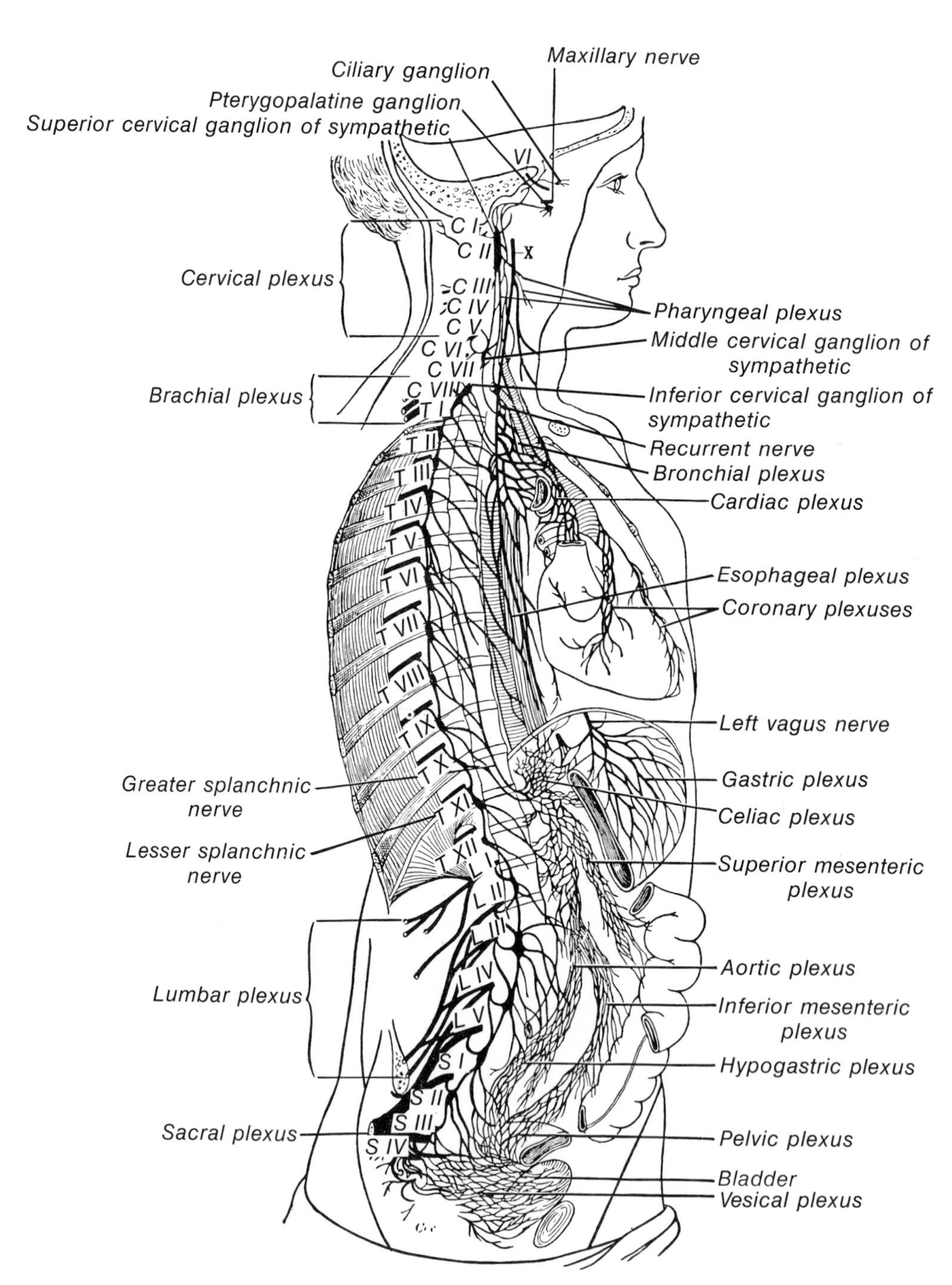

FIG. 4-13. The right sympathetic chain and its connections with thoracic, abdominal, and pelvic plexuses. (After Schwalbe in Gray's Anatomy of the Human Body, 29th ed. C. M. Goss, editor. Philadelphia, Lea & Febiger, 1973.)

All the second order neurons destined for the head and neck arise in the uppermost ganglion of the sympathetic chain. This ganglion has been named the *superior cervical ganglion* and is located at the level of the second cervical vertebra (Figs. 4-13; 4-43). The second order fibers then follow the arterial tree to the various effector organs in the head and neck.

The Parasympathetic Nervous System

The parasympathetic nervous system is termed the *craniosacral* portion of the autonomic nervous system. The first order neurons exit either with the spinal nerves in the sacral region of the vertebral column or with the cranial nerves. This system slows the heart, constricts the pupil of the eye, and increases gastric activity and salivation. Once the fibers exit from the central nervous system, they do not synapse until they are in proximity to the effector organ. A ganglion lies near the effector organ. The second order neurons arise here and by a very short fiber reach the effector organ (Fig. 4-12). There is no parasympathetic ganglion chain as in the sympathetic system.

The Cranial Nerves

The nerves of the body may be classified also by their functions. Those nerves that supply muscles and structures of the body wall are called *somatic.* Nerves to the internal organs (viscera) are termed *visceral.* Sensory fibers that carry impulses toward the brain are designated as *afferent.* Motor (or action-producing) fibers are termed *efferent.* They carry impulses away from brain to an effector.

Cranial nerves carry afferent impulses (taste, pain, proprioception), as well as efferent impulses, to muscles, glands, and blood vessels. Some nerves are highly specialized and do not carry all the components. The optic nerve, for example, carries only afferent fibers from the retina to the brain, to produce vision.

As noted before, no sympathetic fibers have first order neurons in the brain. The sympathetic second order fibers arise in the superior cervical ganglion and reach effector organs by "riding along" with

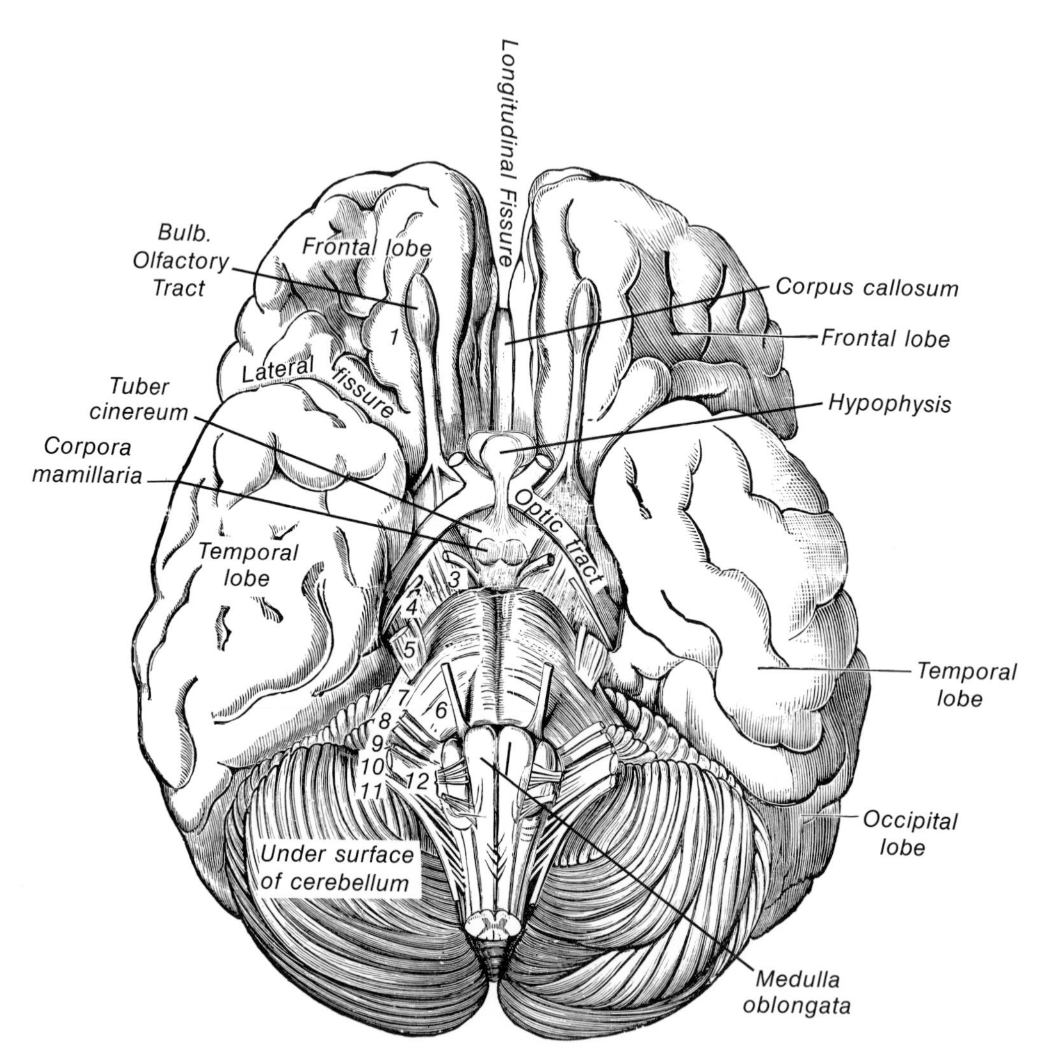

FIG. 4-14. Base of the brain. (From Gray's Anatomy of the Human Body, 29th ed. C. M. Goss, editor. Philadelphia, Lea & Febiger, 1973.)

the arteries. There are, however, several *parasympathetic* fibers originating in the brain. These are carried in 4 of the 12 pairs of cranial nerves.

The cranial nerves are released from the base of the brain (Fig. 4-14) and exit from the cranial cavity through various openings and foramina. They are designated by Roman numerals, and all are paired.

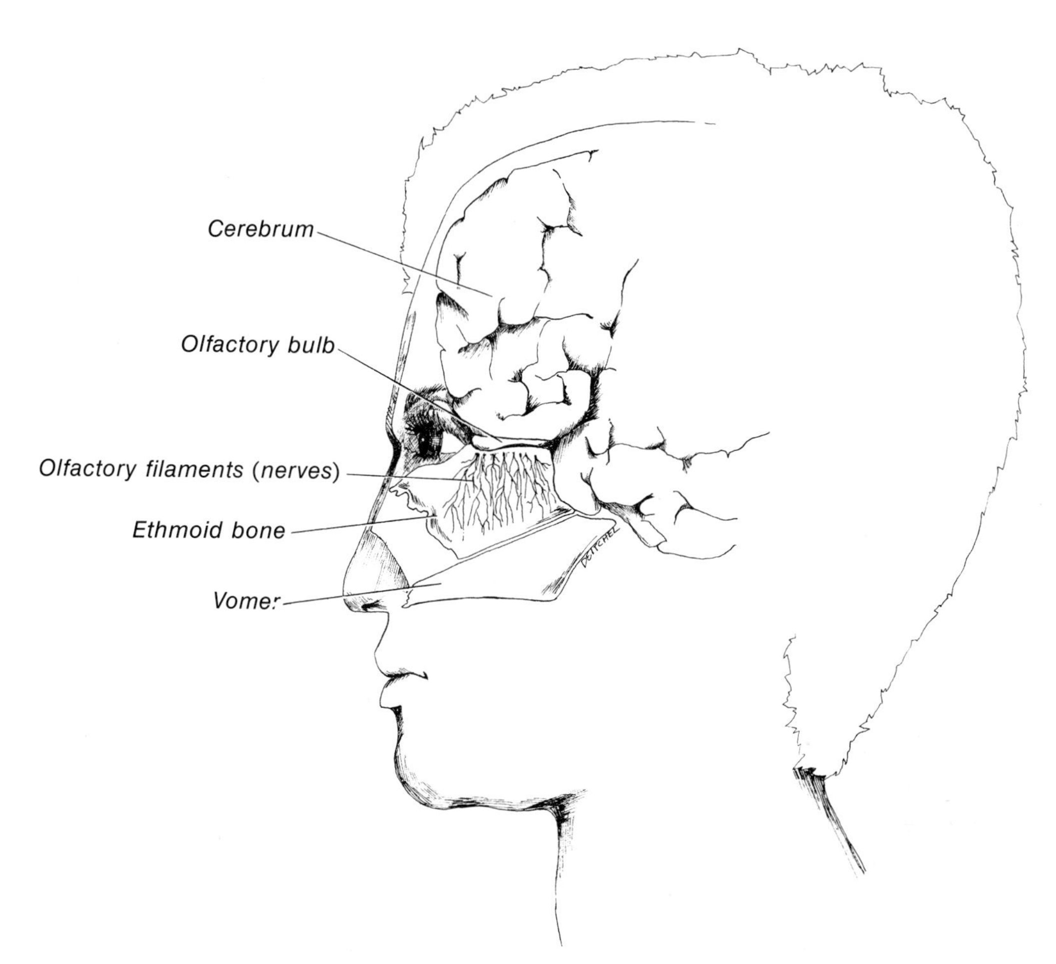

FIG. 4-15. Olfactory nerves. Note: The olfactory filaments leave the olfactory bulb and perforate the foramina of the cribriform plate of the ethmoid bone.

I. *The Olfactory Nerve* (Fig. 4-15)

The olfactory nerve is the nerve providing the sense of smell. It is afferent only. A number of bundles of nerve fibers from the superior choncha and nasal septum pass through the foramina of the cribriform plate of the ethmoid bone. They synapse in the olfactory bulb, which rests on the intracranial surface of the cribriform plate. Although the olfactory bulb tract appears to be a nerve, it more accurately is classified as part of the brain. Impulses are carried back to the brain, and the sense of smell is realized.

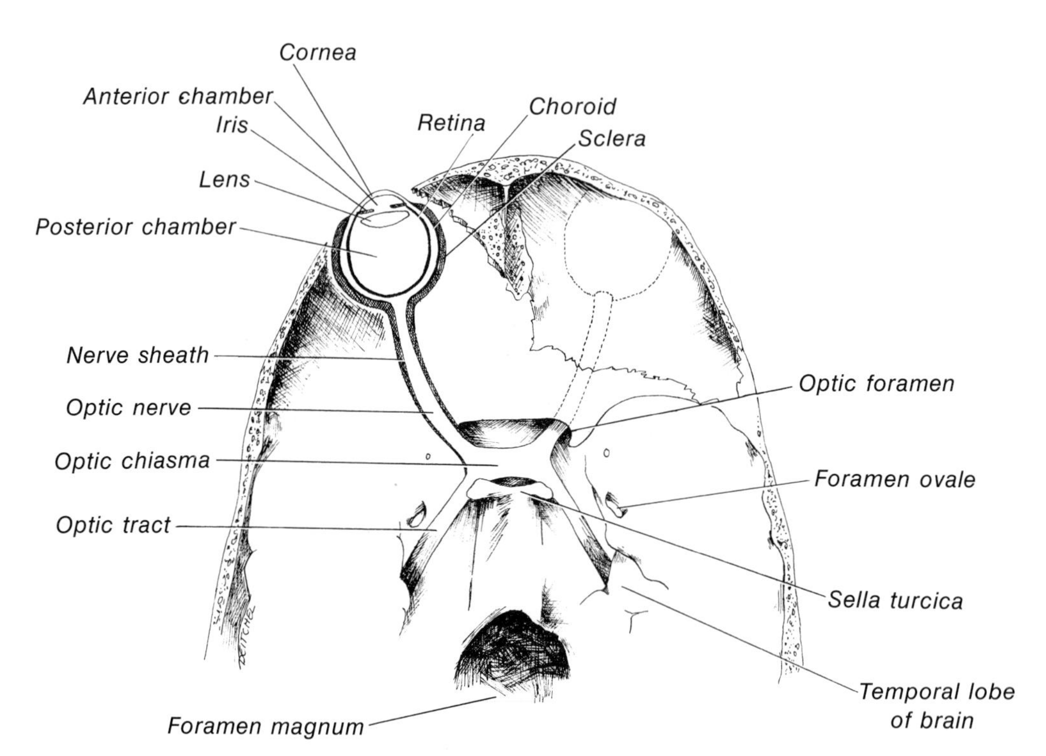

FIG. 4-16. Optic nerve viewed from above. The brain has been removed.

II. *The Optic Nerve* (Fig. 4-16)

The nerve of sight is also an afferent nerve with no efferent component. The flat, cup-shaped *retina* is the internal surface of the back of the eyeball. It converges behind the bulb of the eye and coalesces into a nerve. The nerve passes into the brain case through the optic foramen. It then courses posteriorly to a point in front of the sella turcica. Right and left optic nerves then cross and pass into the brain (Fig. 4-14). The area of crossover is the *optic chiasma.*

III. *The Oculomotor Nerve* (Fig. 4-17)

The third cranial nerve carries both afferent and efferent fibers. It supplies fibers to the extrinsic muscles of the eye with efferent motor fibers and afferent proprioception fibers. *Proprioception* is that sense of awareness of the movement and position of the body and

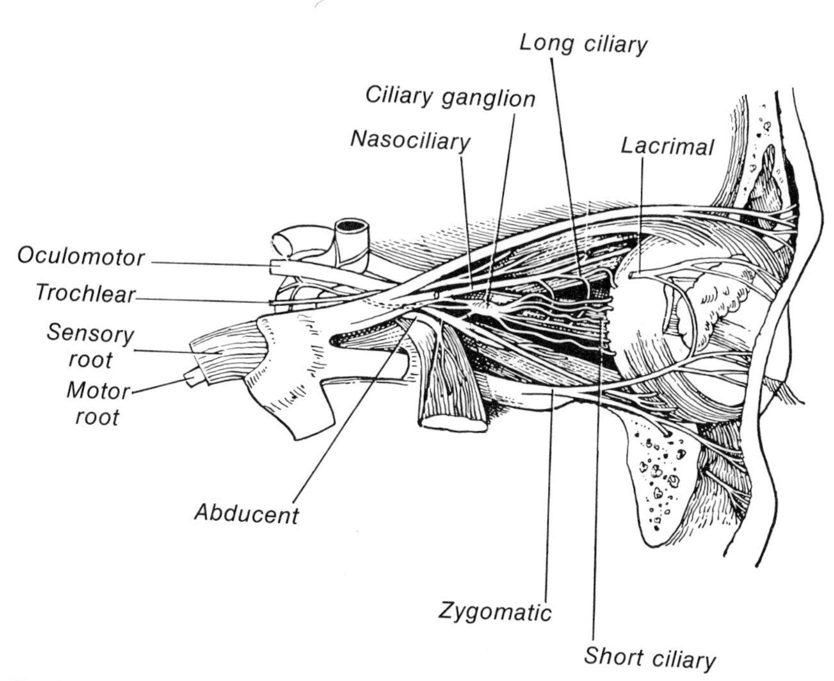

FIG. 4-17. Nerves of the orbit (side view). (From Gray's Anatomy of the Human Body, 29th ed. C. M. Goss, editor. Philadelphia, Lea & Febiger, 1973.)

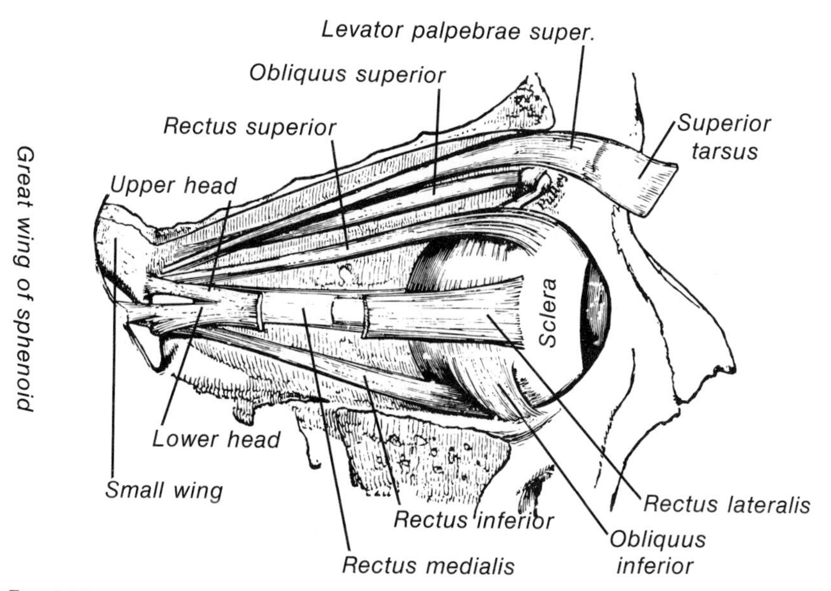

FIG. 4-18. Muscles of the right orbit. (From Gray's Anatomy of the Human Body, 29th ed., C. M. Goss, editor. Philadelphia, Lea & Febiger, 1973.)

its organs. Proprioceptors are located in many parts of the body, particularly in muscles and joints.

The oculomotor nerve also carries first order neurons of the parasympathetic nerves to the intrinsic muscles of the eye. A small ganglion, the *ciliary ganglion,* is located in the orbit, near the posterior wall. This ganglion is the location of synapse between the first order neuron and second order neuron of the parasympathetic portion of the third cranial nerve. First order neurons carried in the oculomotor nerve drop off to synapse in the ciliary ganglion. The second order neurons leave the ganglion via the tiny *short ciliary nerves* to supply the constrictor of the pupil and the ciliary muscle. The ciliary muscle acts to change the shape of the lens, making it more convex.

There are six extrinsic muscles of the eye: superior, medial, lateral, and inferior rectii; the superior and inferior oblique; and the elevator of the upper lid (Fig. 4-18). The lateral rectus and superior oblique muscles are not supplied by the oculomotor nerve. They are supplied by the sixth and fourth cranial nerves, respectively. In order for the oculomotor nerve to reach the muscles and ciliary ganglion, it passes from the cranial cavity into the orbit through the superior orbital fissure.

IV. *The Trochlear Nerve* (Figs. 4-17; 4-18; 4-19)

The trochlear nerve has only one function: to supply the superior oblique muscle of the eye. It provides both proprioceptive afferent fibers and motor (efferent) fibers to this muscle. After passing through the cavernous sinus, it leaves the brain case and enters the orbit via the superior orbital fissure.

V. *The Trigeminal Nerve*

Not only is the fifth cranial nerve the largest of the cranial nerves, but also it is the single most important nerve to members of the dental professions. It supplies almost all the pain and proprioception fibers to the face, jaws, and scalp. The trigeminal nerve also innervates muscles and carries parasympathetic fibers to the salivary glands.

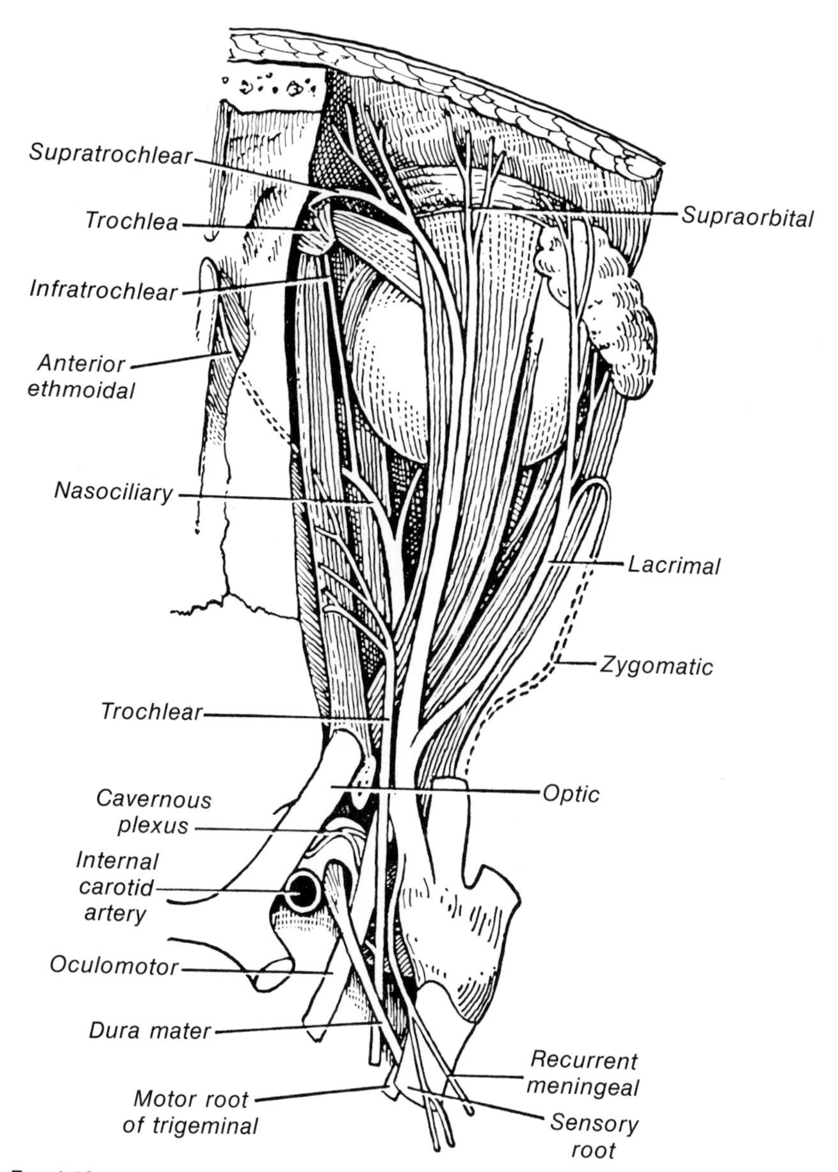

FIG. 4-19. Nerves of the orbit seen from above. (From Gray's Anatomy of the Human Body, 29th ed. C. M. Goss, editor. Philadelphia, Lea & Febiger, 1973.)

The trigeminal nerve arises by a short trunk composed of two closely adapted roots: a thin motor root and a thick sensory root (Fig. 4-14). The two roots pass anteriorly together for a short distance within the cranial cavity. The motor root then exhibits a swelling, the *semilunar (gasserian) ganglion* (Fig. 4-20). In this ganglion lie the cell bodies of the sensory fibers of the fifth cranial nerve. The sensory division then divides into three branches: the ophthalmic, the maxillary, and the mandibular (Figs. 4-17; 4-18; 4-20).

The Ophthalmic Nerve (Fig. 4-19)

The ophthalmic neve is the first division of the trigeminal nerve. It exits from the brain case and enters the orbit through the superior orbital fissure. It supplies sensory fibers to the bulb of the eye, the

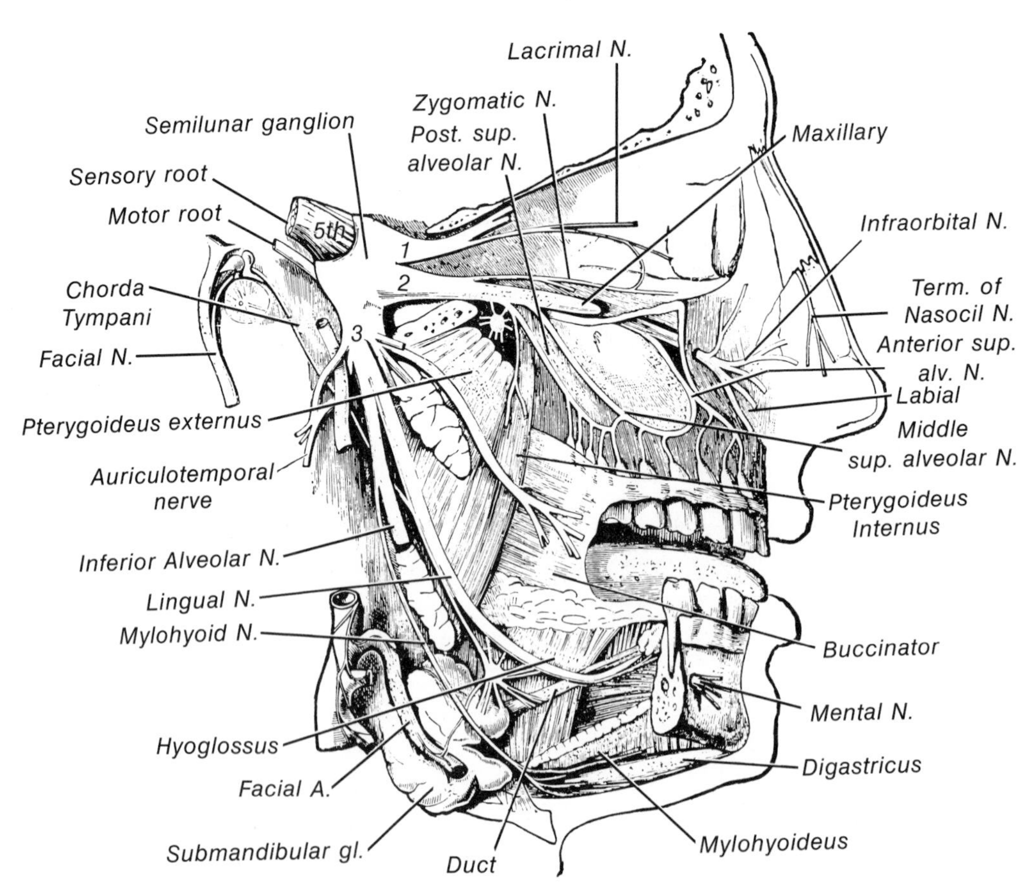

FIG. 4-20. Distribution of the maxillary (2) and mandibular (3) nerves and the submandibular ganglion. (From Gray's Anatomy of the Human Body, 29th ed. C. M. Goss, editor. Philadelphia, Lea & Febiger, 1973.)

conjunctiva, lacrimal gland, inside of the nose, and skin of the eyelids, forehead, and nose. Before entering the orbit, it divides into three branches—the lacrimal, frontal, and nasociliary—all of which pass into the orbit through the superior orbital fissure.

THE LACRIMAL NERVE. This branch supplies the lacrimal gland with sensory fibers. In addition, in the orbit, the lacrimal nerve picks up postganglionic parasympathetic fibers from the zygomaticotemporal nerve, a branch of the *second* division of the trigeminal nerve.

These postganglionic parasympathetic fibers, which are second order neurons from the *pterygopalatine* (sphenopalatine) *ganglion* (page 162), reach the gland via the lacrimal nerve. They innervate the secretory cells of the lacrimal gland, producing tears.

THE FRONTAL NERVE. This branch of the opthalmic nerve divides in the orbit into *supraorbital* and *supratrochlear nerves* to the upper lid and forehead.

THE NASOCILIARY NERVE (Figs. 4-17; 4-19). This branch of the ophthalmic nerve courses toward the medial orbital wall, passes through the ethmoid bone, and reenters the cranial cavity just above the cribriform plate. It then pierces the roof of the nasal cavity, passes into the nose, and supplies the mucosa of the nose. It even sends branches between the cartilage and bone of the nose to supply the skin on the side of the nose (Fig. 4-20).

The Maxillary Nerve (Fig. 4-20)

The second division of the trigeminal nerve leaves the cranial cavity through the foramen rotundum (Figs. 1-12; 4-22). It passes into the pterygopalatine fossa, where it ramifies into several branches. The maxillary nerve is sensory, although some of its branches carry parasympathetic motor fibers.

THE ZYGOMATIC NERVE (Fig. 4-20). This nerve leaves the pterygopalatine fossa and enters the orbit through the inferior orbital fissure. It carries postganglionic parasympathetic motor fibers which it has acquired from the *pterygopalatine ganglion* in the pterygopalatine fossa. While in the orbit, it divides into the *zygomaticofacial* nerve and the *zygomaticotemporal* nerve. The zygomaticofacial nerve pierces the body of the zygoma and supplies sensory fibers to the

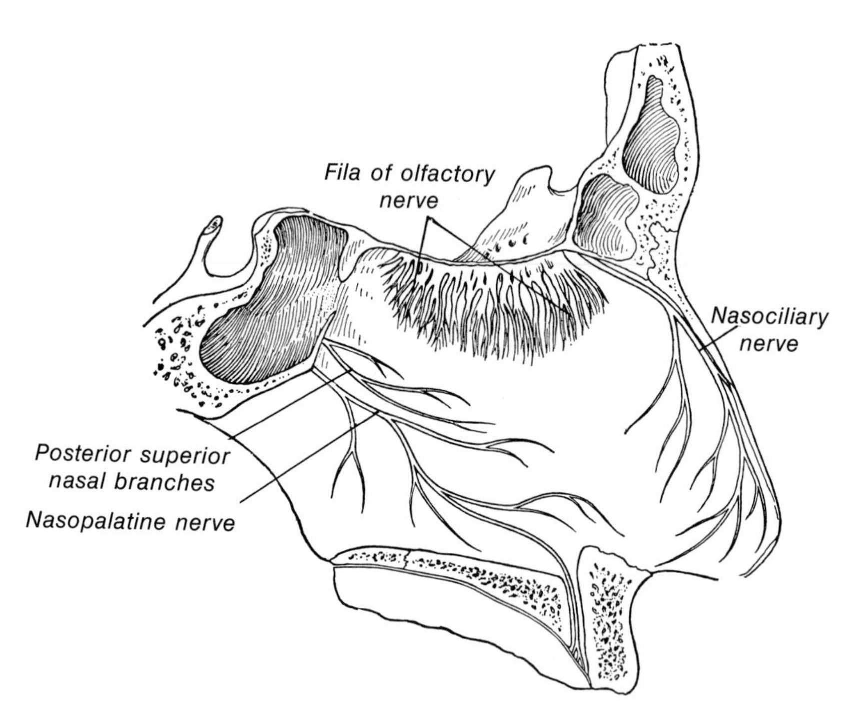

FIG. 4-21. The nerves of the right side of the septum of the nose. (From Gray's Anatomy of the Human Body, 29th ed. C. M. Goss, editor. Philadelphia, Lea & Febiger, 1973.)

skin of the cheek. The zygomaticotemporal nerve reaches the skin and side of the forehead by piercing the lateral orbital wall near the sphenozygomatic suture. The zygomaticotemporal nerve also carries the parasympathetic fibers from the zygomatic nerve from which it branched. While in the orbit, the zygomaticotemporal nerve releases these parasympathetic fibers to the lacrimal nerve. The lacrimal nerve, a branch of the ophthalmic nerve, then carries these parasympathetic fibers to the lacrimal gland (Fig. 4-23).

THE INFRAORBITAL NERVE (Fig. 4-20). The infraorbital nerve in the posterior part of the orbit continues the course of the maxillary nerve. It passes out of the pterygopalatine fossa, laterally and anteriorly, towards the infraorbital fissure. It courses in the floor of the orbit through the infraorbital groove. The groove is roofed over further anteriorly and becomes the infraorbital canal. Just before the infraorbital nerve enters the infraorbital groove, it releases the *posterior superior alveolar nerves.* These branches cross the maxillary tuberosity and supply the molars and buccal gingiva. They also supply the lining of the maxillary sinus.

Within the infraorbital canal, the middle *superior alveolar nerves* may or may not be released. If present, they run downward in the lateral wall of the sinus and supply the premolars and associated buccal gingiva. They communicate with the posterior superior alveolar nerves and the anterior superior alveolar nerves.

Just before the infraorbital nerve exits from the infraorbital canal through the infraorbital foramen, it releases the *anterior superior alveolar* nerves. The anterior superior alveolar nerves drop to the anterior teeth and labial gingiva via the internal surface of the anterior wall of the maxillary sinus. The infraorbital nerve then exits through the infraorbital foramen and supplies the lower lid, the side of the nose, the upper lip, and the anterior surface of the cheek.

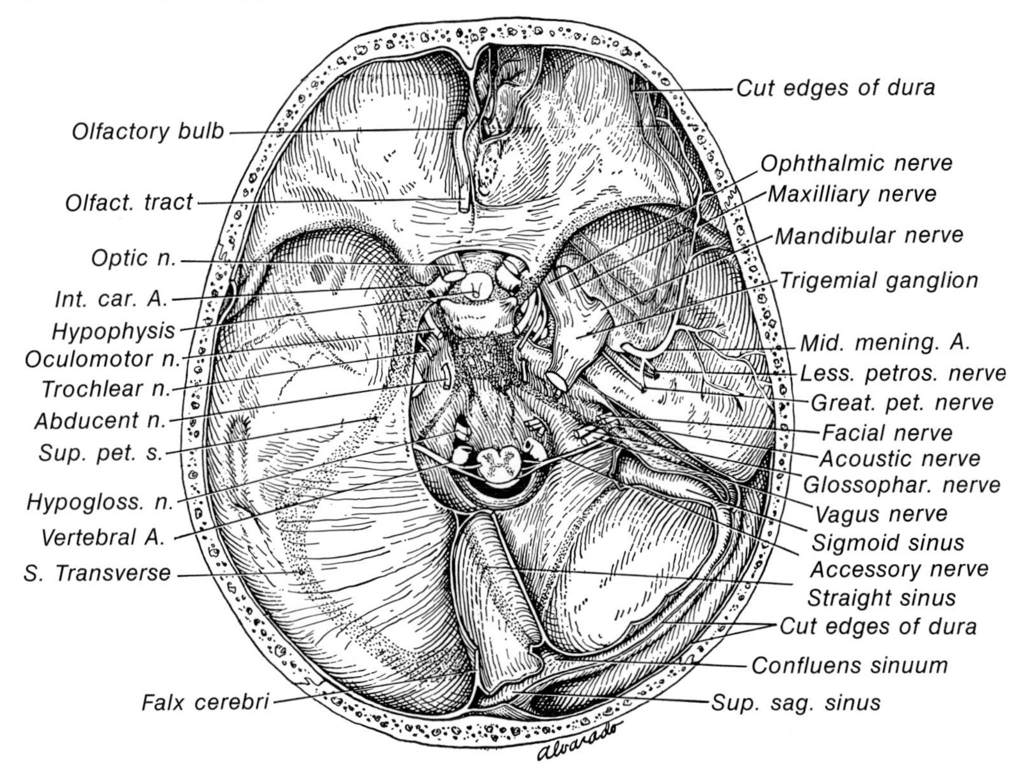

FIG. 4-22. Interior of the base of the skull showing dura mater, dural sinuses, and exit cranial nerves. (Redrawn from Tondury in Gray's Anatomy of the Human Body, 29th ed. C. M. Goss, editor. Philadelphia, Lea & Febiger, 1973.)

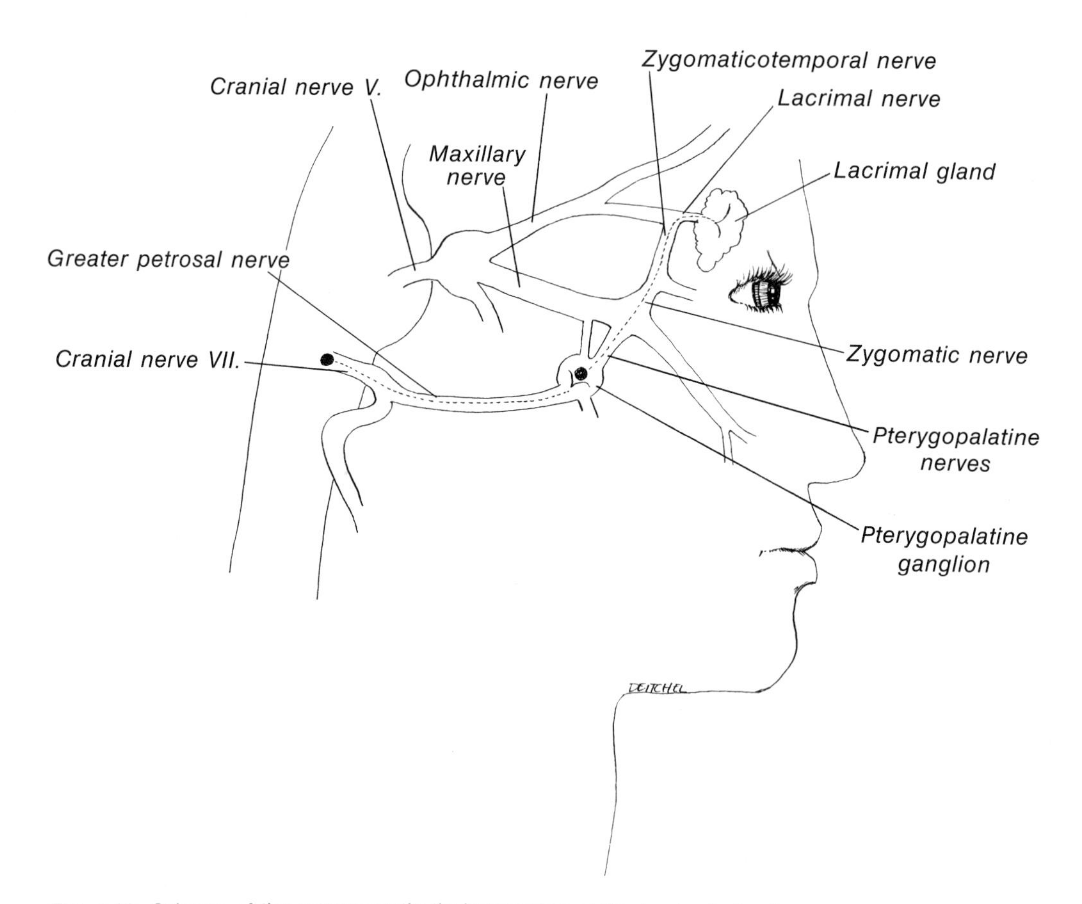

FIG. 4-23. Scheme of the parasympathetic innervation of the lacrimal gland.

PTERYGOPALATINE NERVES (Fig. 4-23; 4-24). In the pterygopalatine fossa, one to five short branches drop off the maxillary nerve to the pterygopalatine (sphenopalatine) ganglion. At the ganglion area, ramification takes place, and the pterygopalatine branches emerge. They pick up postganglionic parasympathetic fibers and carry them to the nasal and oral cavities (Fig. 4-29). Some postganglionic fibers from this ganglion pass into the zygomatic nerve and supply the lacrimal gland (Fig. 4-20). The rest pass into the pterygopalatine branches.

The sensory fibers of the pterygopalatine nerves have only a topographic relation to the pterygopalatine ganglion. No trigeminal sensory fibers synapse in this ganglion.

Nasal Branches (Fig. 4-24). These fibers enter the nasal cavity through the sphenopalatine foramen. They carry sensory fibers from the gasserian ganglion, and they carry second order parasympathetic fibers from the pterygopalatine ganglion to the glands and mucous membrane over the conchae. One long fiber emerges and passes over the septum. It is the *nasopalatine nerve,* which courses as far forward as the nasopalatine canal. It traverses through the canal and enters the mouth through the nasopalatine foramen. The nasopalatine nerve supplies the oral mucosa and minor salivary glands behind the maxillary anterior teeth. (See also Fig. 4-21.)

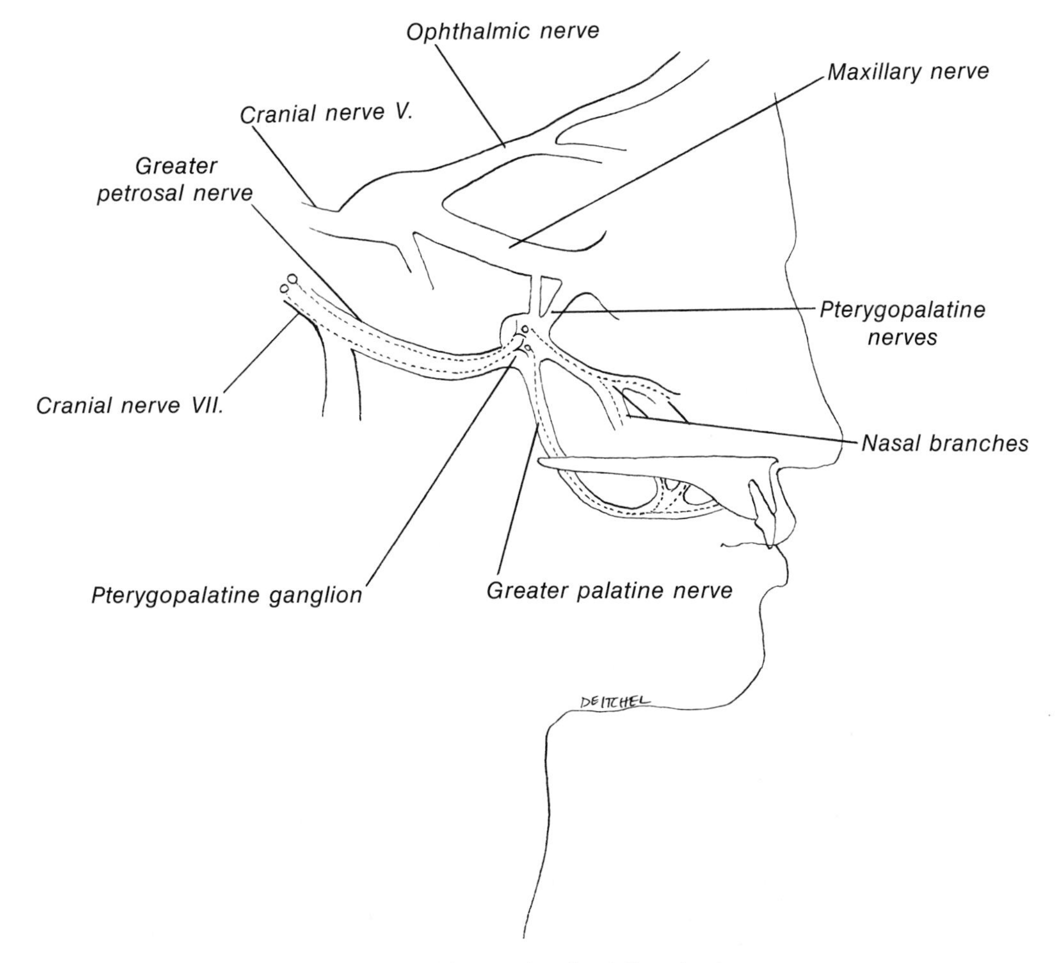

FIG. 4-24. Parasympathetic innervation of the nasal and palatine glands.

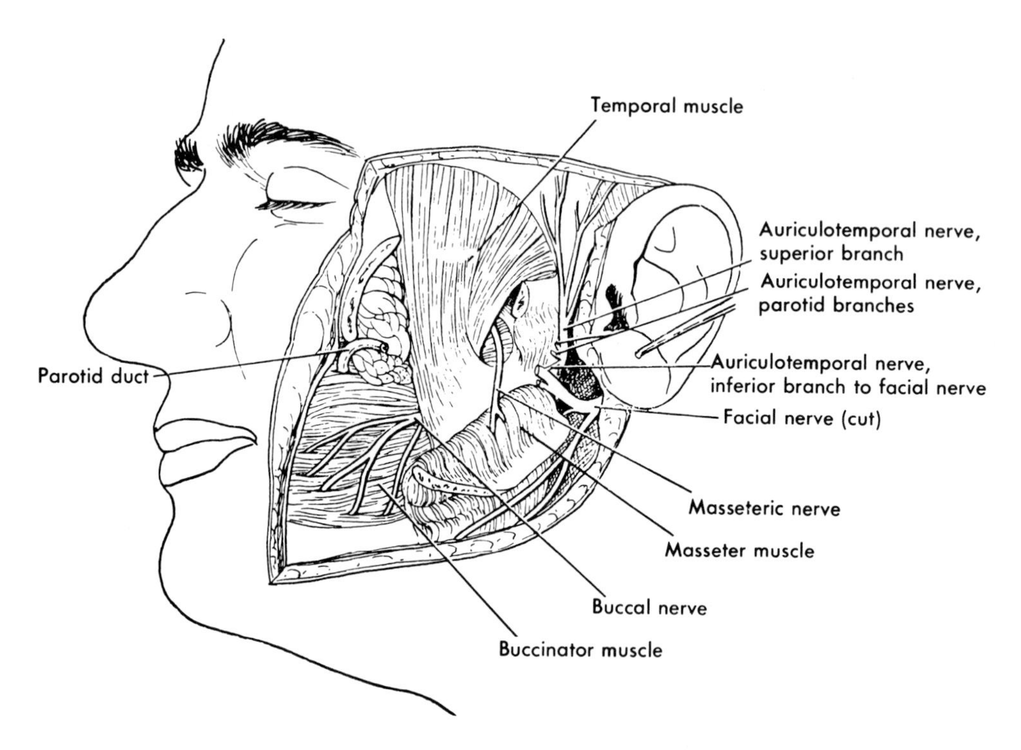

FIG. 4-25. Mandibular nerve, superficial layer. Zygomatic arch and masseter muscle reflected downward. (Modified and redrawn after Sicher and Tandler: Anatomie für Zahnärzte. From Sicher, H., and DuBrul, E. L.: *Oral Anatomy,* 5th ed. St. Louis, C. V. Mosby, 1970.)

Greater Palatine Nerve (Fig. 4-24). This branch emerges from the ganglion and passes downward through the pterygopalatine canal. It carries sensory fibers and postganglionic parasympathetic fibers to the glands of the palate, as well as afferent taste fibers that reach the ganglion on their way to the greater petrosal nerve (page 145). Taste fibers do not synapse in the pterygopalatine ganglion. The greater palatine nerve courses onto the hard palate through the greater palatine foramen and turns horizontally forward. It supplies the glands, taste buds, and mucosa of the palate as far forward as the cuspid.

While in the pterygopalatine canal, the greater palatine nerve releases some *lesser palatine nerves.* These branches emerge on the palate through the *lesser palatine foramina* to supply the glands, taste buds, and mucosa of the soft palate.

The glands of the nasal and oral cavities receive their second order parasympathetic neurons by way of the various branches of the pterygopalatine nerve(s). The first order neurons, which synapse with the second order neurons in the pterygopalatine ganglion, are associated with the facial nerve. Thus, the parasympathetic nerve supply to the glands of the nose and palate has a twofold relation: first order neurons originate with the facial nerve and second order neurons are carried in branches of the trigeminal nerve (Fig. 4-24).

The Mandibular Nerve (Figs. 4-20; 4-29)

The mandibular nerve is the third and largest division of the trigeminal nerve. It also is the only division to carry fibers from the motor root to the skeletal muscle. Both roots leave the brain case through the *foramen ovale* (Figs. 1-9; 1-12; 4-22) and unite into a short main trunk. The nerve then ramifies as it lies in the infratemporal fossa, deep to the lateral pterygoid muscle. Most of the motor nerves to the muscles of mastication are released here (Figs. 4-25; 4-26; 4-27).

MEDIAL PTERYGOID NERVE. This nerve supplies the medial pterygoid muscle and enters it close to the pterygoid process.

MASSETERIC NERVE. It passes laterally through the sigmoid notch and enters the masseter muscle on the deep surface.

DEEP TEMPORAL NERVES. Usually one or two nerves leave the mandibular trunk near the origin of the masseteric nerve. They pass laterally through the two heads of the external pterygoid muscle and enter the temporalis muscle on the deep surface.

LATERAL PTERYGOID NERVE. This nerve arises with the other nerves to the muscles of mastication, passes laterally and inferiorly, and enters the deep surface of the lateral pterygoid muscle.

THE BUCCAL NERVE (Long Buccal Nerve). The buccal nerve sometimes carries a few motor fibers to the temporalis and lateral pterygoid muscles. After releasing these fibers, it passes laterally, between the heads of the lateral pterygoid muscle. It courses down and forward and crosses the anterior border of the vertical ramus of

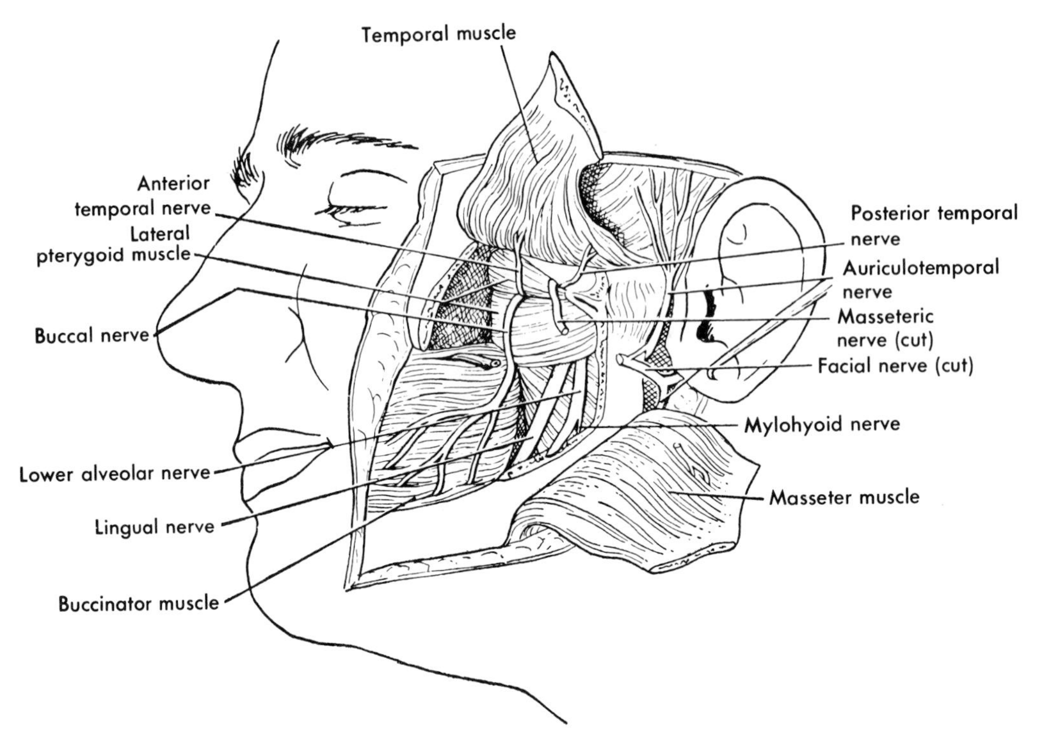

FIG. 4-26. Mandibular nerve, second layer. Temporal muscle and coronoid process reflected upward. (Modified and redrawn after Sicher and Tandler: Anatomie für Zahnärzte. From Sicher, H. and DuBrul, E. L.: Oral Anatomy 5th ed. St. Louis, C. V. Mosby, 1970.)

the mandible. As it reaches the buccinator muscle, it turns forward and lies on the lateral surface. The buccal nerve then ramifies, and the branches pierce the buccinator muscle and supply the buccal mucosa of the cheek, buccal gingiva of the mandibular molars, and sometimes a portion of the lip mucosa. Branches supply the skin of the cheek.

THE LINGUAL NERVE (Fig. 4-28). This sensory nerve branch of the mandibular nerve passes inferiorly, deep to the ramus of the mandible and laterally to the internal pterygoid muscle. It picks up the *chorda tympani nerve* above the level of the inferior alveolar foramen and then passes downward in the pterygomandibular space. When it reaches the level of the occlusal plane of the teeth, the lingual nerve turns forward under the pterygomandibular raphe and lies at the posterior border of the mylohyoid muscle (Fig. 4-29). Here, the chorda tympani fibers drop off to the *submandibular ganglion.* From the ganglion, after synapse, second order neurons pass into the submandibular salivary gland. Other second order neurons reenter the lingual nerve and are carried to the sublingual gland and the tongue. The taste fibers from the tongue to the central nervous system do not synapse in the ganglion.

The chorda tympani nerve originates with the facial nerve and carries taste fibers from the tongue and preganglion parasympathetic fibers to the submandibular ganglion. It is both afferent (taste) and efferent (salivary).

From the area of the submandibular ganglion, the lingual nerve courses anteriorly, in the floor of the mouth to the tip of the tongue. It supplies sensory fibers to the tongue, floor of the mouth, and lingual gingiva.

INFERIOR ALVEOLAR NERVE (Figs. 4-20; 4-28). This nerve descends with the lingual nerve into the pterygomandibular space. It carries

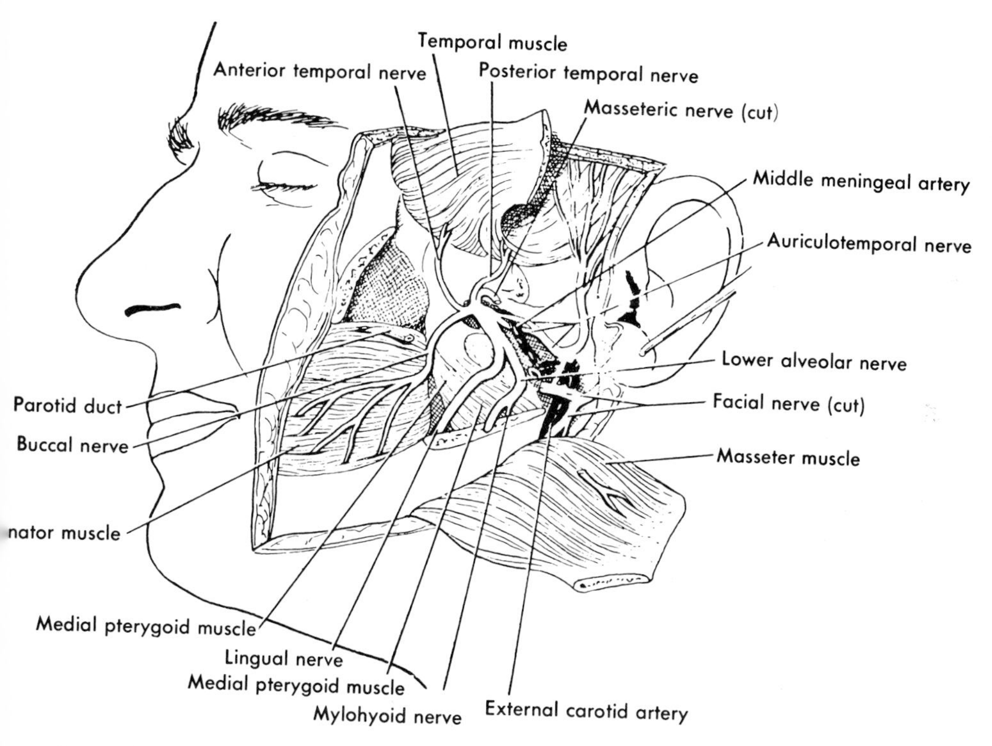

FIG. 4-27. Mandibular nerve, deep layer. Lateral pterygoid muscle and condylar process removed. (Modified and redrawn after Sicher and Tandler: Anatomie für Zahnärzte. From Sicher, H. and DuBrul, E. L.: Oral Anatomy, 5th ed. St. Louis, C. V. Mosby, 1970.)

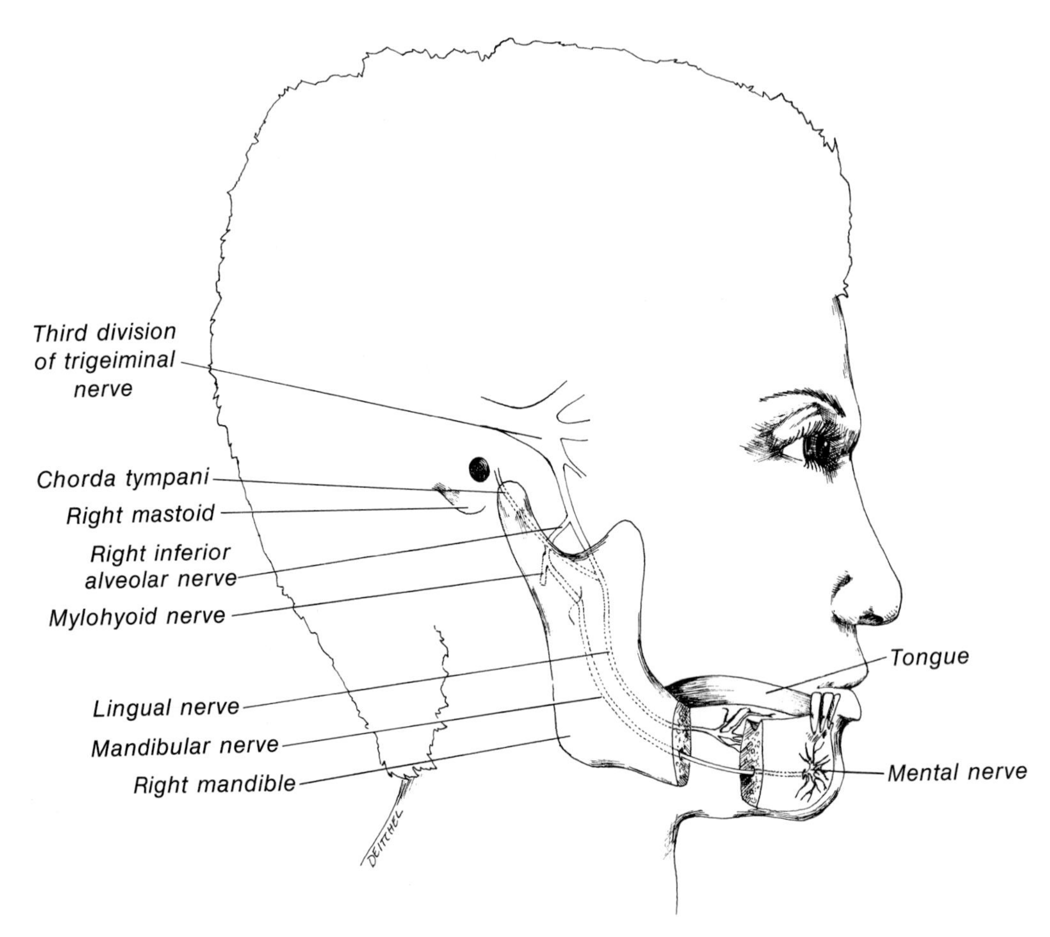

FIG. 4-28. Diagram of inferior alveolar nerve, chorda tympani, and lingual nerve.

some motor fibers, and just before it enters the inferior alveolar foramen it releases these fibers as the *mylohyoid nerve.* The mylohyoid nerve then makes its way to the mylohyoid muscle and the anterior belly of the digastric muscle.

The inferior alveolar nerve, after giving off the mylohyoid nerve, enters the inferior alveolar canal through the inferior alveolar foramen. It passes through the canal, supplying the bone and roots of the mandibular teeth. At the mental foramen, it divides into the *incisive* and *mental nerves.* They follow their associated arteries. The incisive nerve supplies the anterior teeth and labial gingiva; the mental nerve supplies the chin and lower lip.

THE AURICULOTEMPORAL NERVE (Figs. 4-25; 4-26; 4-27). This nerve turns backward from the trunk of the mandibular nerve, arising shortly after the mandibular nerve exits from the foramen ovale. It splits, encircles the middle meningeal artery, fuses, and runs posteriorly (Fig. 4-20). When it reaches the rear of the neck of the mandible, it turns up and accompanies the *superficial temporal artery.* It then passes under the parotid gland and courses over the root of the zygomatic arch. The auriculotemporal nerve supplies the skin of the ear and temporal region with sensory fibers. It also supplies sensation to the temporomandibular joint, sends a few fibers to the zygomatic portion of the cheek, and carries postganglionic secretory fibers from the ninth nerve to the parotid gland (Fig. 4-30).

VI. *The Abducent Nerve*

The sixth cranial nerve has only one function: it supplies the lateral rectus muscle of the eye (Fig. 4-17). Not only are efferent somatic motor fibers contained in this nerve, but also proprioceptive afferent fibers. The nerve arises from the ventral surface of the brain (Fig. 4-14). It passes forward in the floor of the cranial cavity, through the cavernous sinus, and enters the orbit through the superior orbital fissure.

VII. *The Facial Nerve*

The facial nerve is an interesting and complex structure. It contains somatic efferent fibers to all the muscles of facial expression. It also supplies motor fibers to the posterior belly of the digastric muscle, the stylohyoideus muscle, and the stapedius muscle of the middle ear. The facial nerve also contains afferent taste fibers from the tongue, afferent proprioceptive fibers and efferent parasympathetic preganglionic fibers to the mucous glands of the nose and salivary glands of the mouth.

The facial nerve arises by two roots: a large motor root and a small *nervus intermedius* (Fig. 4-14). The latter contains both preganglion parasympathetic secretory fibers and sensory fibers for taste and proprioception. Both roots fuse and pass into the internal

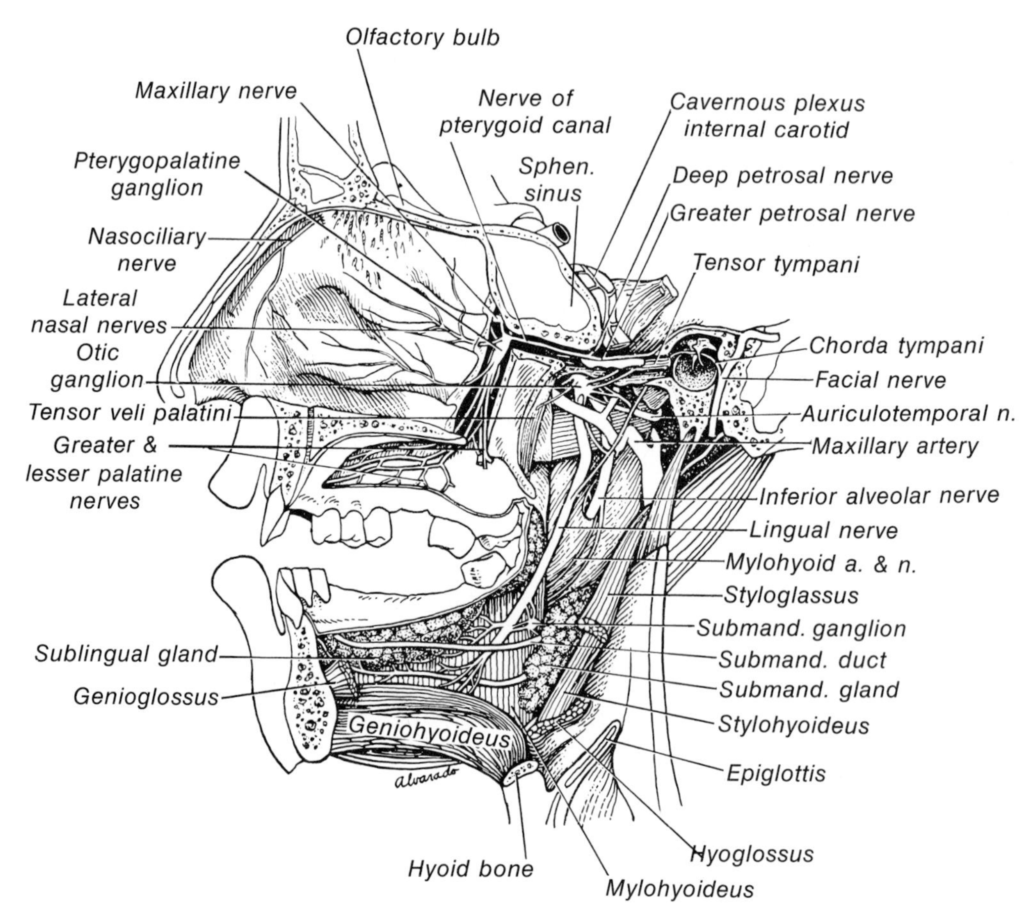

FIG. 4-29. Deep dissection of the region of the face viewed from its medial aspect, showing the pterygopalatine, otic, and submandibular ganglia and associated structures. (Redrawn from Töndury in Gray's Anatomy of the Human Body, 29th ed. C. M. Goss, editor. Philadelphia, Lea & Febiger, 1973.)

auditory meatus and enter the substance of the petrous portion of the temporal bone. While still in the temporal bone, the facial nerve leaves the inner ear and enters the facial canal (Fig. 4-31). The fused nerve takes a tortuous course within the bone in the *facial canal,* a channel inside the temporal bone. While inside the bone, the nerve encounters the sensory ganglion of the facial nerve, the *geniculate ganglion.* No synpase occurs here. This is the location of sensory cell bodies for taste and proprioception fibers entering the brain via the nervus intermedius. Also, while in the facial canal of the temporal bone, the facial nerve releases several branches.

Greater Petrosal Nerve (Fig. 4-32)

The greater petrosal branch contains mixed fibers of sensory and parasympathetic functions. The cell bodies of the parasympathetic fibers are in the brain, and the axons pass through the geniculate ganglion and join the greater petrosal nerve. The cell bodies of the taste fibers and proprioceptive fibers are located in the geniculate ganglion. Their dendrites leave the ganglion with the axons of the parasympathetic fibers and all of these are then incorporated in the greater petrosal nerve. After arising from the geniculate ganglion, this nerve, by a very complex route, finally reaches the pterygopalatine ganglion in the pterygopalatine fossa. The parasympathetic fibers synapse here, and second order neurons are distributed to the various glands of the mouth and nose and the lacrimal gland, via the branches of the trigeminal nerve.

Afferent taste fibers from the palate and posterior third of the tongue are also incorporated in branches of the trigeminal nerves. They reach the geniculate ganglion via the greater superficial petrosal nerve. As noted above, the cell bodies of these sensory nerves are located in the geniculate ganglion, and the axons are carried in the nervus intermedius to the brain.

Nerve to the Stapedius Muscle

This structure supplies the stapedius muscle of the middle ear.

Chorda Tympani Nerve (Figs. 4-28; 4-33)

After arising from the facial nerve in the facial canal and taking a tortuous course, the chorda tympani nerve exits from the skull at the base of the sphenoid bone near the spina angularis. It then courses downward and joins the lingual nerve where the lingual nerve passes between the pterygoid muscles.

The chorda tympani nerve contains preganglionic parasympathetic fibers to the submandibular ganglion. The cell bodies of the first order neurons are located in the brain. The cell bodies of the second order neurons are located in the submandibular ganglion. This nerve also carries taste fibers from the anterior two thirds of the tongue. The cell bodies of the taste fibers are in the geniculate ganglion of cranial nerve VII, and their axons enter the brain via the nervus intermedius along with those of the greater petrosal nerve.

The facial nerve then exits from the petrous portion of the temporal bone through the stylomastoid foramen. Additional branches then arise.

The Posterior Auricular Nerve (Fig. 4-34)

The posterior auricular nerve is released near the stylomastoid foramen. It passes back and up and supplies motor fibers to the auricularis muscles and the occipitalis muscle.

Digastric Branch

The digastric branch also arises close to the stylomastoid foramen. It is the motor nerve to the posterior belly of the digastric.

Stylohyoid Branch

The stylohyoid motor branch to the Stylohyoideus is released next. The facial nerve now runs within the substance of the parotid gland. It divides into two primary branches: a *superior temporofacial* trunk and an inferior *cervicofacial* trunk. From these trunks, five main branches arise: the temporal, zygomatic, buccal, mandibular, and cervical branches.

Temporal Branches

The temporal branches emerge from the parotid gland, cross the zygomatic arch, and supply the auricularis muscles, frontalis, orbicularis oculi, and corrugator muscles.

Zygomatic Branches

Also from the superior trunk, the zygomatic branches pass across the face to the orbicularis oculi muscle.

Buccal Branches

The buccal branches course horizontally forward and supply the procerus, zygomaticus major, quadratus labii superioris, nasal, buccinator, and orbicularis oris muscles.

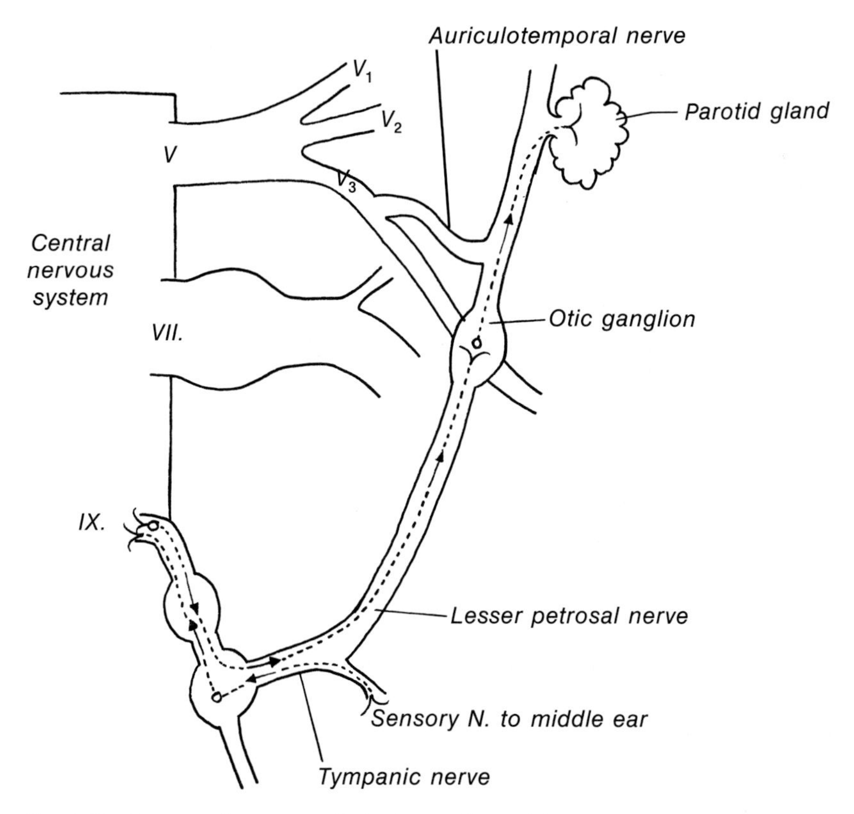

FIG. 4-30. Diagram of parasympathetic innervations of the parotid gland by way of the otic ganglion.

Mandibular Branch

The mandibular branch passes forward and down and curves anteriorly at the level of the inferior border of the mandible. It supplies the muscles of the lower lip and chin.

Cervical Branch

The final branch, the cervical branch, passes downward to supply the platysma muscle.

VIII. *The Acoustic Nerve* (Figs. 4-31; 4-35)

The eighth cranial nerve arises from the brain just behind the two roots of the facial nerve (Fig. 4-14). It consists of two sets of fibers, the *cochlear fibers* (root) and the *vestibular fibers* (root). Both roots, as a common trunk, enter the internal auditory meatus along with the facial nerve. The acoustic nerve remains in the inner ear and divides into the cochlear and vestibular portions. The acoustic nerve is efferent only.

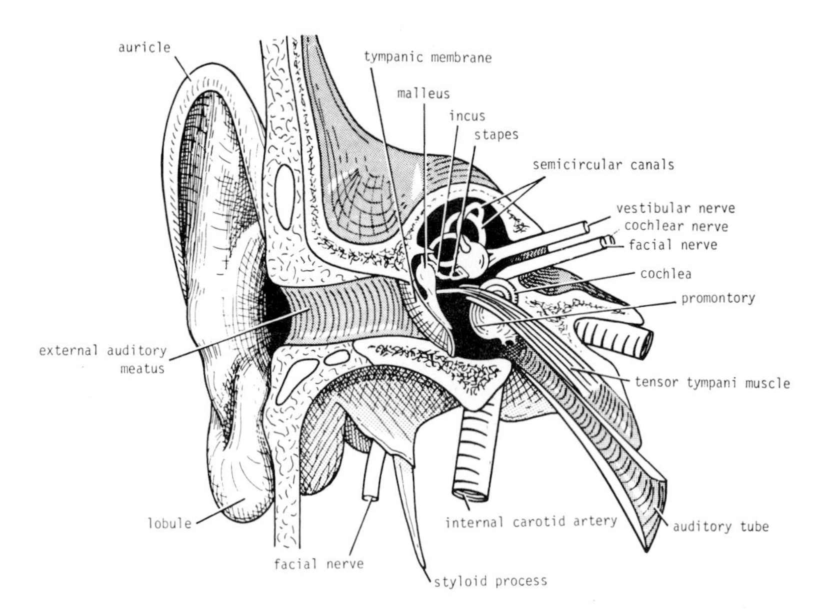

FIG. 4-31. External and middle ear portions of right ear, viewed from front. (From Snell, R. S.: Clinical Anatomy for Medical Students. Boston, Little, Brown and Co., 1973.)

Cochlear Nerve

The cochlear nerve transmits sound impulses to the brain from the inner ear.

Vestibular Nerve

The inner ear contains a sensitive organ of balance made up of three semicircular canals. The vestibular nerve transmits impulses of body and head position to the brain. The semicircular canals contain fluid, and as the position of the head is altered, this fluid moves about in the canals, stimulating the nerve endings of the vestibular nerve.

IX. *The Glossopharyngeal Nerve* (Fig. 4-36)

The glossopharyngeal nerve supplies the tongue and the pharynx. It is composed of a variety of fibers, both afferent and efferent. It leaves the skull through the jugular foramen (Fig. 4-22) at which point its two sensory ganglia are located. These ganglia are known as the *superior petrosal and inferior petrosal ganglia.* The glossopharyngeal nerve has several branches.

The Tympanic Nerve

The tympanic nerve leaves the inferior sensory ganglion. It carries preganglionic parasympathetic secretory fibers and sensory

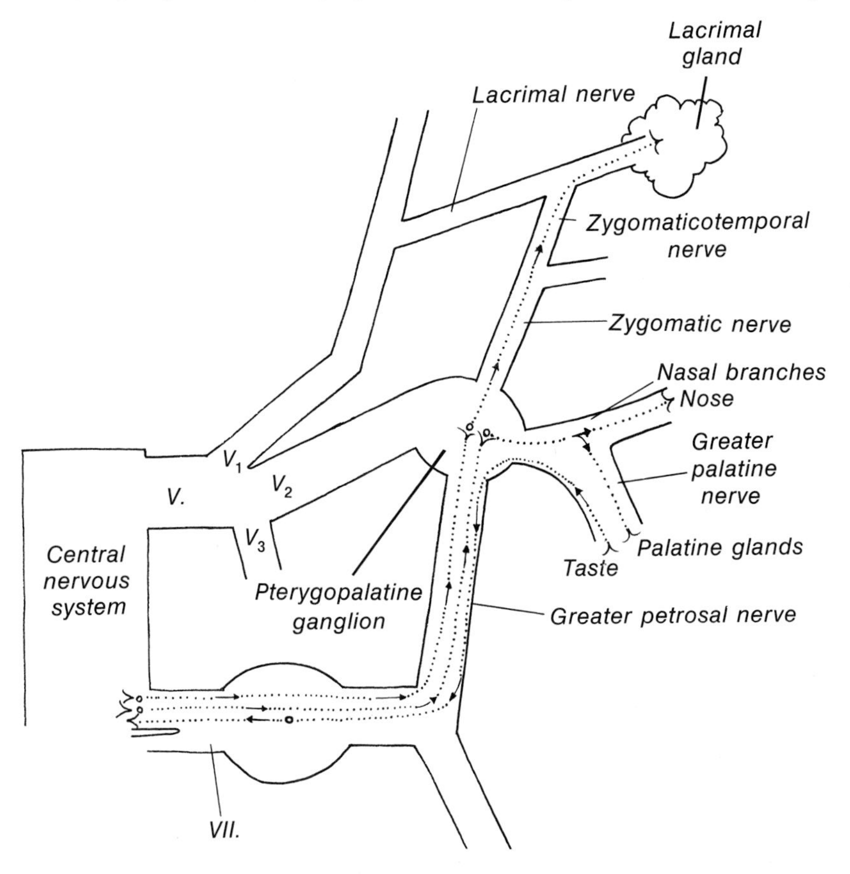

FIG. 4-32. Scheme of the greater petrosal nerve. Arrows denote the direction of impulse flow.

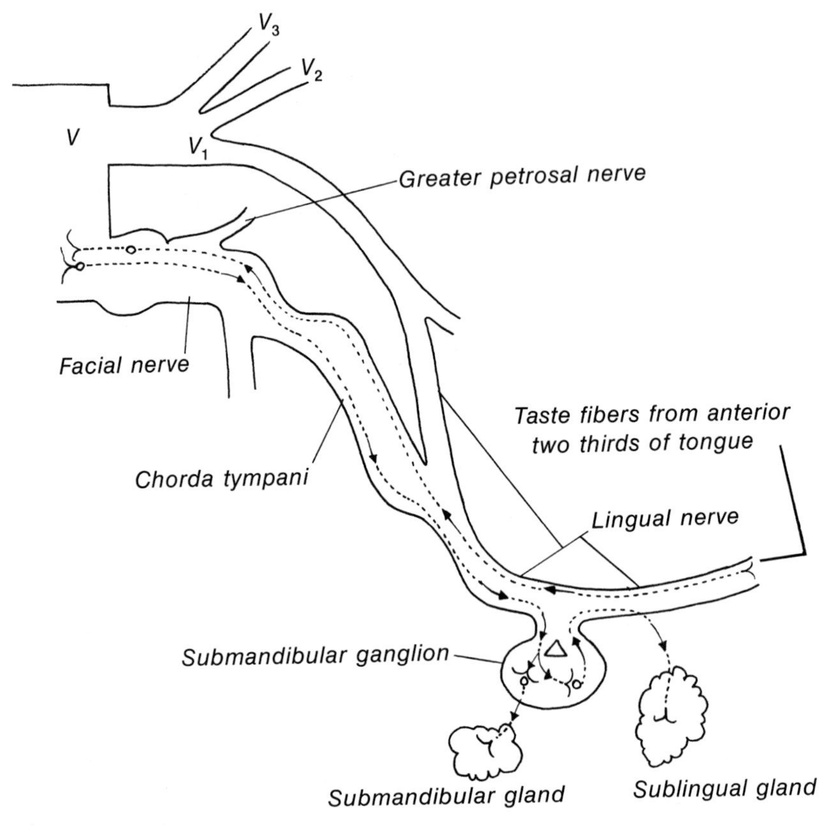

FIG. 4-33. Parasympathetic supply to the sublingual and submandibular salivary glands; scheme of the chorda tympani nerve.

fibers to the mucosa of the middle ear. After leaving the sensory ganglion, it reenters the tympanic cavity. The tympanic nerve then perforates the roof of the cavity and, having lost its sensory fibers, is known as the *lesser petrosal nerve.* Its preganglionic parasympathetic secretory fibers then synapse at the *otic ganglion* (Figs. 4-29; 4-30). Second order neurons then go to the parotid salivary gland after joining the auriculotemporal nerve.

Carotid Branch

At a variable level below the jugular foramen, the carotid branch is released. It supplies afferent fibers to the carotid sinus and carotid bodies. These two tiny organs are blood pressure regulatory mech-

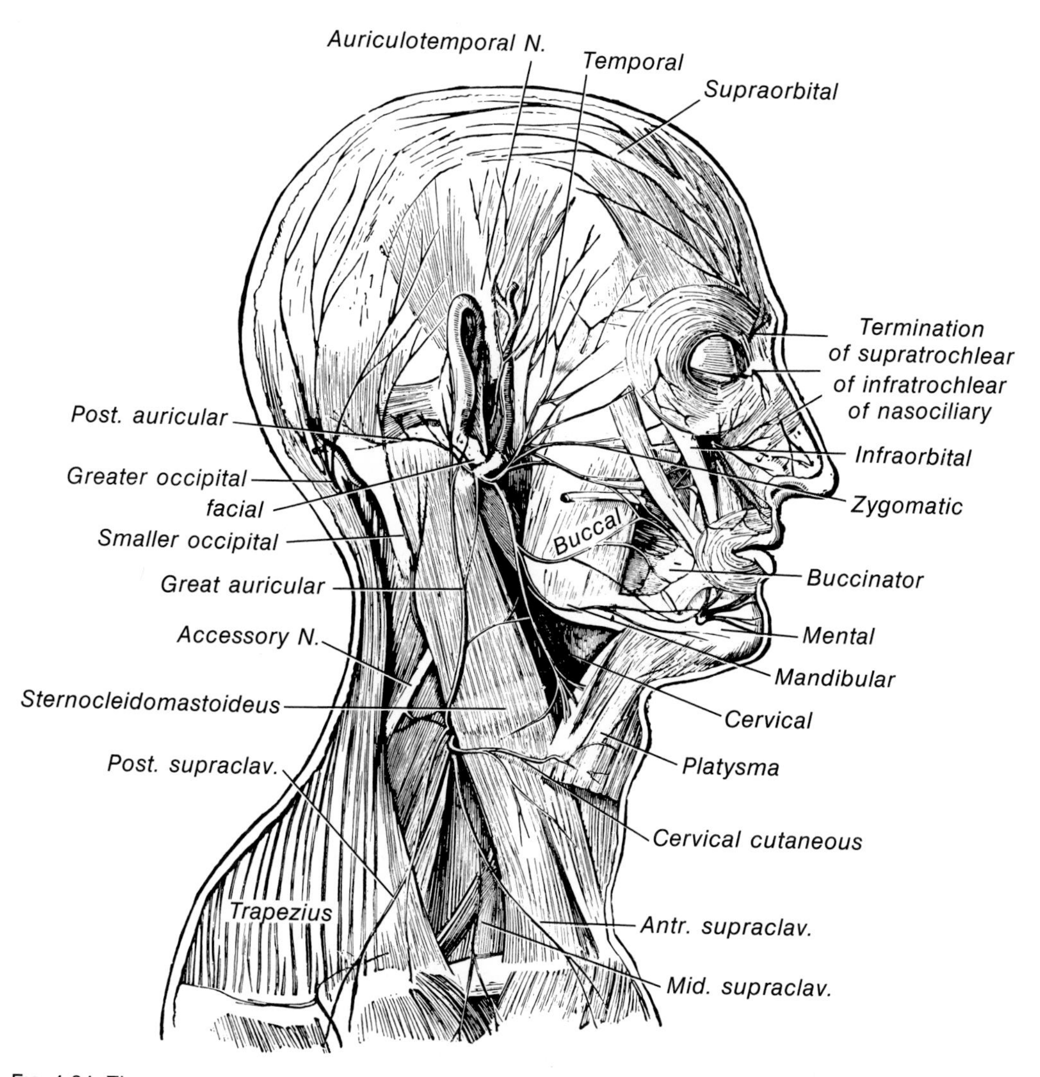

FIG. 4-34. The nerves of the scalp, face, and side of the neck. (From Gray's Anatomy of the Human Body, 29th ed. C. M. Goss, editor. Philadelphia, Lea & Febiger, 1973.)

anisms that are located close to the bifurcation of the common carotid artery.

Stylopharyngeal Nerve

With the stylopharyngeal branch the glossopharyngeal nerve supplies the motor fibers to the stylopharyngeus muscle.

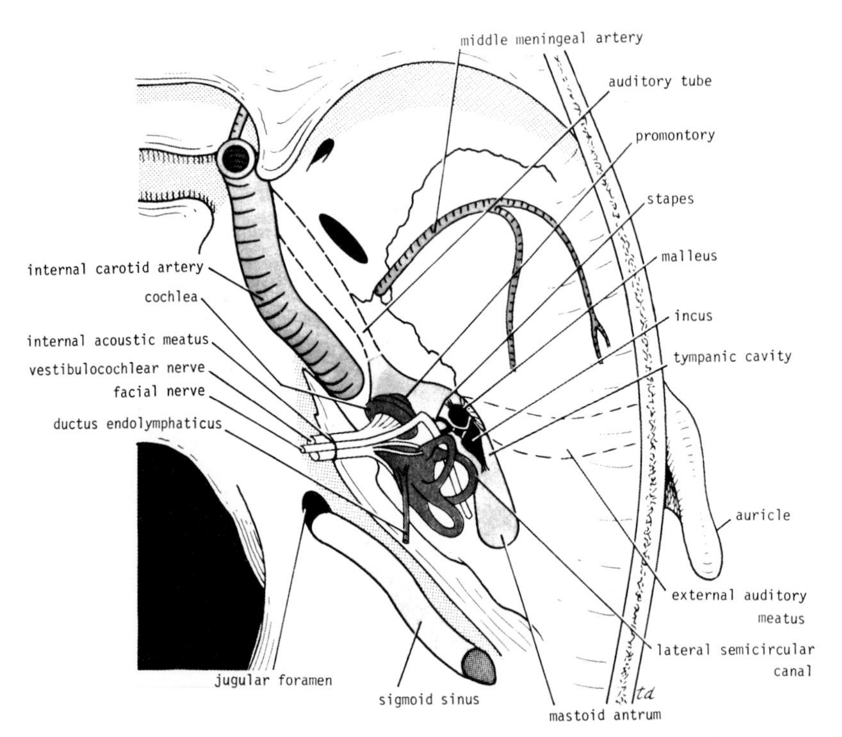

FIG. 4-35. Parts of the right ear in relation to the temporal bone, as viewed from above. (From Snell, R. S.: Clinical Anatomy for Medical Students. Boston, Little, Brown and Co., 1973.)

Pharyngeal Branches

The pharyngeal nerve fibers join with the vagus and spinal accessory nerves to form the pharyngeal plexus. This network supplies the muscles of the pharynx and soft palate, except for the tensor veli platini and the stylopharyngeus muscles. Also, this plexus provides sensory fibers to the mucosa of the soft palate and pharynx.

The glossopharyngeal nerve finally enters the base of the tongue. Here it supplies taste fibers to the posterior one third of the tongue, tonsils, and pillars.

X. *The Vagus Nerve* (Fig. 4-36)

The vagus nerve is the longest of the cranial nerves. It reaches as far down as the abdomen. The nerve exits from the skull through the jugular foramen, along with the ninth and eleventh cranial nerves

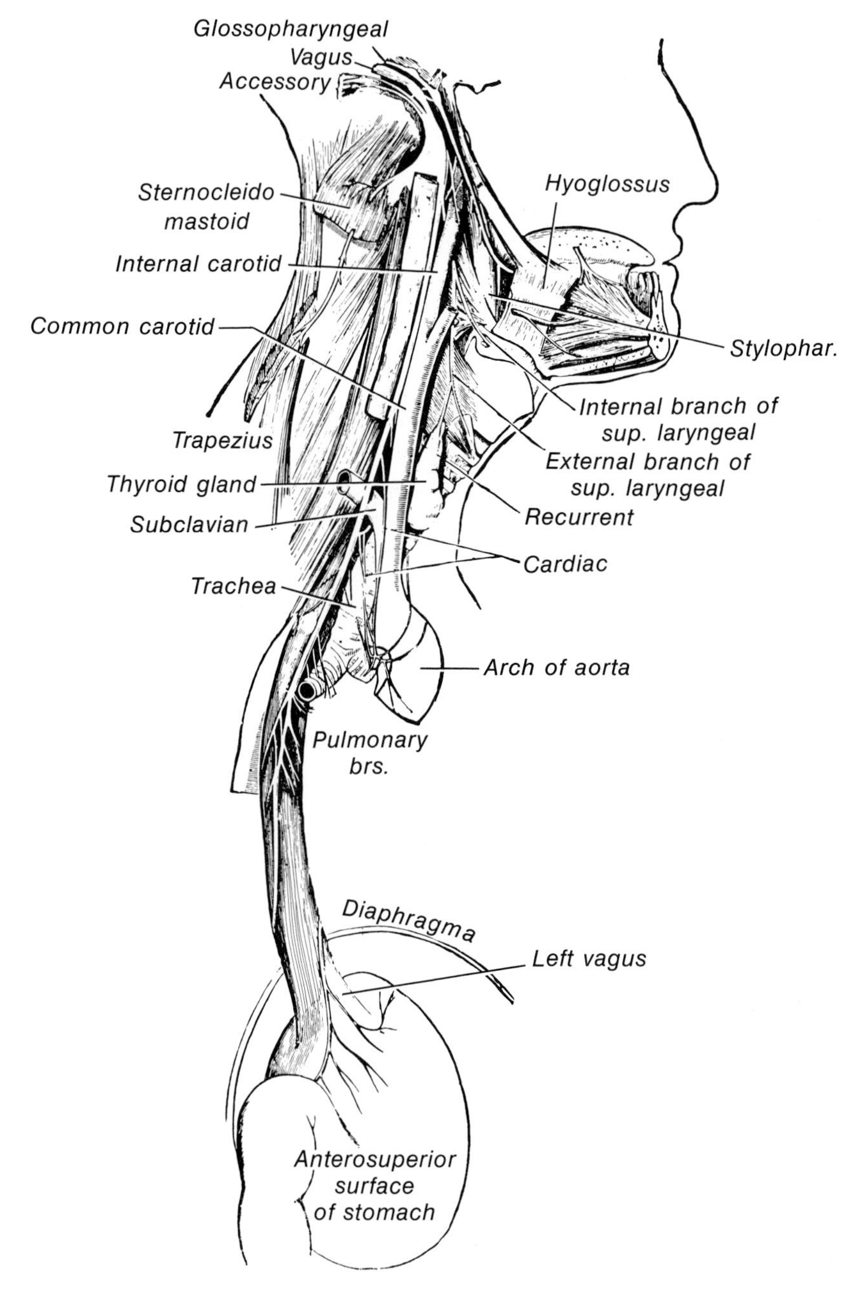

FIG. 4-36. Course and distribution of the glossopharyngeal, vagus, and accessory nerves. (From Gray's Anatomy of the Human Body, 29th ed. C. M. Goss, editor. Philadelphia, Lea & Febiger, 1973.)

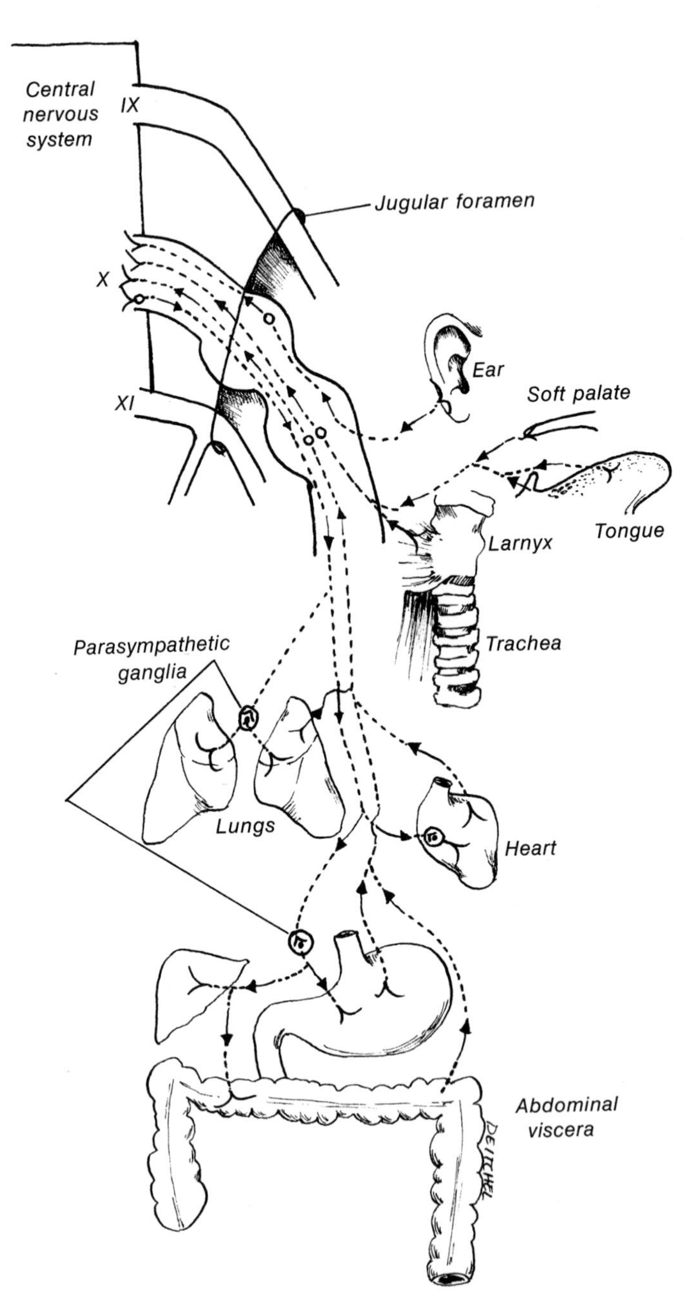

FIG. 4-37. The vagus nerve. The visceral afferent fibers and the parasympathetic divisions are diagrammatically presented. Arrows depict direction of impulse flow. The motor portion of the vagus nerve, which contributes to the pharyngeal plexus, is not illustrated.

(Fig. 4-22). As it descends in the neck, the vagus nerve is enclosed in the carotid sheath with the internal jugular vein and the carotid artery.

Two sensory ganglia—the *jugular* (*superior*) *ganglion* and the *nodose* (*inferior*) *ganglion*—are located in the portion near the jugular foramen. The jugular ganglion is superior and smaller. The fibers whose cell bodies lie in these ganglia supply sensory innervation to the skin of the ear, the pharynx, larynx, trachea, bronchi, esophagus, and the thoracic and abdominal viscera.

Both somatic and visceral fibers and afferent and efferent fibers are contained in the vagus nerve. It also carries parasympathetic preganglionic fibers. Primarily, the vagus is a visceral nerve. The visceral afferents come from the gut, the lungs, the heart, and the mucous membranes of the pharynx and larynx. This nerve also has a few taste fibers arising in the region of the epiglottis. Parasympathetic preganglionic efferent fibers originate in the brain and go to the heart, lungs, and abdominal viscera. The second order neurons are located in or near the respective organs.

The vagus nerve contributes to the pharyngeal plexus and the muscles of the pharynx. It also provides innervation to the muscles of the larynx. One of the most important functions of the vagus is its parasympathetic control of heart rate. Stimulation of the parasympathetic portion results in bradycardia, decreased heart rate. The parasympathetic portion also supplies the glands of the gastrointestinal tract (Fig. 4-37).

XI. *The Accessory* (*Spinal*) *Nerve*

The twelfth cranial nerve is only partly a cranial nerve. It consists of two roots, one root arising from the brain and the other arising from the spinal cord. The nerve carries efferent and afferent impulses and is intimately associated with the vagus nerve.

Cranial Root

The cranial root portion of the accessory nerve arises by several small rootlets just behind the root of the vagus (Fig. 4-14). It passes into the jugular foramen after joining the spinal part.

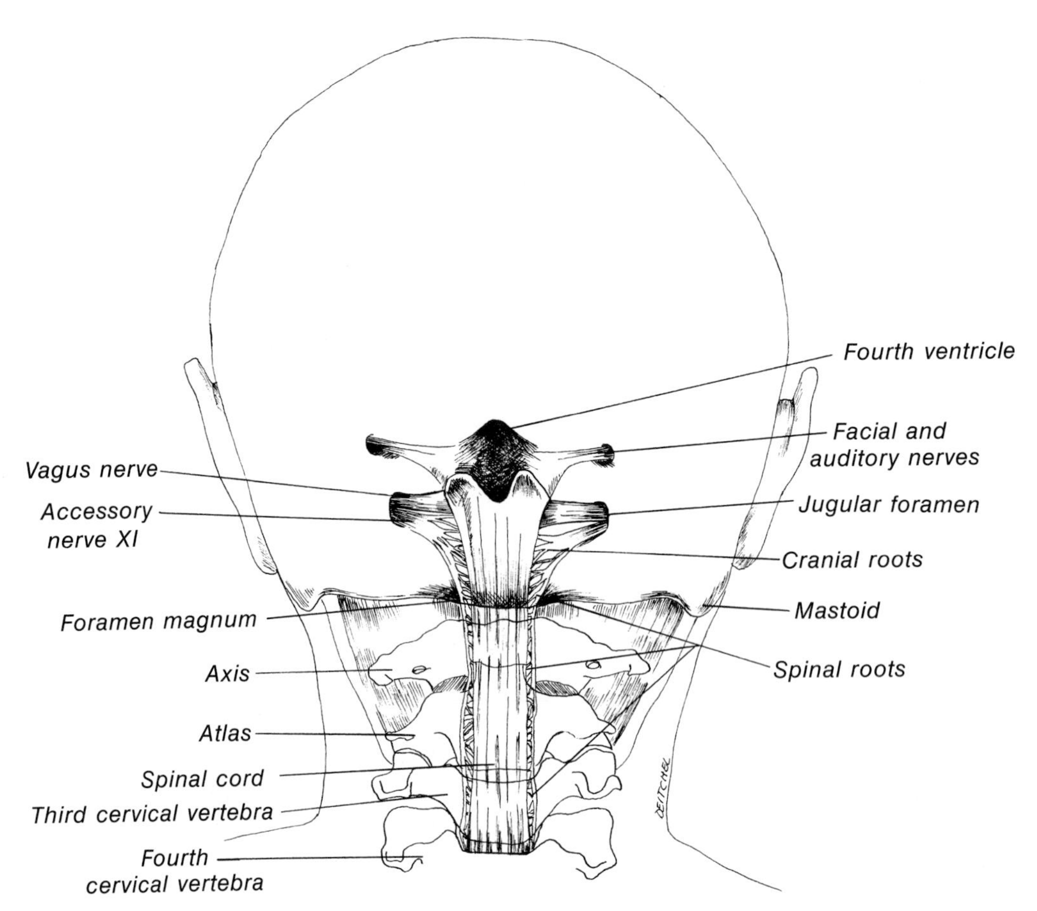

FIG. 4-38. Cranial nerves X and XI, exposed from behind. Note: Cerebellum has been removed and vertebrae have been ghosted over to expose spinal cord.

The Spinal Portion

The spinal portion of the accessory nerve arises from several rootlets of the first five cervical divisions of the spinal cord (Fig. 4-38). These spinal roots fuse and pass upward through foramen magnum into the cranial cavity. The nerve then passes into the jugular foramen and joins the cranial part. The cranial and spinal fibers intermix and pass through the foramen as one nerve, the spinal accessory nerve.

As the spinal accessory nerve exits from the foramen, it divides into two nerves again, each carrying representatives of both roots but still containing a majority of either spinal fibers or cranial fibers. The cranial fibers join the vagus and contribute to the pharyngeal plexus. The spinal fibers pass backward and down to supply the trapezius and sternocleidomastoideus muscles (Fig. 4-36).

The spinal accessory nerve not only contributes to the pharyngeal plexus and supplies motor impulses to the Trapezius and Sternocleidomastoideus, but also it carries proprioceptive afferent fibers from the muscles it supplies. Loss of function of this nerve results primarily in paralysis of the trapezius and sternocleidomastoideus muscles.

XII. *The Hypoglossal Nerve* (Fig. 4-39)

The twelfth cranial nerve supplies motor fibers to all the intrinsic and extrinsic muscles of the tongue. No other nerve supplies these muscles with motor fibers. In addition, the hypoglossal nerve carries proprioceptive impulses from the muscles of the tongue to the brain.

The course of this nerve is extremely interesting. It arises by several rootlets at the base of the brain, just inferior to the origin of cranial nerves IX, X, and XI (Fig. 4-14). These rootlets merge into a trunk, which passes out from the brain case via the *hypoglossal canal* (Fig. 4-22). Here, it is close to the vagus nerve. About the level of the mandibular foramen, as the nerve is passing downward, it turns anteriorly and superficially (Fig. 4-39). It courses forward and is almost horizontal as it reaches a level deep to the angle of the mandible. Passing deep to the posterior belly of the digastric muscle, it ramifies to supply the muscles of the tongue.

As the nerve descends from the hypoglossal canal, it picks up fibers from the first and second cervical nerves (Fig. 4-40). It carries these fibers for a short distance and then drops them off to the suprahyoid muscles. Also it drops off a branch called the *descendens hypoglossi.* The descendens hypoglossi, passing down to the infrahoid muscles, picks up branches from the third cervical nerve. The loop thus formed is known as the *ansa hypoglossi.* From the ansa, fibers composed of C 1, C 2, and C 3 are distributed to the infrahyoid muscles.

Loss of function due to lacerations or tumor destruction of the

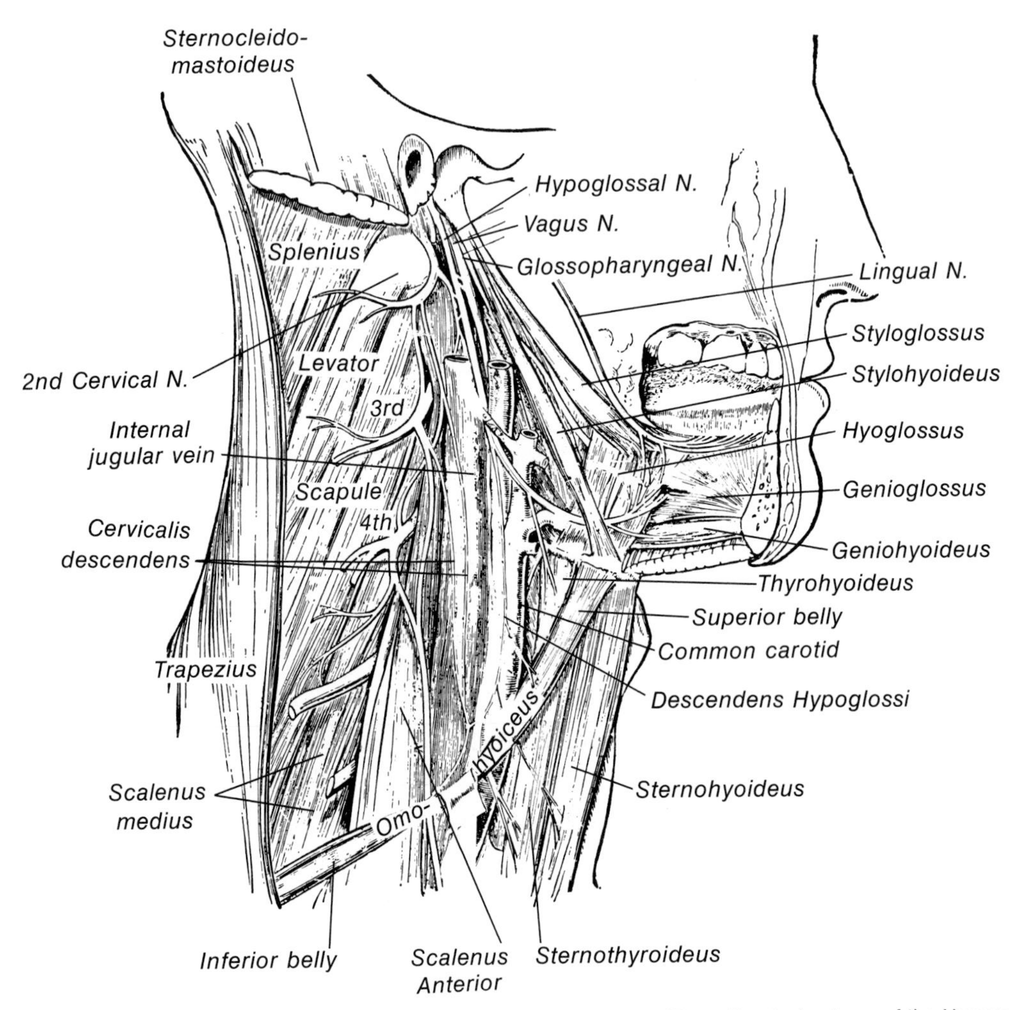

FIG. 4-39. Hypoglossal nerve, cervical plexus, and their branches. (From Gray's Anatomy of the Human Body, 29th ed. C. M. Goss, editor. Philadelphia, Lea & Febiger, 1973.)

hypoglossal nerve will result in inability of the patient to move one side of the tongue. When the patient attempts to protrude the tongue, the tongue will deviate toward the affected side.

Autonomic Nervous System of the Head

Many organs of the head are controlled partially or completely by the autonomic nervous system. The eye, salivary glands, and nasal glands all have autonomic innervation. The autonomic fibers are carried to the various organs by "hitching a ride" on the cranial nerves and arteries supplying that area. For example, the parasympathetic supply to the salivary glands is carried in the trigeminal nerve branches.

Sympathetic Nerves of the Head

All the sympathetic fibers in the head are postganglionic fibers. The short dendrites and the cell bodies of the preganglionic fibers lie in the thoracic portion of the spinal cord (Fig. 4-13). The axon fibers pass out to the sympathetic chain and synapse with second order neurons. This synapse takes place in the superior cervical ganglion which is the uppermost ganglion in the sympathetic chain. It is located about the level of the axis. In this ganglion lie the cell bodies and dendrites of second order sympathetic neurons whose fibers (axons) supply sympathetic impulses to the head (Fig. 4-12). The axons leave the ganglion and are carried to various areas by attaching themselves to arteries. The sympathetic fibers do not synapse again. If and when they reach various parasympathetic ganglia in the head, they pass through without synapsing. Only the parasympathetic fibers synapse in the ciliary, pterygopalatine, and otic ganglia. Sympathetic fibers supply the dilator of the pupil and sweat glands of the skin. Reduced salivation, dilation of the pupil, and sweating are examples of sympathetic impulses reaching various areas of the head.

Parasympathetic Nerves of the Head

The parasympathetic fibers are carried in various cranial nerves rather than along arteries. The parasympathetic supply is complex but extremely interesting. It is not difficult to understand if the system is organized by the various parasympathetic ganglia (Fig. 4-41).

The Otic Ganglion (Fig. 4-29)

The small otic ganglion is attached to the medial surface of the third division of the fifth nerve just as the nerve exits the foramen ovale. This is purely an anatomic attachment. The otic ganglion

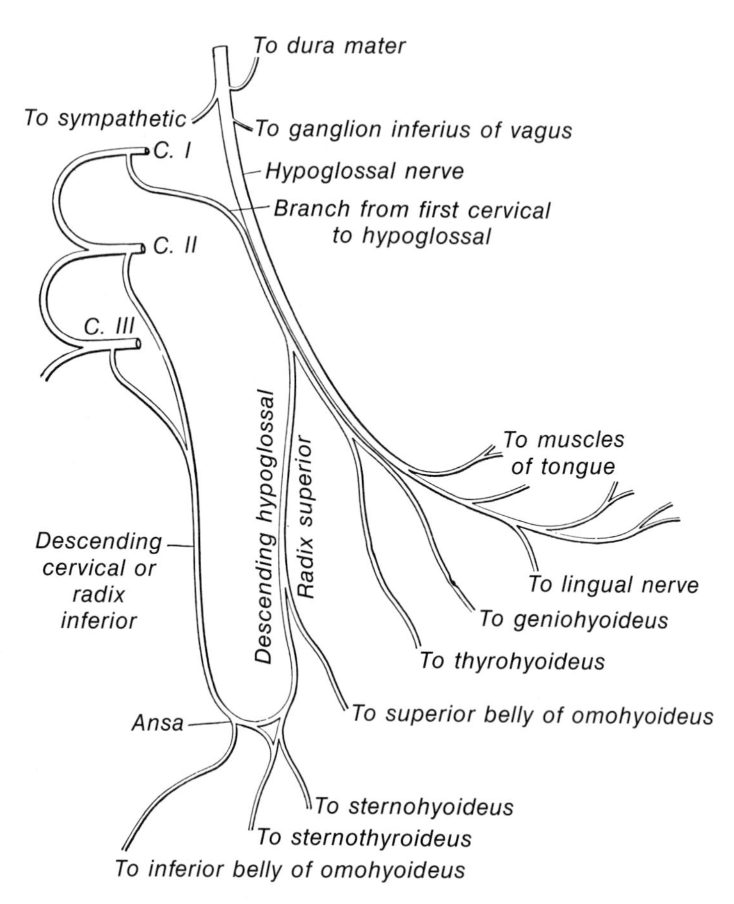

FIG. 4-40. Plan of hypoglossal and first three cranial nerves. (From Gray's Anatomy of the Human Body, 29th ed. C. M. Goss, editor. Philadelphia, Lea & Febiger, 1973.)

contains only cell bodies of second order parasympathetic fibers, and only parasympathetic fibers synapse here. All other fibers pass directly through the ganglion without synapsing.

Preganglionic parasympathetic fibers destined for the otic ganglion originate in the brain and exit with the glossopharyngeal nerve through the jugular foramen. They branch from the glossopharyngeal nerve as the *tympanic nerve* which becomes the *lesser petrosal nerve* after emerging from the tympanic cavity. The fibers of the lesser petrosal nerve synapse in the otic ganglion. Postganglionic fibers join the auriculotemporal branch of the mandibular nerve and are carried to the parotid gland (Fig. 4-30). The function of these parasympathetic fibers is to activate production and secretion of saliva from the parotid gland.

The sympathetic fibers to the parotid gland pass up with the external carotid artery and the internal maxillary artery and finally reach the parotid gland via the middle meningeal artery.

Submandibular Ganglion (Fig. 4-33)

The submandibular ganglion lies above the submandibular gland at the posterior border of the mylohyoid muscle (Fig. 4-29). It is suspended from the lingual nerve by several preganglionic parasympathetic fibers of the chorda tympani nerve, which joins the lingual nerve before the lingual nerve reaches the submandibular ganglion.

Preganglionic parasympathetic fibers destined for the submandibular ganglion begin in the brain and exit as the nervus intermedius

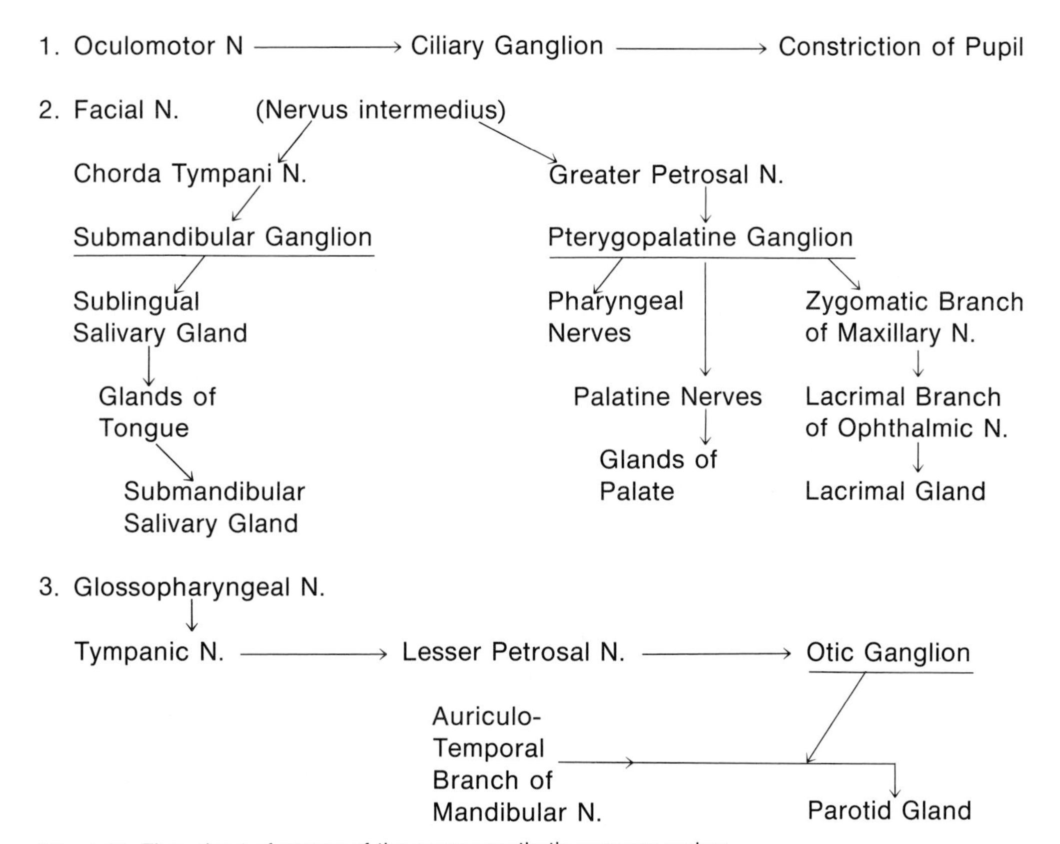

FIG. 4-41. Flow sheet of nerves of the parasympathetic nervous system.

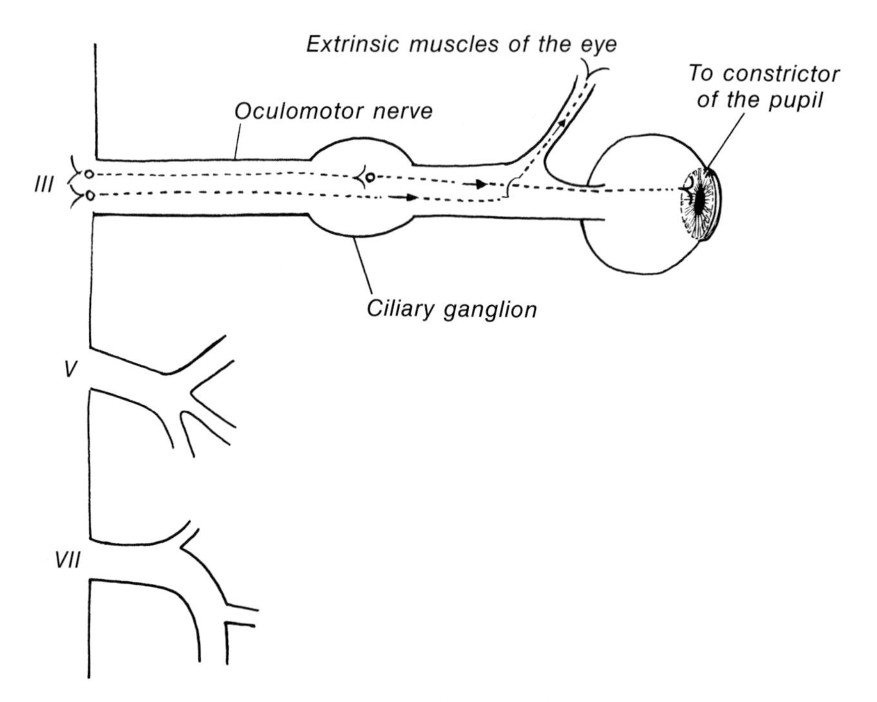

FIG. 4-42. Parasympathetic innervations of the constrictor of the pupil by way of the ciliary ganglion.

of the facial nerve. They leave the intermediate nerve as the chorda tympani, which joins the lingual nerve. Preganglionic fibers drop off the lingual nerve and synapse with postganglionic fibers at the submandibular ganglion. Some postganglionic secretomotor fibers rejoin the lingual nerve and are carried to the sublingual salivary gland under the tongue. Other postganglionic secretomotor fibers leave the ganglion and pass down into the submandibular salivary gland which lies below the mylohyoid muscle.

The sympathetic fibers to these two salivary glands arrive by following the facial artery.

Pterygopalatine (Sphenopalatine) Ganglion (Figs. 4-24; 4-29)

The pterygopalatine ganglion is located in the pterygopalatine fossa. It is positioned quite close to the maxillary nerve as the nerve exits from the foramen rotundum. Preganglionic parasympathetic fibers destined for this ganglion also originate in the brain as the intermediate nerve, a branch of the facial nerve. They leave the intermediate nerve as the *greater petrosal nerve.* The greater petrosal nerve passes through a small canal, enters the pterygopalatine fossa, and synapses with postganglionic parasympathetic neurons. From the ganglion, postganglionic fibers easily join various branches of the first and second divisions of the trigeminal nerve.

Fibers to the lacrimal gland leave the ganglion and join the maxillary nerve, course through the zygomatic nerve, and its branch, the zygomaticotemporal nerve, switch to the lacrimal nerve (branch of the ophthalmic nerve), and innervate the lacrimal gland (Fig. 4-23).

The postganglionic fibers to the salivary and mucous glands of the palate leave the ganglion and join the greater and lesser palatine branches of the maxillary nerve (Fig. 4-24).

The fibers to the nasal mucous glands reach that area from the ganglion by way of the nasal branches of the maxillary nerve (Fig. 4-24). Those to the pharynx are carried in small pharyngeal branches of the maxillary nerve. The greatest parasympathetic activity of the pterygopalatine ganglion, however, is to supply innervation for production and secretion of tears from the lacrimal gland.

Ciliary Ganglion (Fig. 4-42)

No secretory activity is associated with the ciliary ganglion. The ciliary ganglion serves as a synaptic center for parasympathetic supply to the constrictor of the pupil. Preganglion fibers originate in the brain and are carried to the ganglion via the oculomotor nerve. The ciliary ganglion is located in the posterior of the orbit (Fig. 4-17). Postganglionic fibers leave the ganglion and supply the constrictor muscle of the eye.

Sympathetic fibers enter the orbit and dilate the pupil. They are carried into the cranial vault on the internal carotid artery and pass into the orbit through the superior orbital fissure.

5

Lymphatic Drainage of the Head and Neck

In addition to the intricate vascular tree of arteries and veins, there is another system of vessels known as the lymphatic system. The function of this system is anatomically and physiologically related to that of the circulatory system.

Tissue Fluid

All of the tissues of the body are bathed in *tissue fluid.* The tissue fluid that is not within the blood vessels is termed extravascular fluid. It serves as a medium through which nutrients, electrolytes, and waste products can pass to and from cells and vessels. The fluid within blood vessels is termed intravascular fluid. Intravascular fluid is composed of plasma, blood cells, electrolytes, and protein. When intravascular fluid moves from the intravascular area to the extravascular area by the process of osmosis, it is then termed extravascular fluid.

Extravascular fluid is different from blood or intravascular fluid. Normally, it contains no red blood cells and a different percentage of white cells, electrolytes, and proteins. Extravascular fluid returns to the intravascular space by the same process, osmosis. In normal circumstances, the volume of the fluid within the vascular tree and the volume of fluid in the tissue spaces remain unchanged. The two fluid "compartments" are in balance and physiologically dependent on one another.

The fluid in the tissue spaces that is returning to the circulatory system returns primarily by way of the veins and lymphatic system. Physical activity and the milking action of the muscles promote fluid flow in these two types of vessels.

Venous return to the heart is enhanced by a strong heart. Aside from those in the head and neck, veins have valves at various intervals to prevent backflow of blood as it is returning to the heart. Anything that reduces flow of fluid in the veins and lymphatic vessels will reduce their ability to carry as much fluid in a given unit of time. In such a situation, tissue fluid flow is decreased, and less is picked up from the tissues. Thus, more fluid will accumulate in the tissues. Swelling or *edema* will be the result. People with failing hearts or clots in the leg veins (thrombophlebitis) will have thick, edematous ankles and legs. This ankle edema is merely an accumulation of tissue fluid.

Components of the Lymphatic System

The lymphatic system consists of lymphatic capillaries, vessels, nodes, lymphatic organs, and lymph.

Lymphatic Capillaries

The lymphatic capillaries are small lymph vessels located throughout the body. They collect the tissue fluid and coalesce into larger vessels much the same as the venous system is arranged. Small vessels flow into larger vessels.

Lymphatic Vessels

The collecting system that ultimately returns its contents to the blood stream is composed of lymphatic vessels. Most of the

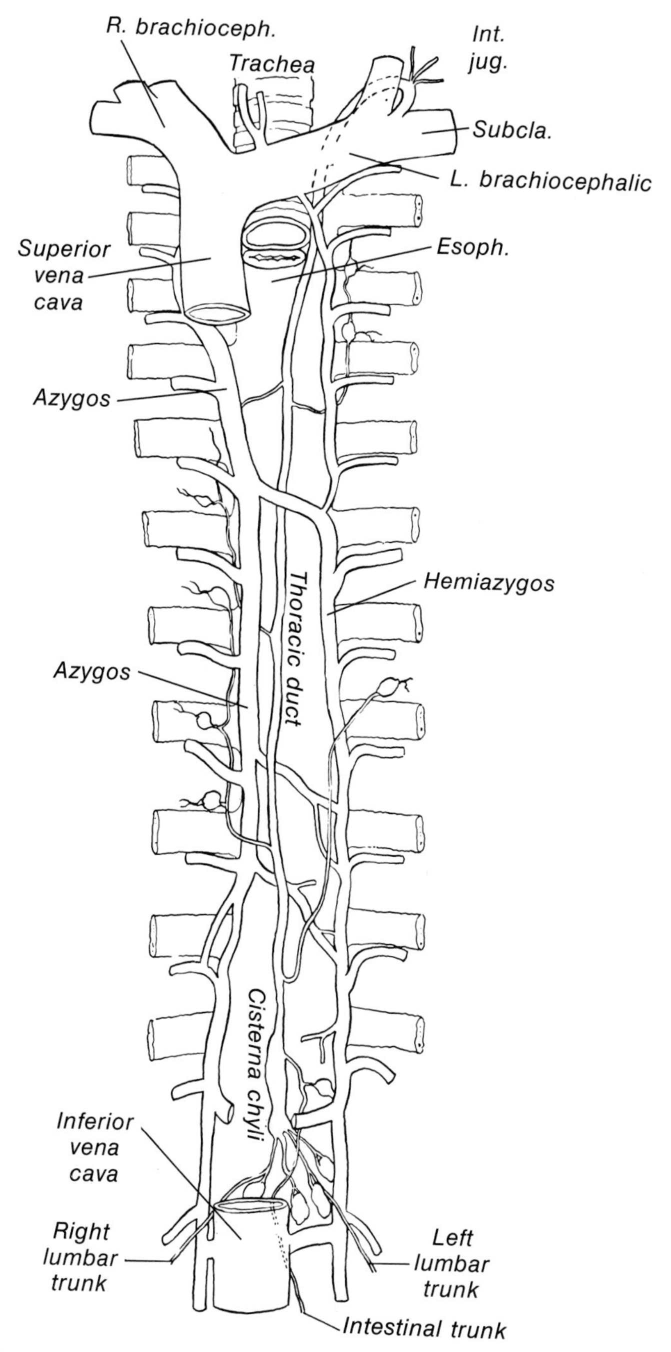

FIG. 5-1. The thoracic duct with lymphatic vessels and lymph nodes. (From Gray's Anatomy of the Human Body, 29th ed. C. M. Goss, editor. Philadelphia, Lea & Febiger, 1973.)

lymphatic vessels flow into the *thoracic duct* which empties into the junction of the left internal jugular vein with the left subclavian vein (Fig. 5-1). The thoracic duct carries all the lymph except that from the right head, right neck, right thorax, right arm, right lung, and right side of the heart. It is a thin-walled tube extending from the lumbar vertebrae to the root of the neck. It is situated close to the spine as it courses superiorly.

Lymph not drained by the thoracic duct is drained by the *right lymphatic duct.* The right lymphatic duct empties into the junction of the right subclavian vein with the right internal jugular vein. It is a short vessel, approximately 1 to 1.5 cm long.

Nodes

Small bean-shaped bodies, usually in groups but sometimes alone, are positioned along the course of the lymph vessels. Their function is to filter the lymphatic fluid flowing through them. Also, they produce and discharge lymphocytes, one type of white blood cell used in body defense against infection. Even though these assorted groups of nodes have wide and varied communications, the lymph drains in a relatively specific pattern toward the main collecting ducts.

Lymphatic Organs

Certain organs in the body resemble the lymph nodes and serve much the same function. The tonsils and adenoids are examples. The spleen contains lymphatic tissue. The thymus gland also has lymphatic connections, but no function for this organ has been definitely determined.

Lymph

By definition, lymph is the fluid found only in the closed lymphatic vessels. It is a transparent, colorless, watery fluid closely resembling blood plasma. Lymph is more dilute than plasma and lacks some of the protein and other elements found in plasma.

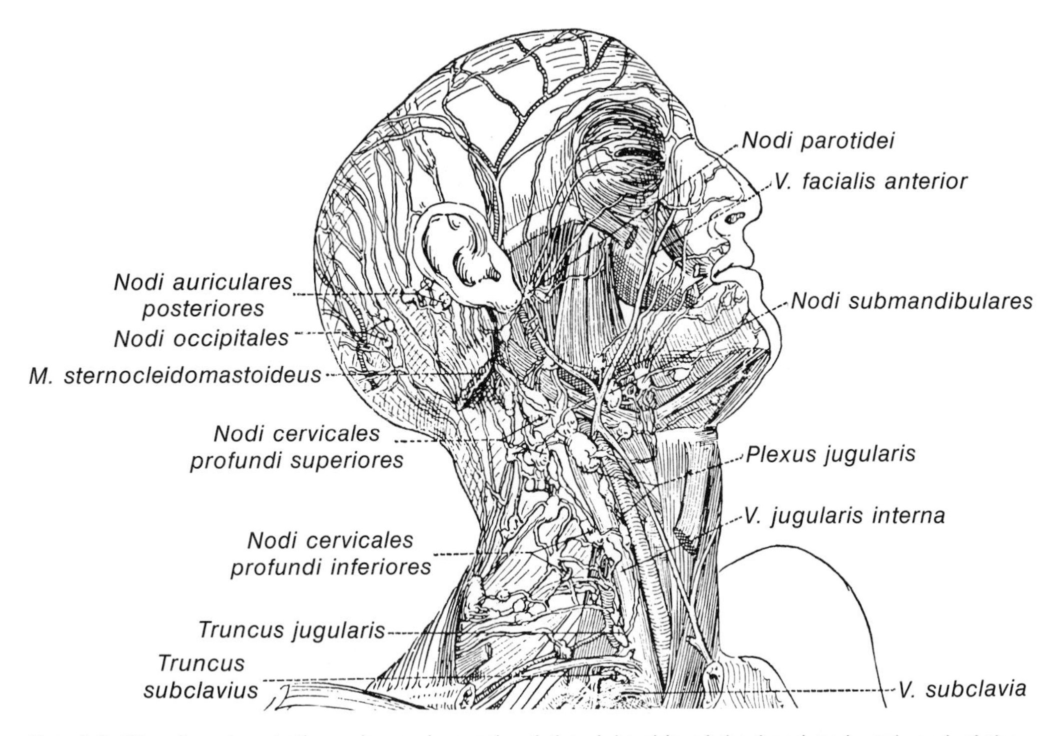

FIG. 5-2. The deep lymphatic nodes and vessels of the right side of the head and neck and of the mammary and axillary regions. (From Gray's Anatomy of the Human Body, 29th ed. C. M. Goss, editor. Philadelphia, Lea & Febiger, 1973.)

The Lymphatic System of the Head and Neck

As in other parts of the body, lymph to the head and neck is carried by the very thin-walled lymph vessels to aggregations of nodes. From the nodes it drains to other areas and ultimately into the thoracic duct or right lymphatic duct. Various groups of nodes located in specific areas drain the local tissues (Fig. 5-2).

For ease of learning, the nodal system of the head and neck can be divided into groups of nodes according to location.

Occipital Nodes (Fig. 5-3)

Located near the occipital protuberance, the occipital nodes drain the occipital portion of the scalp and empty into the cervical nodes.

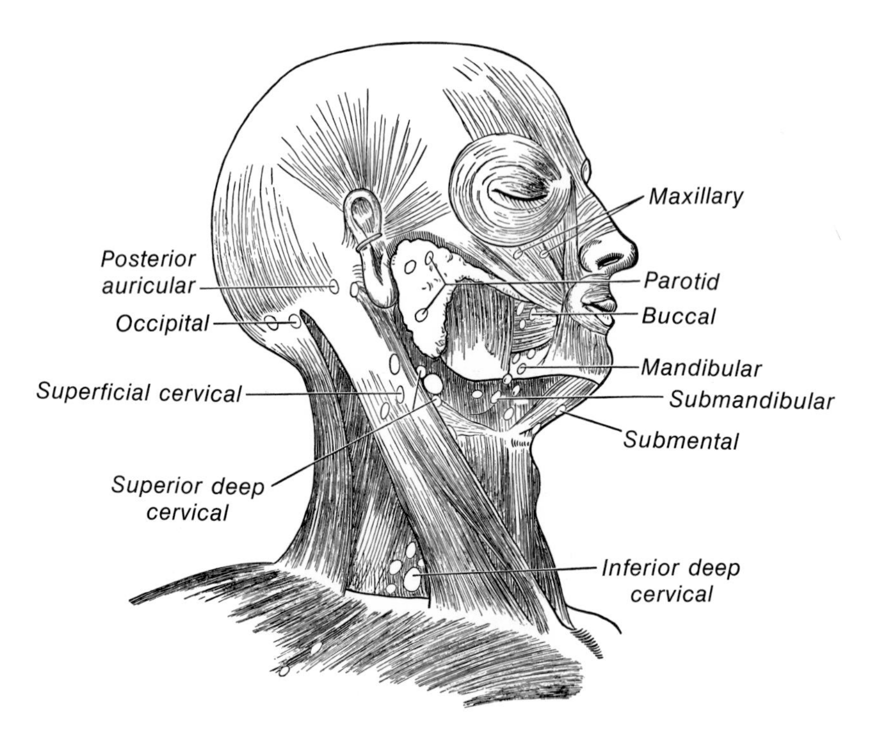

FIG. 5-3. Superficial lymph nodes and lymphatic vessels of head and neck. (From Gray's Anatomy of the Human Body, 29th ed. C. M. Goss, editor. Philadelphia, Lea & Febiger, 1973.)

Posterior Auricular Nodes (Fig. 5-3)

The posterior auricular nodes drain the mastoid region.

Anterior Auricular Nodes (Fig. 5-3)

The anterior auricular group drains the temporal region and skin of the ear.

Parotid Nodes (Fig. 5-3)

Several nodes are located around and in the parotid gland. They drain the nasal cavities, eyelids, frontotemporal region, posterior palate, anterior auricular nodes, and parotid region.

Facial Nodes (Fig. 5-3)

Several groups of nodes are located in the facial structures. Most of these drain into the *submandibular nodes.*

Submandibular Nodes (Fig. 5-3)

The submandibular nodes lie near the inferior border of the mandible. They collect lymph from the submental region, upper and lower teeth, tongue, lips, and jaws and drain into the *deep cervical nodes.*

Submental Nodes (Fig. 5-3)

The unpaired submental group lies in the submental triangle between the anterior bellies of the digastric muscles. These nodes drain the lower incisors and empty into the submandibular and *deep cervical nodes.*

Cervical Nodes (Fig. 5-4)

Various groups of nodes are located along the internal and external jugular veins. They are divided into *superficial* and *deep cervical nodes.* The superficial nodes are usually found in the upper region of the neck (Fig. 5-3). They receive lymph from the ear and

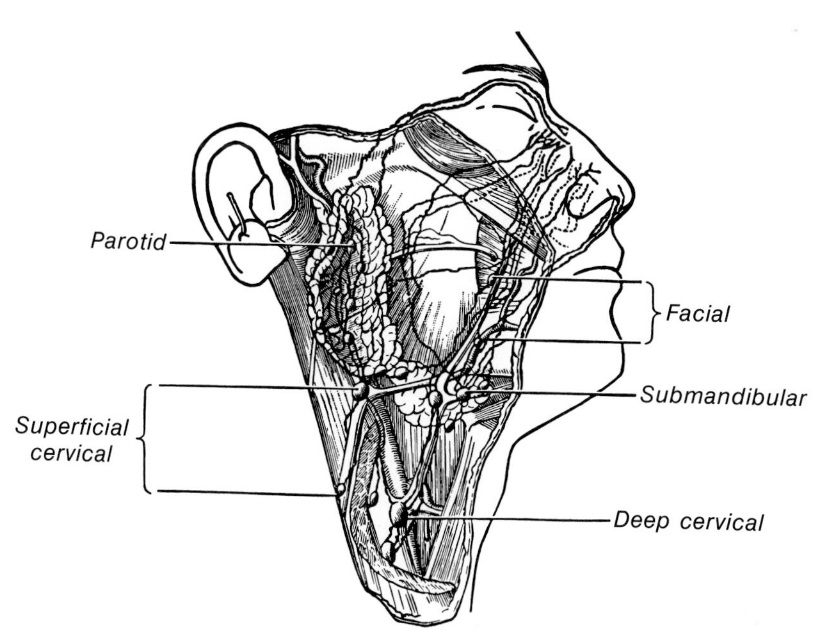

FIG. 5-4. The lymphatics and lymph nodes of the face. (After Küttner in Gray's Anatomy of the Human Body, 29th ed. C. M. Goss, editor. Philadelphia, Lea & Febiger, 1973.)

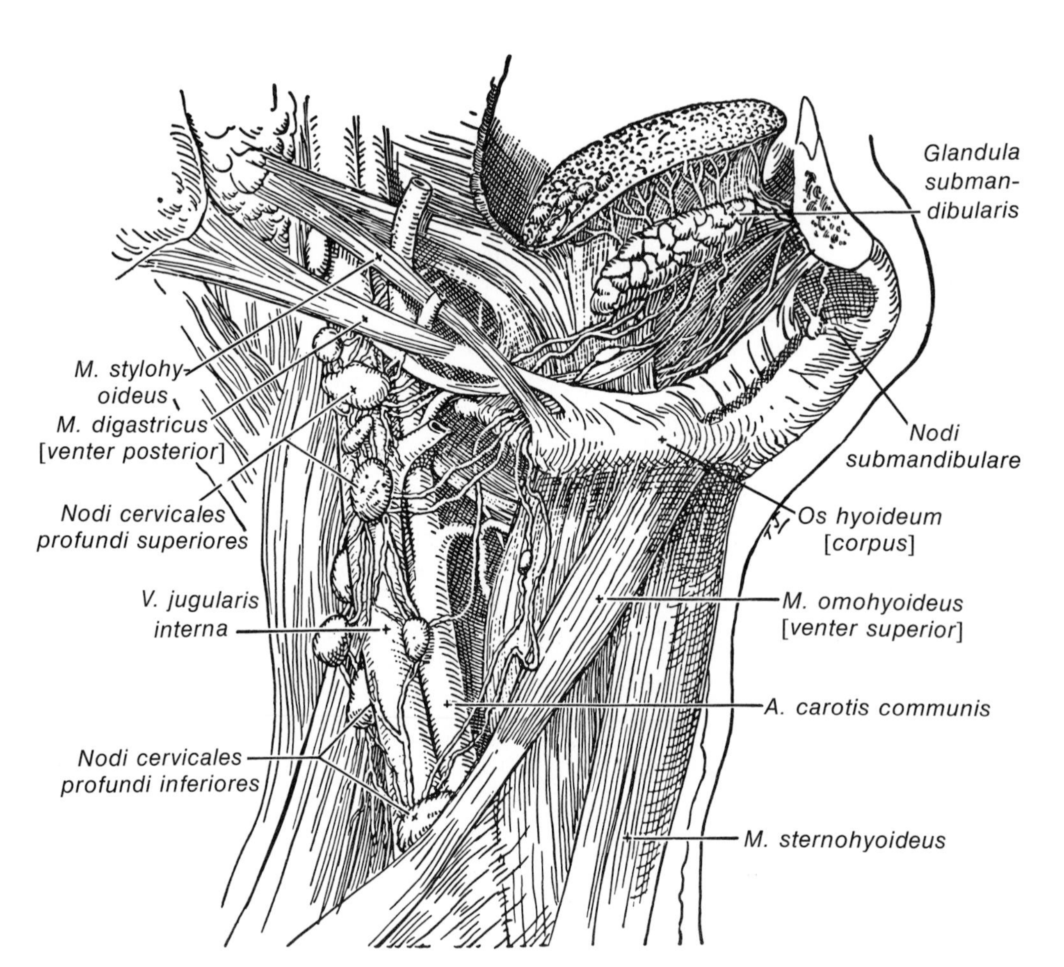

FIG. 5-5. The deep cervical lymphatic nodes and vessels of the right upper cervical triangle. The lymphatic drainage of the tongue is shown. (After Eycleshymer and Jones in Gray's Anatomy of the Human Body, 29th ed. C. M. Goss, editor. Philadelphia, Lea & Febiger, 1973.)

adjacent regions. The deep cervical nodes are further subdivided into *superior* and *inferior* deep cervical nodes (Fig. 5-5). Lymph from the base of the tongue and posterior floor of the mouth drains directly into these nodes. From the auricular nodes, submandibular nodes, submental nodes, facial nodes, occipital nodes, and viscera of the neck, lymph drains into the deep cervical nodes (Fig. 5-2). On the right, lymph from the deep cervical nodes empties into the right lymphatic duct. The thoracic duct collects lymph from the left deep cervical nodes.

Clinical Notes

Knowledge of the lymph system and the geography of the various groups of nodes is important in diagnosis. In the resting normal state lymph nodes are not palpable. Infection or cancer in an area drained by the vessels to these nodes causes the nodes to become very active, and they become quite firm and palpable. Palpable nodes are a most important sign of clinical disease.

The cause of palpable lymph nodes should always be investigated. Pain and swelling of the submandibular nodes are clearly indicative of oral pathology. In addition, we can predict that the pathologic condition, if left untreated, will spread and involve the cervical nodes. Infection in the area of the scalp may be the cause of palpable occipital nodes. Malignancy is also a common cause of palpable nodes. Cancer cells often spread to regional lymph nodes and produce seeding and growth of the tumor within the nodes. Identification of palpable lymph nodes can result in early diagnosis and thereby improve the prognosis.

6 Temporomandibular Joint

Most human bones are joined with one another. These connections of bones to each other are termed *articulations* or *joints.* Some joints are immovable as in the articulations of the bones of the skull (Fig. 1-4). This type of joint is known as a *suture.* In movable joints, the surfaces of the bones are slightly separated (Fig. 6-1). The articulating surfaces are covered by cartilage, and the entire joint is enveloped by a fibrous tissue capsule. The lining cells of the interior of the capsule form a *synovial* membrane which secretes an intracapsular fluid. This fluid acts as a lubricant for the joint. The bones forming the joint are supported, strengthened, and connected by fibrous bands called *ligaments.*

Ligaments are strong bands of fibrous tissue. They appear white and silvery and are somewhat pliant and flexible. Thus, they allow freedom of movement, but still protect the joint against over-excursion. They normally lie outside the capsule and are attached to the involved bones. Sometimes one end of the ligament is attached

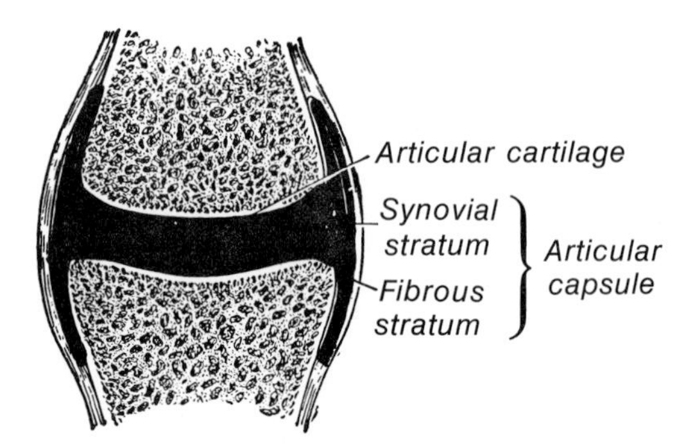

FIG. 6-1. Diagrammatic section of a synovial (diarthrodial) joint. (From Gray's Anatomy of the Human Body, 29th ed. C. M. Goss, editor. Philadelphia, Lea & Febiger, 1973.)

to a more distant bone. The *sphenomandibular ligament* is an example. Often the ligaments are incorporated in and fused to the capsule (Figs. 6-2; 6-3).

Classification of Joints

There are three main classes of joints: synarthrosis, amphiarthrosis, and diarthrosis.

SYNARTHROSIS. Synarthrosis designates an immovable joint such as the sutures of the bones of the skull (Fig. 6-4).

AMPHIARTHROSIS. Joints that are only slightly movable fall into this category. The articulations between the vertebrae are a good example for this class (Fig. 1-2).

DIARTHROSIS. Joints that are freely movable are members of this group. The joints of the arms, legs, and fingers are diarthroidal joints. Also, the temporomandibular joint is diarthroidal (Fig. 6-5).

Diarthrodial joints may be subclassified according to the various types of movements of which they may be capable. The two important divisions relative to the study of head anatomy are the ginglymus (hinge) and the arthrodia (gliding) types.

In a diarthrosis, the articulating surfaces are covered by cartilage. Also, the joint is divided into an upper and a lower compartment by an interposed piece of cartilage known as the *meniscus* or articular disk. The edges of the disk blend with the capsule, and the surfaces of the disk are covered with the synovial membrane (Fig. 6-6).

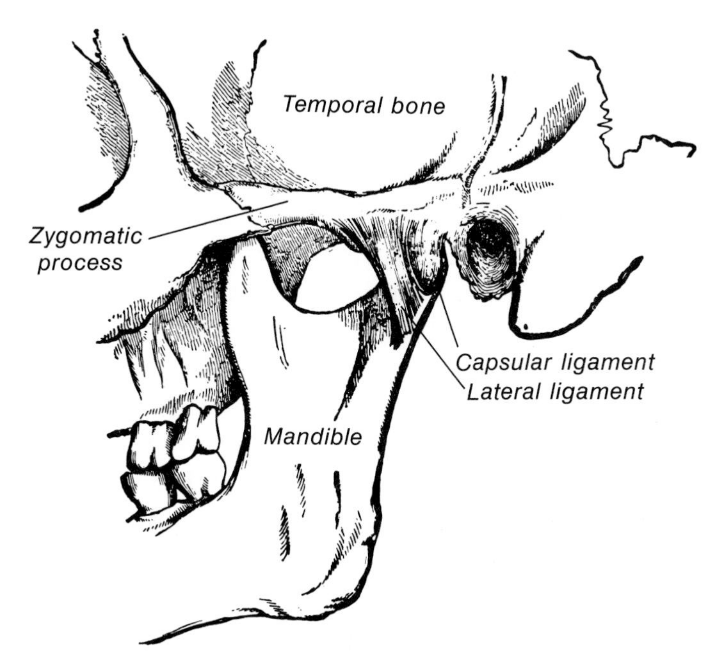

FIG. 6-2. Articulation of the mandible. Lateral aspect. (From Gray's Anatomy of the Human Body, 29th ed. C. M. Goss, editor. Philadelphia, Lea & Febiger, 1973.)

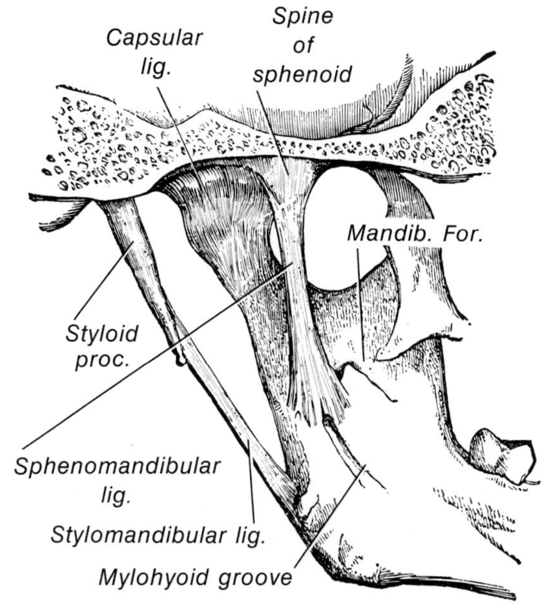

FIG. 6-3. Articulation of the mandible. Medial aspect. (From Gray's Anatomy of the Human Body, 29th ed. C. M. Goss, editor. Philadelphia, Lea & Febiger, 1973.)

Structure of the Temporomandibular Joint (TMJ)

The temporomandibular joint is the articulation between the mandible and the cranium. It is a diarthrosis, but is unique in several aspects from other diarthrodial joints. First of all, both the articulating complexes of this joint house teeth. The shapes and positions of these teeth influence the function of the temporomandibular joint. Secondly, the articulating surfaces are covered with a nonvascular fibrous tissue that contains some cartilage cells and is designated as a *fibrocartilage.* Thirdly, the right and left articulations are functionally intercoupled and have a restricting influence on each other.

The condylar head of the mandible conforms to the anterior slope of the glenoid fossa of the temporal bone (Fig. 6-7). Anteroposteriorly, the condyle is convex. Mediolaterally, it is also convex but is wider than the anteroposterior dimensions. Below the head of the condyle is the neck which is quite thin. The neck is a frequent area of fracture when a strong force is abruptly applied to the lower jaw.

The glenoid fossa of the temporal bone is a smooth concavity covered with a fibrocartilage (Fig. 6-8). Anteriorly, a transverse bony ridge, the *articular eminence,* forms the boundary. The thin bone of the fossa separates it from the cranial cavity. It is interesting that this thin bone is rarely fractured or perforated. The reason is twofold. The articular disk protects the fossa from any trauma the condyle might exert. Also, the head of the condyle is angled forward on the neck of the ramus (Fig. 6-5). Posteriorly, the glenoid fossa is bounded by a bony ridge, the *postglenoid process.*

Between the glenoid fossa and the head of the mandibular

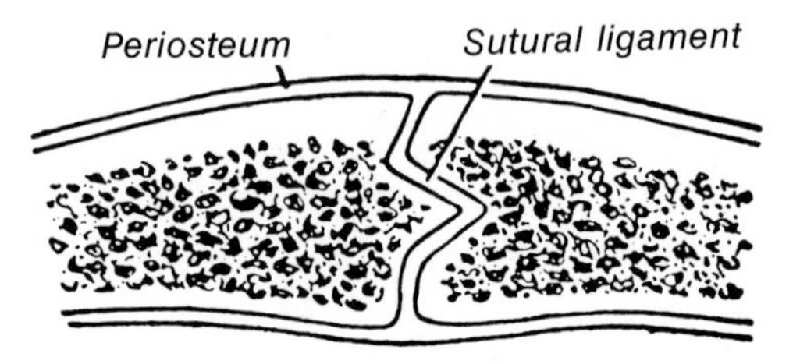

FIG. 6-4. Section across the sagittal suture. (From Gray's Anatomy of the Human Body, 29th ed., C. M. Goss, editor. Philadelphia, Lea & Febiger, 1973.)

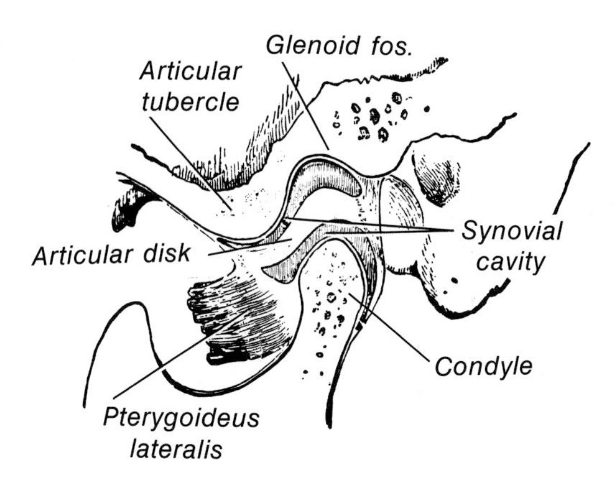

FIG. 6-5. Sagittal section of the articulation of the mandible. (From Gray's Anatomy of the Human Body, 29th ed. C. M. Goss, editor. Philadelphia, Lea & Febiger, 1973.)

condyle lies a fibrous oval disk, the *meniscus* (Fig. 6-5). It is thinner in the central part than at the periphery. Posterior to the disk is a thick layer of loose, vascular connective tissue which fuses with the capsule of the temporomandibular joint.

The capsule encloses the joint. Superiorly, it encircles the glenoid fossa. Below, it attaches to the circumference of the neck of the condyle just beneath the head of the condyle.

The disk is fused to the capsule anteriorly. Some fibers of the external pterygoid muscle attach to the disk here (Fig. 6-5). Laterally and medially, the disk is attached to the medial and lateral poles of the condylar neck. The capsule is too, but somewhat lower than the attachments of the disk.

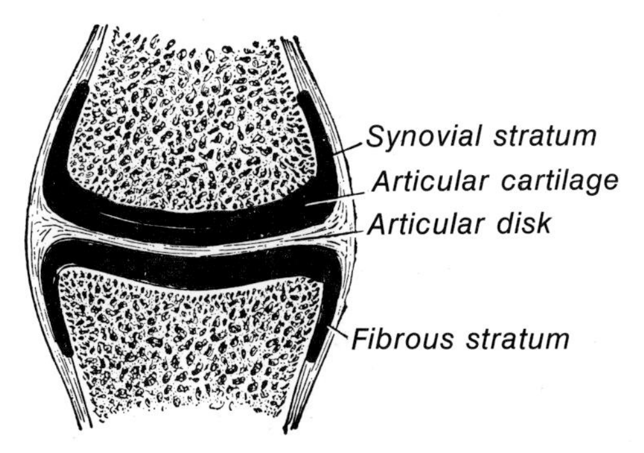

FIG. 6-6. Diagrammatic section of a synovial joint with an articular disk. (From Gray's Anatomy of the Human Body, 29th ed. C. M. Goss, editor. Philadelphia, Lea & Febiger, 1973.)

Laterally, the capsule is thickened, reinforced, and strengthened by the *temporomandibular* (lateral) ligament (Fig. 6-2). The *sphenomandibular* ligament is a band of fibrous tissue arising from the spine of the sphenoid bone and attaching at the lingula of the mandibular foramen. It is located on the medial aspect (Fig. 6-3). The *stylomandibular ligament* is another fibrous band. It extends from the styloid process to the mandibular angle. Only the temporomandibular ligament limits the joint. Neither the stylomandibular ligament nor the sphenomandibular ligament has any influence upon the movement of the lower jaw.

The inner surface of the temporomandibular capsule is lined by a synovial membrane which reflects onto the superior and inferior surfaces of the meniscus. Synovial fluid lubricates the joint.

The temporomandibular joint is divided into two compartments, an upper and a lower, by the articular disk. In theory, these two compartments may move in conjunction with one another or independently. The upper compartment is a sliding joint and the lower compartment is a hinge joint. Because of this dual action the temporomandibular joint is classified as a ginglymoarthrodial joint.

Movements of the Temporomandibular Joint

Under normal circumstances, mandibular movement is a result of upper and lower compartments of the temporomandibular joint acting together. In addition, the condyle, disk, and fossa are in contact at rest and in all movements. Moreover, the head of the condyle, being angled slightly forward, actively holds the disk against the posterior slope of the articular eminence, *not in the depths* of the fossa.

In general, the movements of the temporomandibular joint include opening, closing, protrusion, and lateral displacement (Fig. 6-9). The disk slides on the articular tubercle, and the condyle rotates on the disk when the jaw is opened and closed. In protruding the mandible, both disks glide forward; rotation of the condyles is prevented by contraction of the elevating muscles of the mandible. When the mandible goes into lateral positions, one disk glides

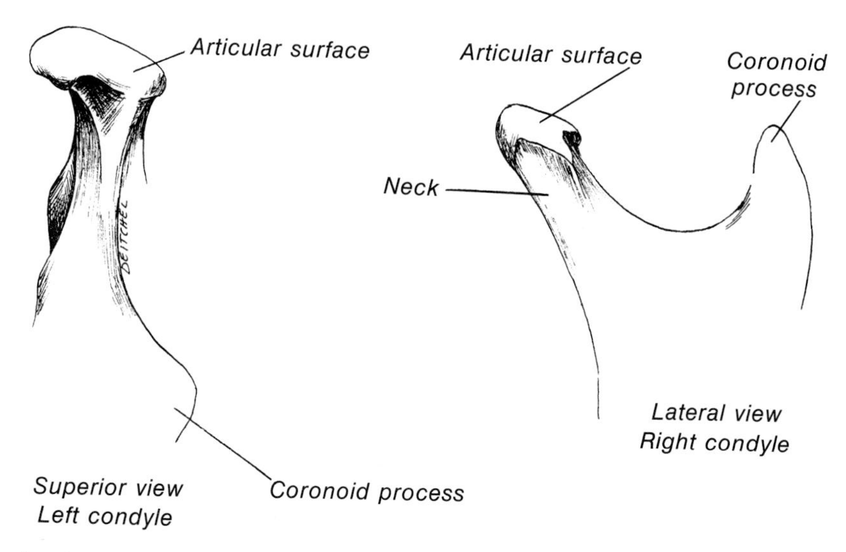

FIG. 6-7. Mandibular condyle.

forward while the other remains stable in position. In chewing, first there is lateral movement; then the mandible is returned to position by the closing muscles.

Opening is accomplished by contraction of the external pterygoid, digastric, mylohyoideus, and geniohyoideus muscles. Elevation is produced by the masseter, temporalis, and internal pterygoid muscles.

The mandible is protruded by simultaneous action of both external pterygoids and the closing muscles. It is retracted by the posterior fibers of the temporalis muscle. Lateral movement is accomplished by contractions of the pterygoid muscles. When the left pterygoid muscles are contracted, the left condyle glides forward, the right condyle stays in position, and the mandible shifts to the right.

Mandibular Positions

There are much research and controversy regarding positions of the mandible, definitions of terms, and relative functions of the positions. However, a general discussion of the subject is necessary to acquaint the student with a basic knowledge of maxillomandibular relations.

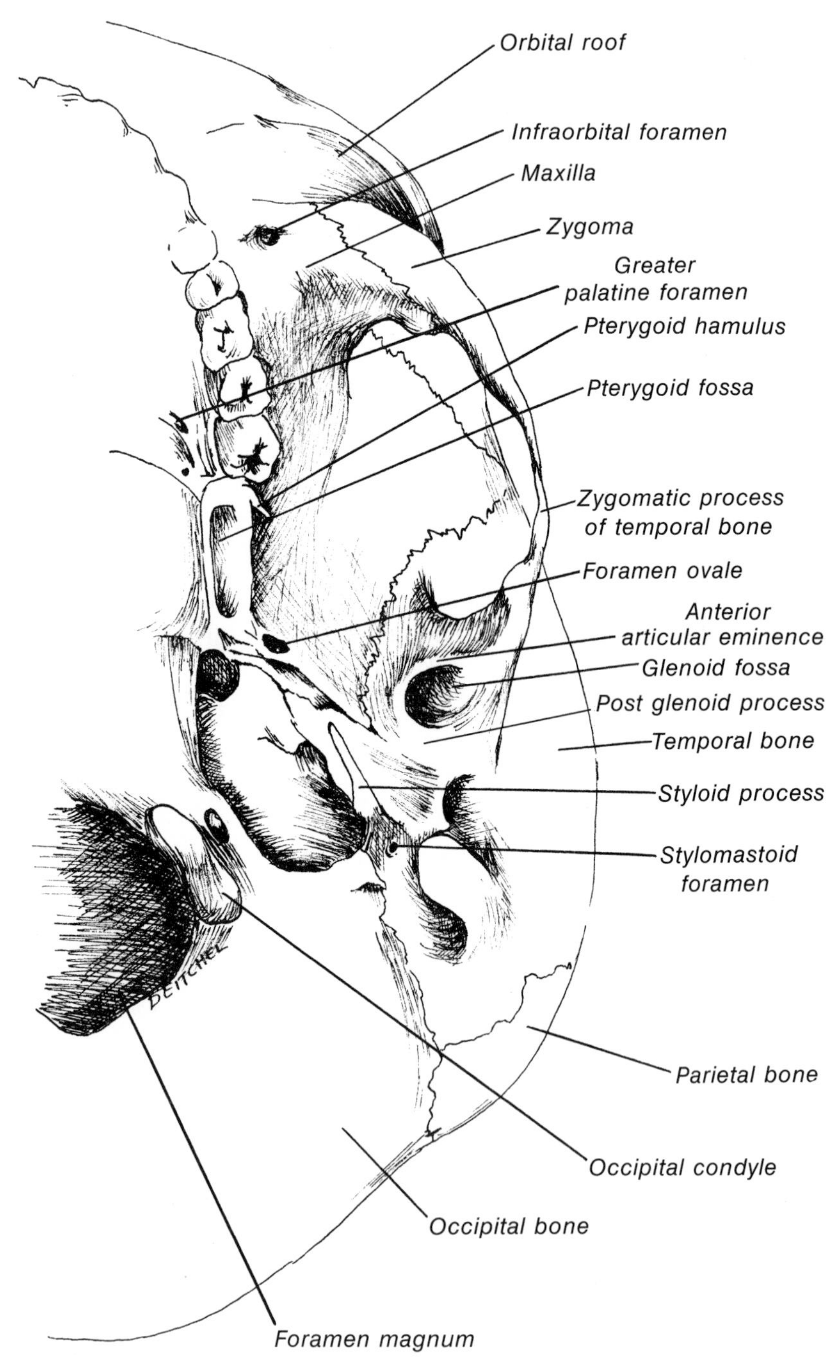

FIG. 6-8. Glenoid fossa and base of skull.

Rest position is herein defined as that position of the mandible when the muscles of mastication are in repose. The individual is sitting or standing, looking straight ahead. The teeth are not in contact. In rest position, approximately 3 millimeters of space are between the upper and lower bicuspids.

When the lower teeth and upper teeth are occluded with the most complete and comfortable intercuspation, the position is termed *centric occlusion.*

Centric relation may be defined as the teeth positioned in occlusion with the condyles in the most posterior location possible.

It is from these various neutral positions that the mandibular excursions take place.

Clinical Notes

The temporomandibular joint, because of its anatomy and unique features, presents several interesting clinical problems. There are proprioceptive fibers in the temporomandibular joint, the periodontal membranes of the teeth, and muscles of mastication. Those in the periodontal membranes are more sensitive and exacting. Patients without their natural dentition, therefore, often have difficulty positioning their jaws in the neutral positions which are based on occlusion.

Pain

Pain in the temporomandibular joint may be a result of a pathologic condition in the joint or of a pathologic condition elsewhere that is radiating pain to the joint. Infection in the joint is rare. Degenerative disease of the various structures of the joint, unfortunately, is not so rare. *Arthritis* is inflammation of a joint. The temporomandibular joint may become inflamed either as an extension of generalized arthritis of all joints or as a result of malfunction of the masticatory machine. Malocclusion can result in undue stress and abnormal pressures on the protective fibrocartilages. The cushion effect of the cartilages is reduced and inflammation results. Tumors of the

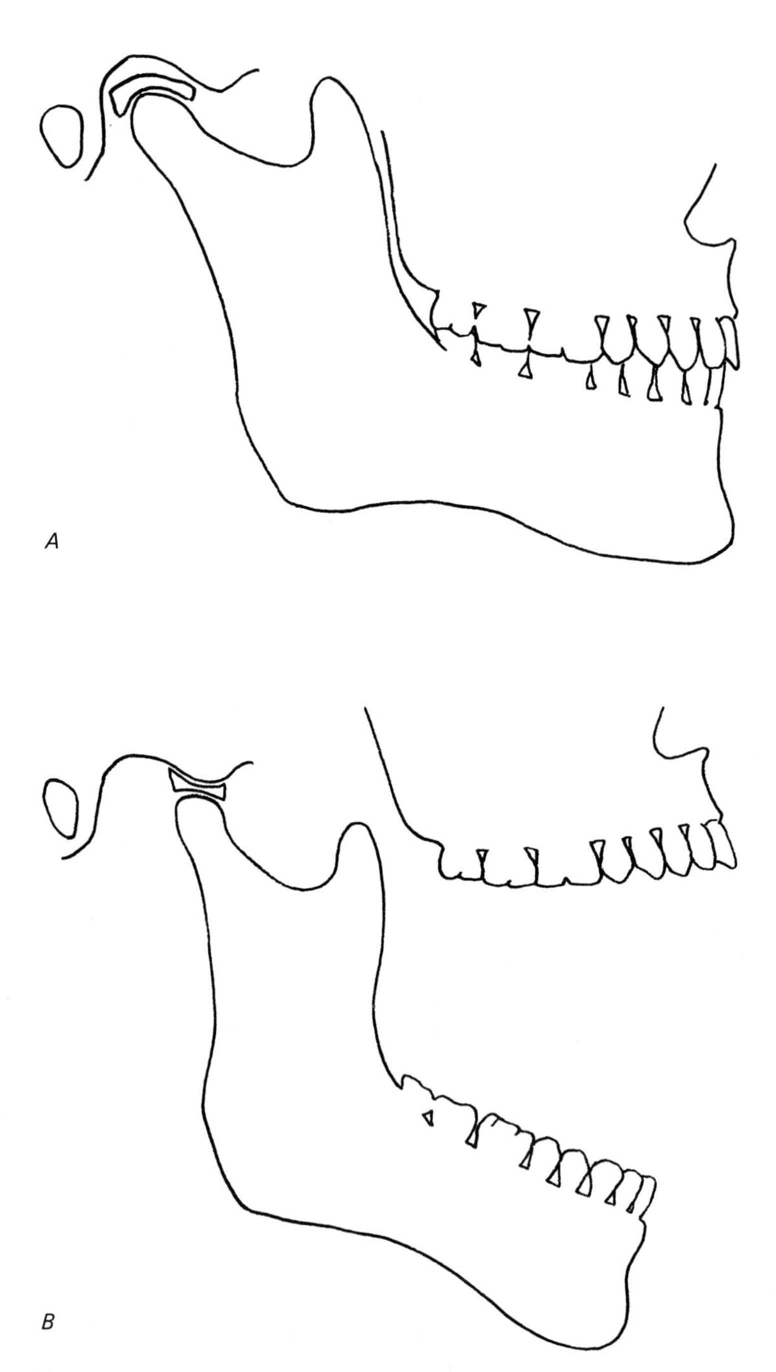

FIG. 6-9. A. Normal occlusion. B. Mouth wide open. Note position of condyle and its relation to meniscus and anterior articular eminence.

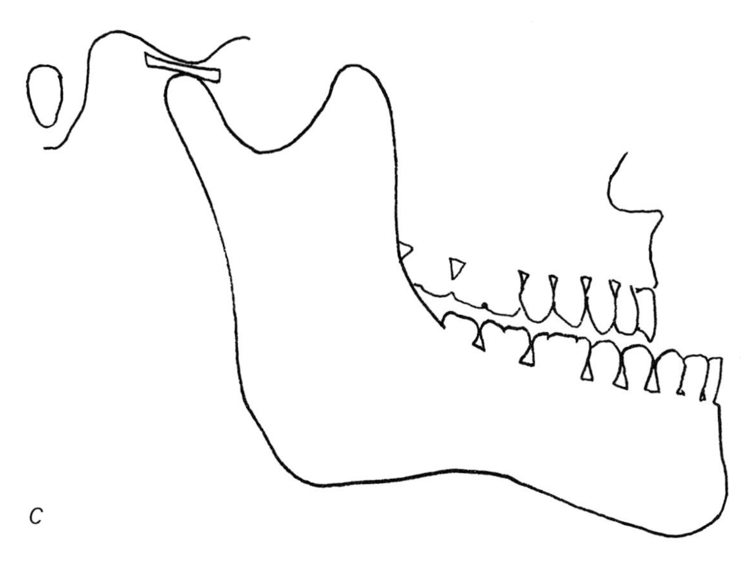

FIG. 6-9. C. Protrusion.

condyle or other associated parts may cause painful pressure on nerves.

The patient with an infection or a pathologic condition of the ear, may complain of temporomandibular joint pain. Pain in the teeth or jaws, primarily in the lower molars or mandible, often radiates to the temporomandibular joint. The sensory nerve supply to the temporomandibular joint and ear is via the auriculotemporal nerve. Pain in the jaws, radiating to the ear and the temporomandibular joint is thought to be based on central nervous system synapses and connections of the various branches of the fifth nerve. Thus, pain stimuli to the inferior alveolar nerve from a lower molar may be interpreted as ear or joint pain via the auriculotemporal nerve. A similar situation occasionally occurs with pain from the upper teeth radiating to the lower teeth or ear.

Clicking

Many patients hear or feel clicking in the temporomandibular joint during various excursions of the jaw. If the meniscus becomes more mobile than normal, clicking or *crepitus* is often the result. Hyper-

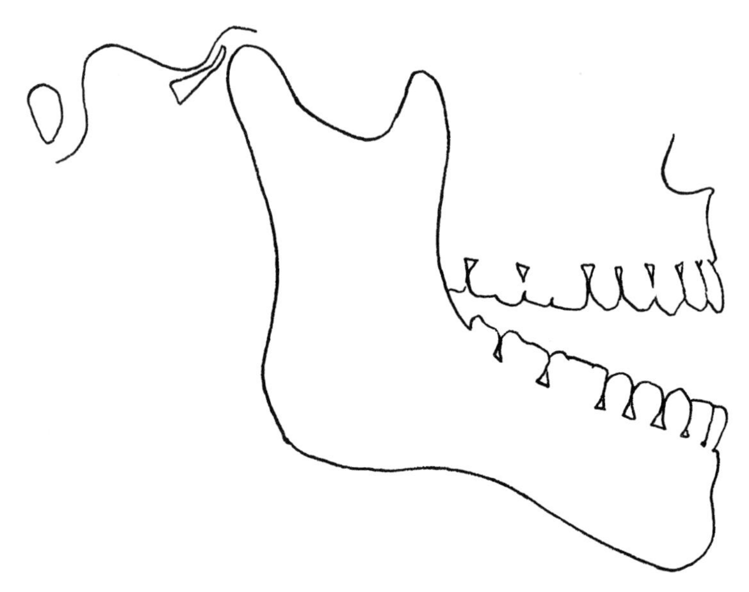

FIG. 6-10. Dislocation.

mobility of the meniscus may result from degeneration and reduction in thickness of the disk. It may also result from a "sloppy" joint caused by overstretching and loss of resiliency of the temporomandibular ligament and capsule. Stretching of the fibrous attachments of the meniscus to the condyle may also cause hypermobility of the meniscus and result in crepitus.

Dislocation

When the condyle comes to rest in *front* of the anterior articular eminence, it is dislocated (Fig. 6-10). The patient is unable to close his mouth. Normally, this situation is painful. Even under normal conditions, when the jaw is opened wide, the condyle and disk pass beyond the summit of the anterior articular eminence (Fig. 6-9B). The closing movement begins with retraction of the mandible, bringing the condyle into position behind the height of the eminence. Then the elevators replace the condyle and eminence into the fossa. This movement must be made with extreme precision and coordination. If, however, this coordinated motion is disturbed and the *elevators* contract *before* the retractors, the condyle may be moved superiorly before it is moved posteriorly. Dislocation will occur. Overexcursion upon opening may stretch the capsule or muscles, causing spasm of the muscles, loss of coordination, and dislocation. This painful malposition anterior to the eminence is then maintained by the continued spasm of the elevators.

Reduction of the dislocation is aimed at overcoming the muscle spasm. This may be accomplished by general anesthesia, which is usually not necessary. The dislocation can often be reduced by pressing downward on the lower molars. Since the muscle spasm may be a result of increased reflex stimuli to the muscles from capsule damage, injection of a local anesthetic directly into the capsule sometimes reduces the dislocation by interrupting these reflex stimuli.

If a patient is able to reduce the dislocation for himself, the reduction is termed *subluxation.* If help from another person is required, the term is *luxation.* Patients who have frequent luxation of the temporomandibular joint may be treated by surgically reducing the height of the articular eminence (eminectomy). Also, surgical tightening of the temporomandibular ligament may reduce joint mobility and prevent dislocation from overexcursion.

7

Salivary System

Fluid-producing glands and ducts surrounding and within the oral cavity comprise the salivary system. The smaller glands are found in the submucosa of the mouth. The larger, major glands are located further from the oral cavity and open into the mouth through a large duct.

Saliva

The oral cavity is kept continually moist by a mucoserous fluid known as *saliva.* This secretion is composed of mucus and serous fluid. The serous material contains ptyalin, an enzyme for digesting starches. The pH of saliva is between 6.0 and 7.0 (acid). The total quantity of saliva produced in one day is approximately 1500 cc.

Saliva is produced and secreted by the various major and minor salivary glands located in or near the oral cavity. The secretion of saliva is stimulated by several sources. Taste stimuli and tactile stimuli increase salivation. It is interesting to note that smooth objects in the mouth produce increased salivation while rough objects reduce the secretion of saliva. Moreover, pleasant foods increase salivation, whereas unpleasant foods reduce salivation.

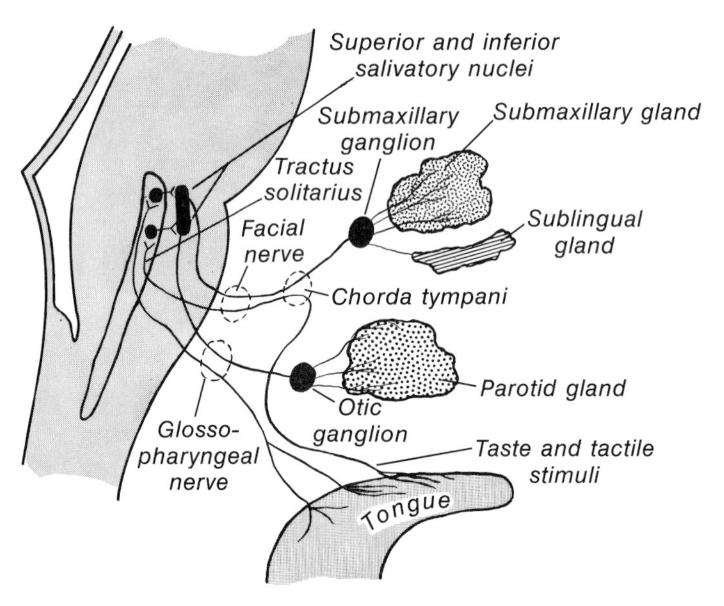

FIG. 7-1. Nerves regulating salivary secretion. (From Guyton, A. C.: Textbook of Medical Physiology, 4th ed. Philadelphia, W. B. Saunders Co., 1971.)

Hunger and the smell of food also increase salivation. Figure 7-1 diagramatically illustrates a few of the various nerves and pathways which neurologically control salivation.

Saliva not only lubricates the oral cavity and digests starches, but it also has other very important functions. Constant lavage of the oral cavity and cleansing of the teeth and surrounding tissues is an extremely important function of saliva. Patients who have reduced amounts of saliva for extended periods of time suffer from increased dental caries and poor oral hygiene, with fetor oris. Some authorities also believe that saliva has some antibacterial properties.

Major Salivary Glands

The major salivary glands may be divided into groups based on their anatomic size and location.

Parotid Gland (Fig. 7-2)

The right and left parotid glands lie in the right and left retromandibular fossae. Normally, the gland is divided into a superficial

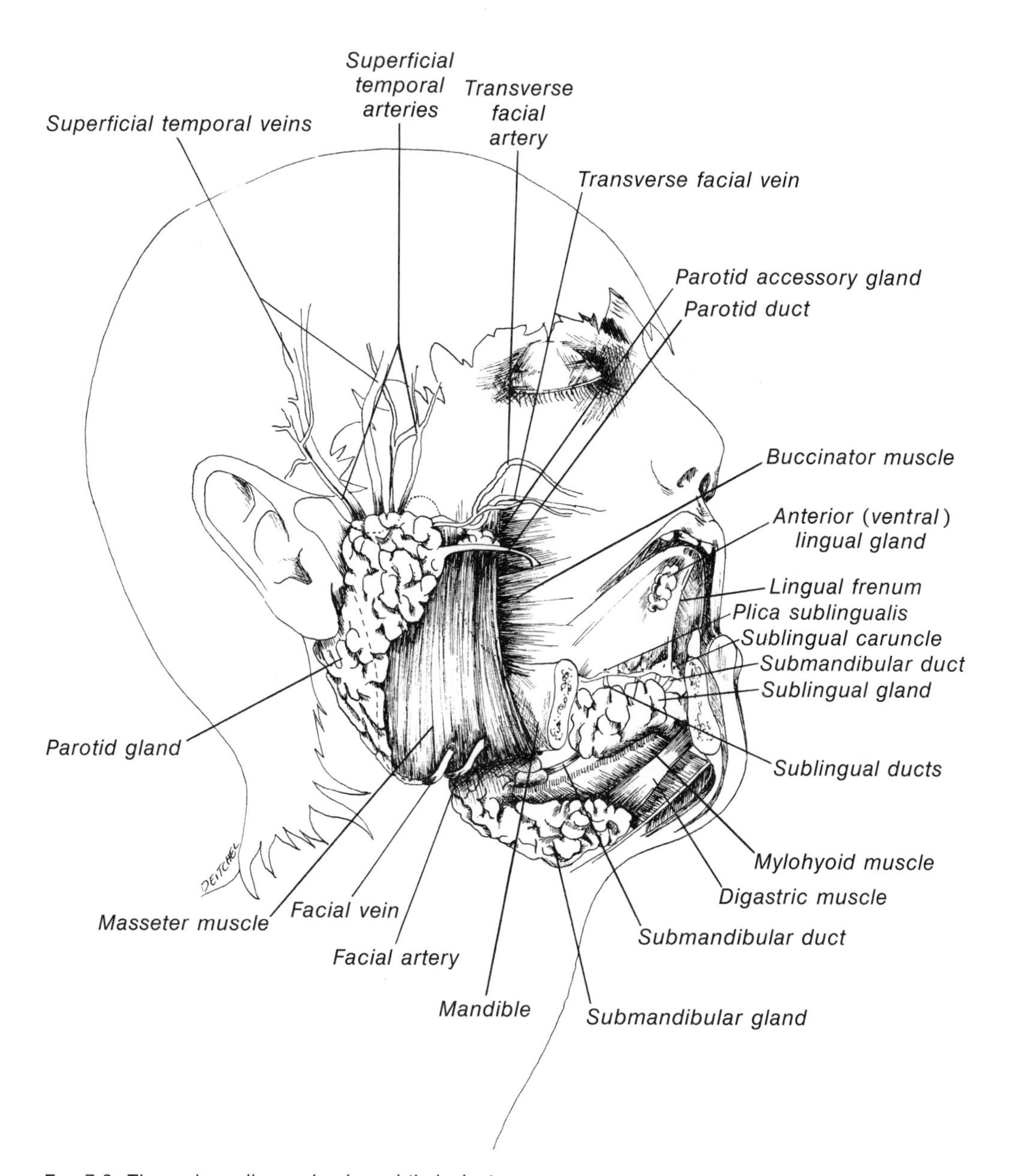

FIG. 7-2. The major salivary glands and their ducts.

lobe and a deep lobe. Both lobes are connected by a variable thick *isthmus.* The superficial lobe extends anteriorly on the lateral surface of the mandibular ramus. Medially, the deep lobe reaches the styloid process and associated muscle and ligaments. It surrounds the inferior surface of the auditory canal and is in relation to the mastoid process and sternocleidomastoid muscle posteriorly.

Emerging from the gland, near the anterior border, is the *parotid (Stensen's) duct.* It passes horizontally, anteriorly on the lateral surface of the masseter muscle. Once the duct reaches the anterior border of the masseter muscle, it turns sharply medialward, pierces the buccinator muscle, and opens into the oral cavity opposite the maxillary molars. Usually *the parotid papilla,* a small papilla protruding from the cheek mucosa opposite the maxillary second molar, marks the opening of the parotid duct. In many patients, a small accessory parotid gland lies just superior to the duct on the surface of the masseter muscle.

The parotid gland is enclosed in a fibrous capsule that is continuous with the masseteric and investing layer of deep cervical fascia. Laterally, the capsule is more dense and stronger than it is medially. Medially, the capsule is continuous with the pterygoid fascia and stylomandibular ligament. Posteriorly, the capsule is closed and fuses with fascia of the sternocleidomastoid muscle. The lobes of the parotid gland are subdivided into lobules by strong fibrous septa which are continuous with the capsule.

Various important structures are close to the parotid gland. The external carotid artery enters the substance of the gland and exits as the superficial temporal artery. The temporal vein and retromandibular vein are also embedded in the substance of the gland. During surgical procedures, blood vessels can be identified, ligated, and divided. Blood supply is then provided by collateral circulation and surgery of the parotid region may be continued without the hazard of excessive bleeding.

One other important group of structures is related to the parotid gland. This is the ramification of the facial nerve. This nerve *must not* be cut and must be indentified, retracted, and protected during surgical procedures because nerves regenerate slowly, if at all. The facial nerve enters the parotid gland after leaving the stylomastoid foramen. The nerve ramifies within the gland and emerges as the five main branches. These branches leave the gland on the deep surface of the superficial lobe and pass to the various muscles of the facial expression (Fig. 4-34).

The saliva elaborated by the parotid salivary gland is collected in small ducts within the gland which ultimately empty into the main parotid duct. The parotid duct then carries the saliva into the mouth through the opening opposite the maxillary molars. Saliva produced by this gland is purely serous and contains no mucus except in the newborn.

Briefly, the arterial supply is via branches of the external carotid arteries. Veins empty into the external jugular vein. The nerve supply for salivation arises from the glossopharyngeal parasympathetic nerves. Afferents of pain and touch are from the auriculotemporal branch of the trigeminal nerve. (Details are in Chapters 3 and 4.)

The Submandibular Gland (Fig. 7-2)

The submandibular salivary gland occupies a large portion of the submandibular space. It lies primarily in the posterior portion of this area with its upper pole extending around the free border of the mylohyoid muscle and lying on the superior surface of this muscle. From the upper pole extends the *submandibular duct* (Wharton's duct). The duct courses anteriorly in the floor of the mouth. It begins by numerous branches, from the deep surface of the gland, and passes forward between the sublingual gland and the genioglossus muscle. When the duct reaches the lingual frenum, it opens at a small papilla, the *sublingual caruncle.*

The saliva elaborated by the submandibular gland is mixed, primarily (80%) serous and 20% mucus. By raising the tongue and massaging the gland extraorally, one can easily see saliva exuding from the sublingual caruncle.

A capsule of investing fascia encloses the gland. This capsule, however, is only loosely attached to the gland. Thus the gland is easily "shelled out" during dissection. On the inner surface of the submandibular gland, one sees the facial artery which may even be embedded in the gland.

Nerve supply is via the chorda tympani and lingual nerves. Branches of the facial and lingual arteries supply the blood. Venous drainage is via the sublingual, submental, and facial veins. The submandibular lymph nodes drain lymph from this area.

The Sublingual Gland (Fig. 7-2)

The sublingual gland is a diffuse, poorly encapsulated gland. The saliva it secretes is mixed also, but the mucous portion predominates. It is the smallest of the major glands and is located beneath the mucous membrane of the floor of the mouth. Below is the mylohyoid muscle; laterally is the body of the mandibule; behind is the deep part of the submandibular gland; and medially is the genioglossus muscle.

The sublingual gland empties by several (8 to 20) small ducts (*ducts of Rivinus*). Some of these ducts open directly into the mouth; others empty into the submandibular duct. Along the lingual frenum and under the tongue is the *plica sublingualis,* an elevated crest of

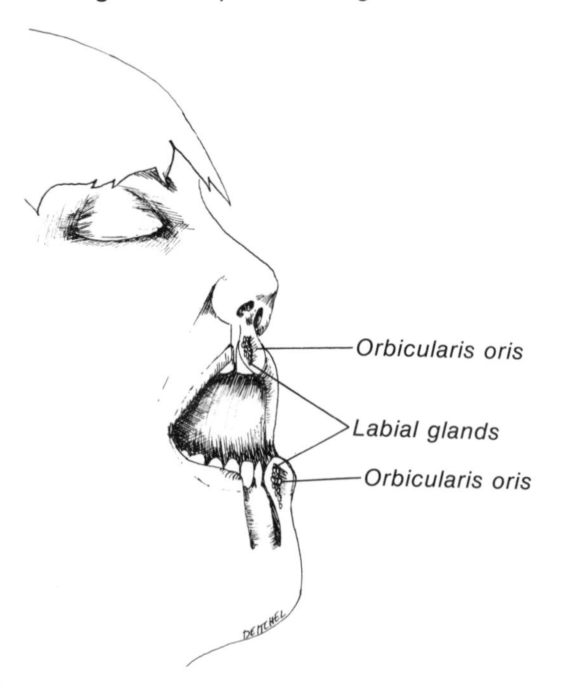

FIG. 7-3. Labial glands.

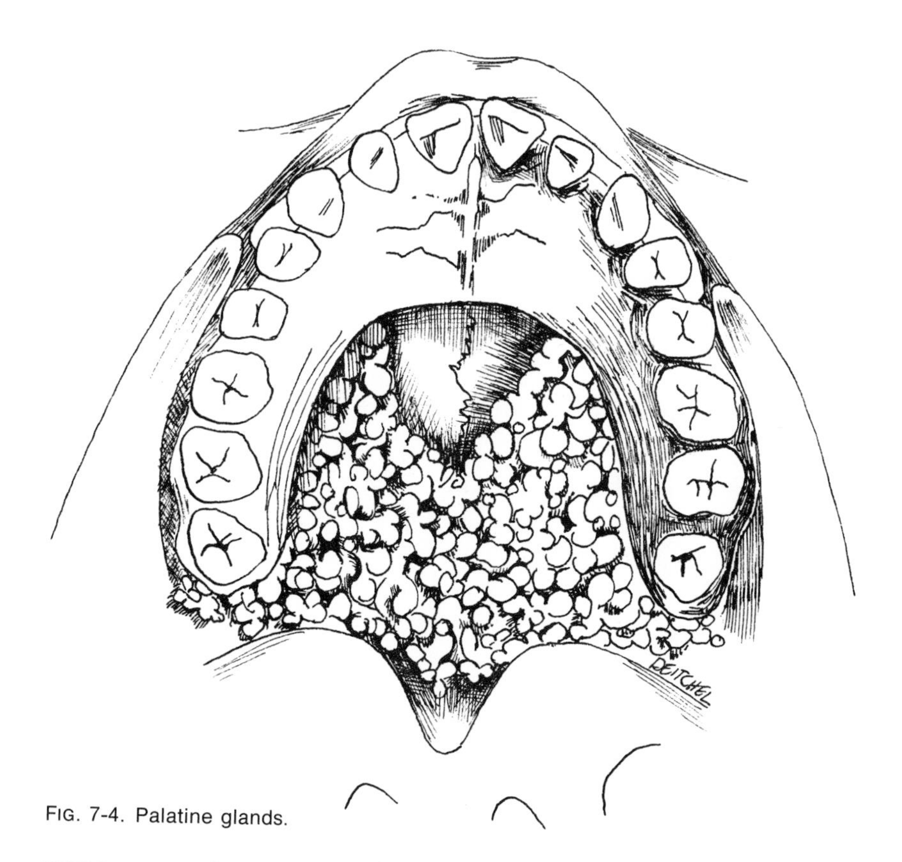

FIG. 7-4. Palatine glands.

mucous membrane caused by the projection of the gland and openings of the sublingual ducts.

The blood supply to the sublingual gland is via the sublingual and submental arteries. The nerve supply is the same as that for the submandibular gland.

Minor Salivary Glands

The saliva from minor salivary glands is usually of the mixed type, with one component predominating. The labial, buccal, and palatal glands are primarily mucus secreting, as is the anterior ventral lingual gland. The posterior dorsal lingual glands are predominately serous in nature and probably serve to keep the convolutions free of debris. Moreover, the dorsal lingual glands assist in the distribution of the substance to be tasted.

The labial, buccal, and lingual glands are supplied by the chorda tympani nerve. The palatal glands are supplied by the greater petrosal branch of the facial nerve.

Classification of the minor salivary glands is also based on their location. They are poorly encapsulated and normally open by many ducts directly into the mouth.

Labial Glands

A large percentage of the substance of the lips is glandular tissue (Figs. 7-3; 10-2). Here multiple small glands lie between the mucosa of the lip and the orbicularis muscle. They open by many small ducts directly onto the lip mucosa.

Buccal Glands

In a similar, diffuse fashion, to that of the labial glands, the buccal glands are located in the cheek, between the mucosa and the buccinator muscle.

Palatine Glands (Fig. 7-4)

The posterior one third of the hard palate is covered with the palatine glands. These glands are sparsely found anterior to the bicuspid region of the palate. The are also found in the soft palate. Indeed, they are quite numerous here.

Lingual Glands

There are two groups of lingual glands. Anteriorly, on the ventral surface of the tongue, is the *anterior lingual gland* (*of Blandin and Nuhn*) (Fig. 7-2). On the dorsal surface of the tongue are the glands surrounding the trough of the circumvallate papillae. Also, on the dorsal surface are glands around the lingual crypts at the base of the tongue (Fig. 10-7).

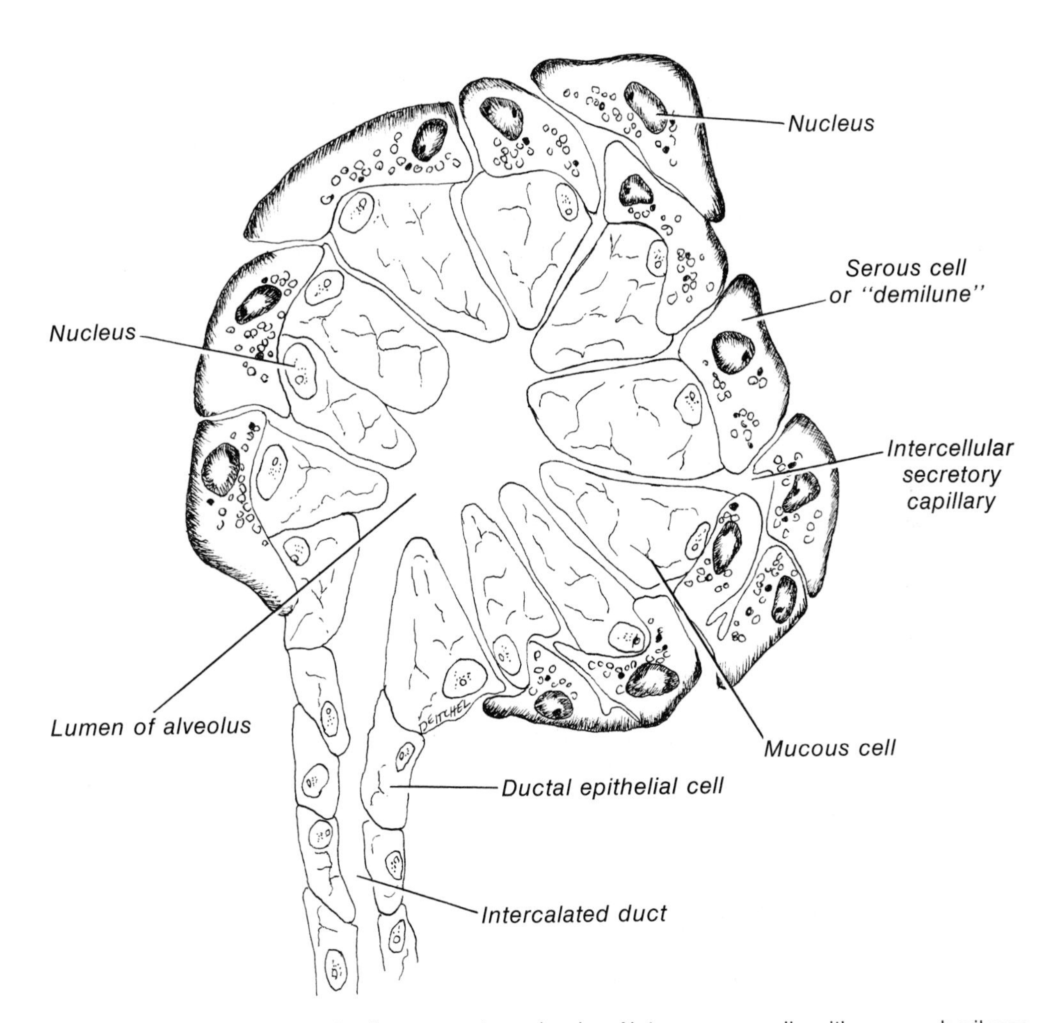

FIG. 7-5. Microscopic view of salivary secretory alveolus. Note mucous cells with serous demilunes.

Histophysiology

The microscopic anatomy of the salivary system explains the function of these glands. The serous cells appear different from the mucous cells (Fig. 7-5).

The *serous (albuminous) cells* are polyhedral in shape and form globular alveoli (acini). An *acinus* is a saclike dilation forming the end of the microscopic duct. Serous secretions of these cells are expelled into the duct, carried to larger ducts, and finally flow into the main excretory duct of the gland.

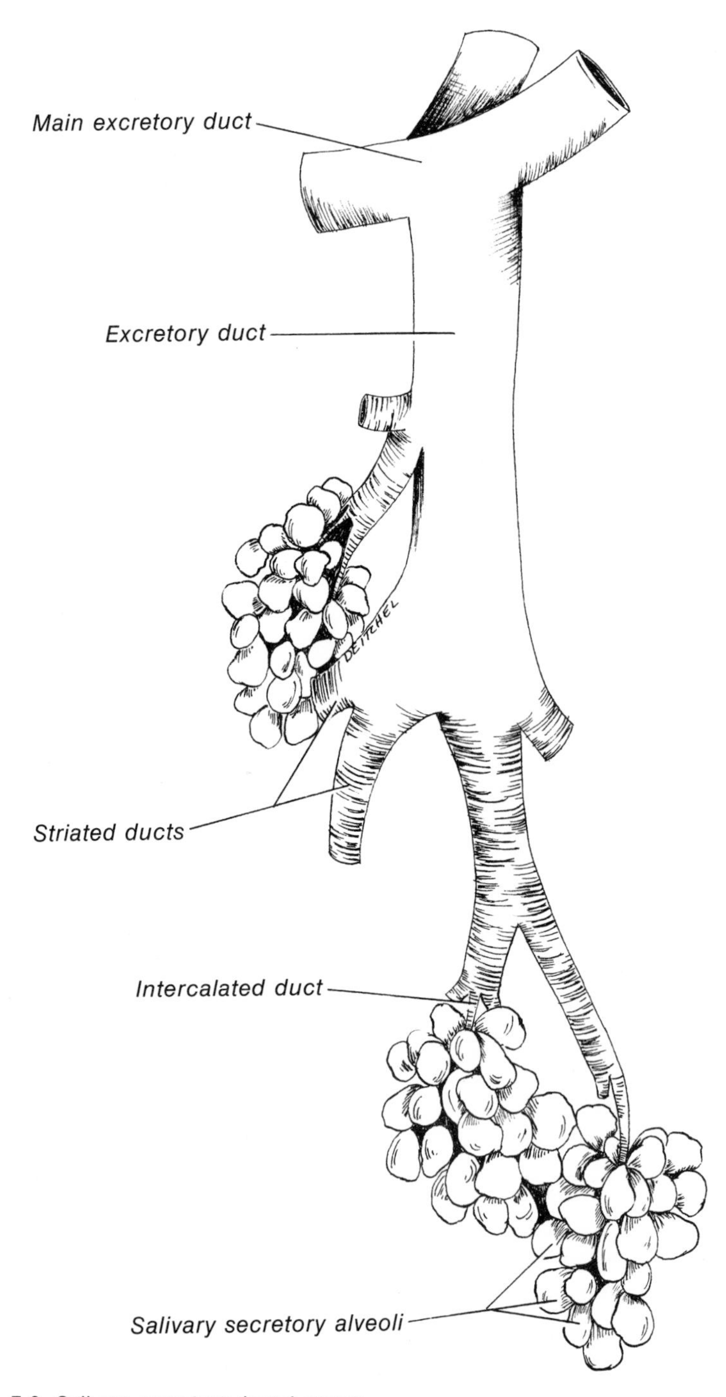

FIG. 7-6. Salivary excretory ductal system.

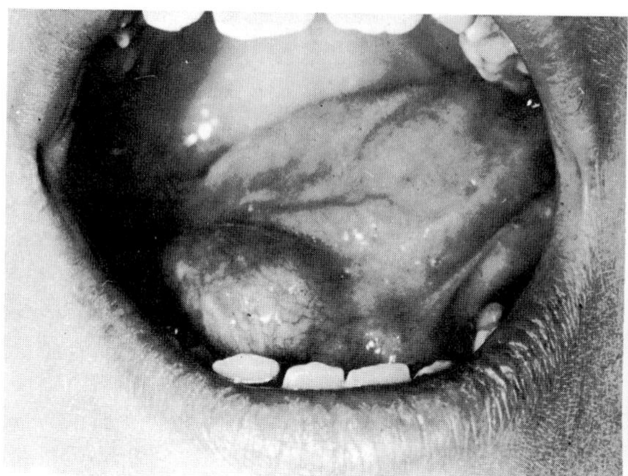
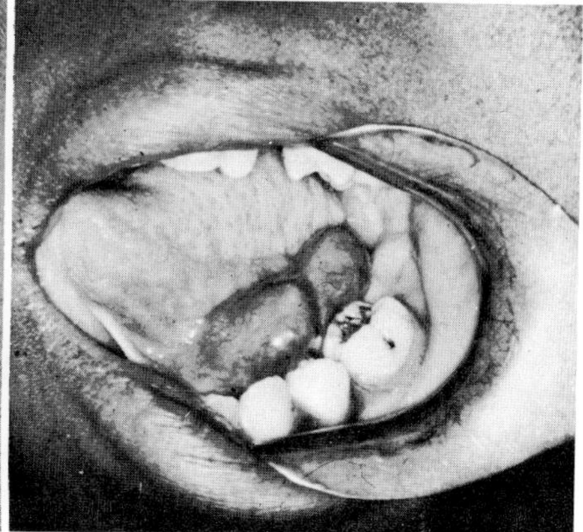

FIG. 7-7. Ranula. Note the swelling in the floor of the mouth opposite the bicuspids. (From Zegarelli, E. V., Kutscher, A. H., and Hyman, G. A.: Diagnosis of Diseases of the Mouth and Jaws. Philadelphia, Lea & Febiger, 1969.)

Mucous cells secrete mucin, which is more viscous than the watery material elaborated by serous cells. Mucous cells are cuboidal in shape, and the lumen formed by a mucous acinus is wider. Once mucin is secreted into the lumen of the acinus, it is carried by increasingly larger ducts to the main excretory duct (Fig. 7-6).

The duct system is complex and branching. In the smaller minor glands, the system is not as intricate as in the larger major glands. *Intercalated ducts* are the smallest channels, and they connect the alveoli to the rest of the ductal system. The less serous-producing the gland, the shorter and less apparent are the intercalated ducts. In purely mucus-secreting glands, intercalated ducts are absent. *Striated ducts* are the next larger sized ducts. They empty into the *excretory ducts* which ultimately release their contents into the main duct of the gland.

In mixed glands, the arrangement of cells of the alveoli takes several patterns. All the cells of some alveoli are serous; some are purely mucous, and some alveoli are mixed. In mixed alveoli, the serous cells are placed at the periphery of a clump of mucous cells. These serous cells appear as crescent-shaped caps known as *demilunes* (Fig. 7-5). Small separations between the acinar cells, the *intercellular secretory capillaries,* carry the serous material to the lumen of the alveolus where it mixes with the mucin and empties into the intercalated ducts.

The nerve supply to the salivary glands is described in detail in Chapter 4. Chapter 3 covers the details of various associated arteries and veins.

Clinical Notes

The salivary glands are often involved in pathologic processes that result in clinical symptoms.

Mucocele

Usually found on the lower lip, a mucocele is the pooling of saliva in the tissues surrounding a minor salivary gland. It usually results from trauma to the excretory duct allowing extravasation of saliva into the tissues beneath the mucosa. It appears as a variably sized vesicle and is normally painless.

Ranula

The ranula is similar to a mucocele, but the sublingual salivary gland is involved rather than a minor gland (Fig. 7-7). Also, a duct may be blocked by a mucous plug or small salivary stone (*sialolith*). If left untreated,this saliva-filled sac may grow and eventually occupy the entire floor of the mouth.

Sialolith

Calcareous concretions form in various ducts of the body. Gallbladder and kidney stones are of common occurrence. The pain associated with such phenomena may be quite severe.

When a stone is blocking a major salivary gland, the patient complains of pain in the area of the gland. The pain increases during mealtime, owing to the blockage of increased salivary flow during these periods. Swelling and retrograde infection are common. Often,

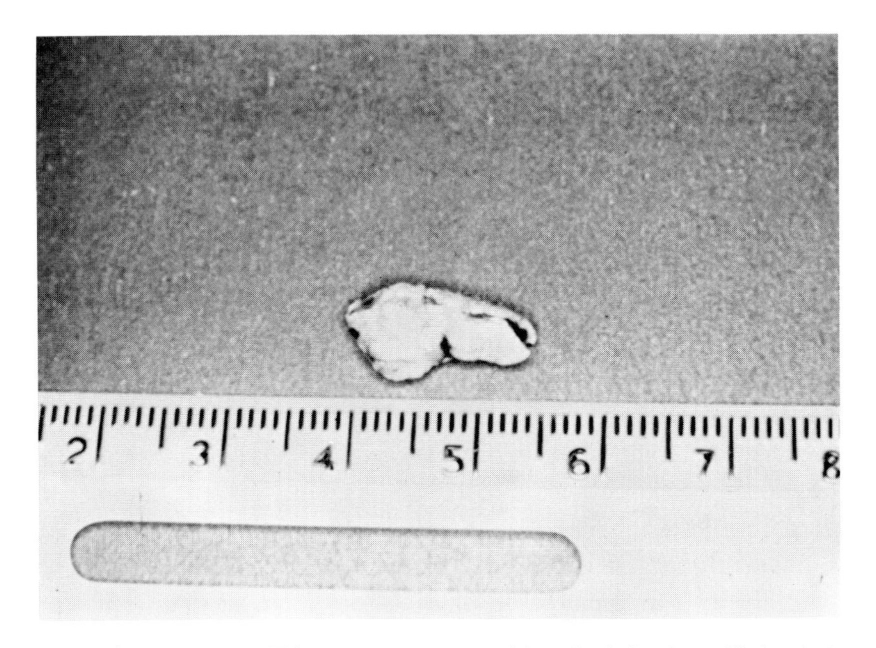

FIG. 7-8. Salivary stone. This stone was recovered from the left submandibular duct.

the stone can be radiographically demonstrated. Treatment centers around removal of the stone (Fig. 7-8). Removing a deep stone of the parotid gland can be most challenging. Fortunately, sialoliths are more frequently found in the submandibular gland because of the greater mucin content of the saliva of this gland.

Infections

Infections usually cause swelling and pain. Unless saliva is blocked, however, primary bacterial infections of salivary glands are rare. The glands are self-irrigating. Infections may be disseminated from other sites. Masticator space infections often spread to the parotid space and cause purulent parotid saliva, a good diagnostic sign of infection. Similarly, dental infection of the submandibular space often involves the submandibular salivary gland secondarily.

Probably the most common primary infection of the salivary glands is mumps, a viral infection that usually involves the parotid gland. Mumps may spread to other salivary glands, and in the adult, to other organs of the body (prostate, ovaries, pancreas). This could lead to rather severe complications.

Tumors

Tumors of salivary glands may be either benign or malignant. They may involve either major or minor glands.

Benign tumors can usually be easily removed and the prognosis is excellent, especially when they involve minor glands. Parotid surgery is always challenging because of the intimacy of the facial nerve.

Malignant tumors vary widely in their configuration and type. The treatment and prognosis depend on the type, size, location, and duration of the lesion.

An important and ominous diagnostic sign of tumors of the parotid gland is facial nerve palsy. Benign tumors of the parotid gland rarely cause paraylsis of the facial nerve.

Sialorrhea

Increased salivation beyond normal is usually of psychic origin. It may be drug-induced. Some patients taking parasympathetic-type drugs complain of sialorrhea.

Xerostomia

Dry mouth, or xerostomia, results from reduced function of the glands. This also may be psychic. Removal of as many as four of the major glands, however, may not result in xerostomia. Radiation for the treatment of cancer of the head and neck will often produce xerostomia. Elderly patients whose glands have atrophied may present with xerostomia. Patients on parasympathetic blocking drugs for the treatment of ulcers often complain of dry mouth. Vitamin deficiencies have also been implicated.

8

Fascial Planes of the Head and Neck

The entire body is enclosed in a large envelope, the skin. The organs, glands, muscles, blood vessels, and other structures are enclosed in individual envelopes deep to the skin. These deep envelopes are composed of fibrous connective tissue.

Connective Tissue

The connective tissues are all the connecting and supporting tissues in the body. Even bone and cartilage are considered connective tissues. These tissues may be classified according to the texture or character of the intercellular matrix.

The *intercellular matrix* is the material or substance that comprises the bulk of the tissue and serves to bind the entire mass into a unit. It is found between the connective tissue cells and incorporates them into the tissue as a whole (Fig. 8-1).

The matrix in bone is tough and fibrous. It is made rigid by the presence of mineral salts. The matrix in cartilage is also firm, but it is more elastic. Cartilage tolerates compression well and is therefore frequently found between articulating bones, where it serves as a cushion.

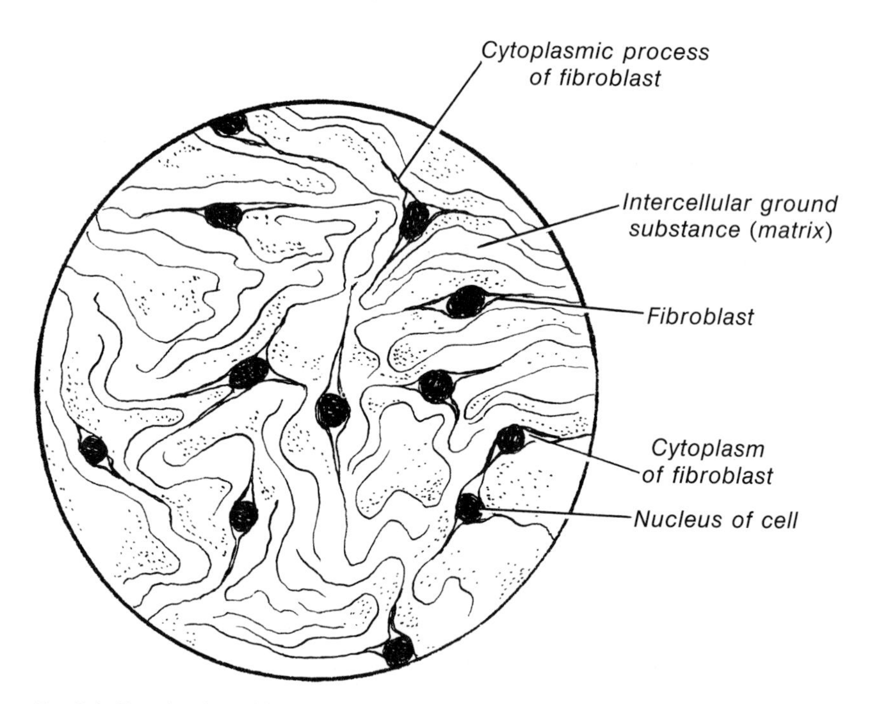

FIG. 8-1. Drawing from high power magnification of loose alveolar connective tissue.

Fibrous connective tissue is another type of connective tissue. It can also be further classified according to its texture. This texture depends on the type of microscopic fibers and substance comprising the matrix. Tendons and ligaments are made of *dense fibrous* connective tissue (Fig. 8-2), which exhibits a great tensile strength.

Loose fibrous connective tissue has a strong binding power also, but it is more loosely interwoven. Between the fibers of the matrix are interstices that are filled with tissue fluid or ground substance. Loose fibrous connective tissue is found in various degress of denseness, and its tensile strength is somewhat less than that of dense fibrous connective tissue (Fig. 8-1).

The capsules of joints and the capsules of some glands and organs are strong and closely woven structures. They are composed of a type of fibrous connective tissue known as *fibroelastic tissue. Fibroareolar* or *areolar* connective tissue implies that the tissue is *loosely* woven, fairly weak, and friable.

Fibrous tissue cells are called *fibroblasts.* They are widely sepa-

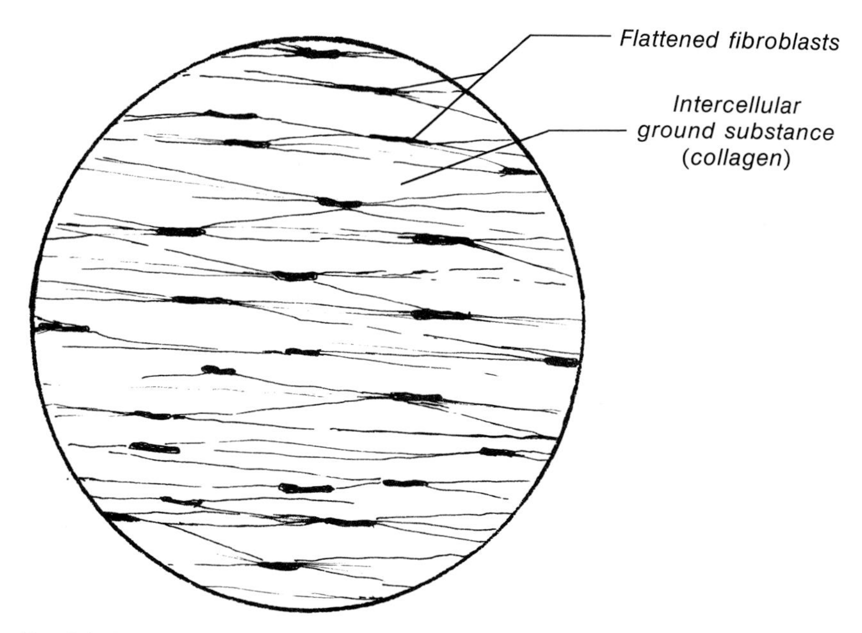

FIG. 8-2. Drawing from high power magnification of dense fibrous connective tissue (tendon). Note flattened nuclei and cells.

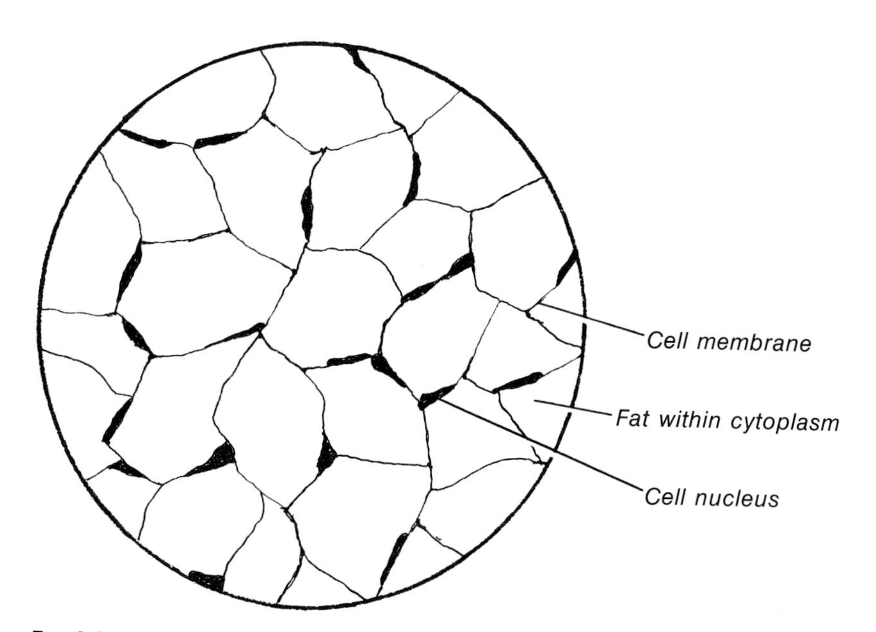

FIG. 8-3. Drawing from high power magnification of adipose tissue (fat).

rated except in *adipose tissue* where the cells are clumped together, and each cell contains a large glob of fat (Fig. 8-3). The *buccal fat pad* (Fig. 10-4) is a mass of adipose tissue, lying superficial to the buccinator muscle at the anterior border of the masseter muscle.

Fascia

Other than the specifically organized structures, tendons, and ligaments, the fibrous connective tissue envelopes are called *fascia.* The fascia of the head and neck may be divided into *superficial fascia* and *deep fascia.*

Superficial Fascia

The superficial fascia covers the entire body and lies just under the skin. It is composed of two layers. The outer layer contains varying amounts of fat (Fig. 8-4). A thin membrane serves as the inner layer which has a large amount of elastic fibers. The superficial fascia attaches the skin to the deep fascia which covers and invests the structures lying deep to the skin (Fig. 8-4). Although the superficial fascia attaches to the deep fascia, it glides freely over the deep fascia in most areas, producing the characteristic movability of the skin. In areas where the skin is movable, the superficial and deep fascia can be separated by blunt dissection, using a probing finger or blunt instrument. Thus, the interspace between these two levels of fascia is a *potential cleft* or *space.*

The muscles of facial expression are invested in the superficial fascia. Under the superficial fascia of the face and scalp, no deep fascia is found. The superficial fascia of the face and scalp is continuous with the superficial fascia of the neck, the *superficial cervical fascia.*

Deep Fascia

In the head and neck region, the deep fascia begins at the anterior border of the masseter muscle and is attached to the superior temporal and nuchal lines. The deep fascia is found only posterior and inferior to these margins.

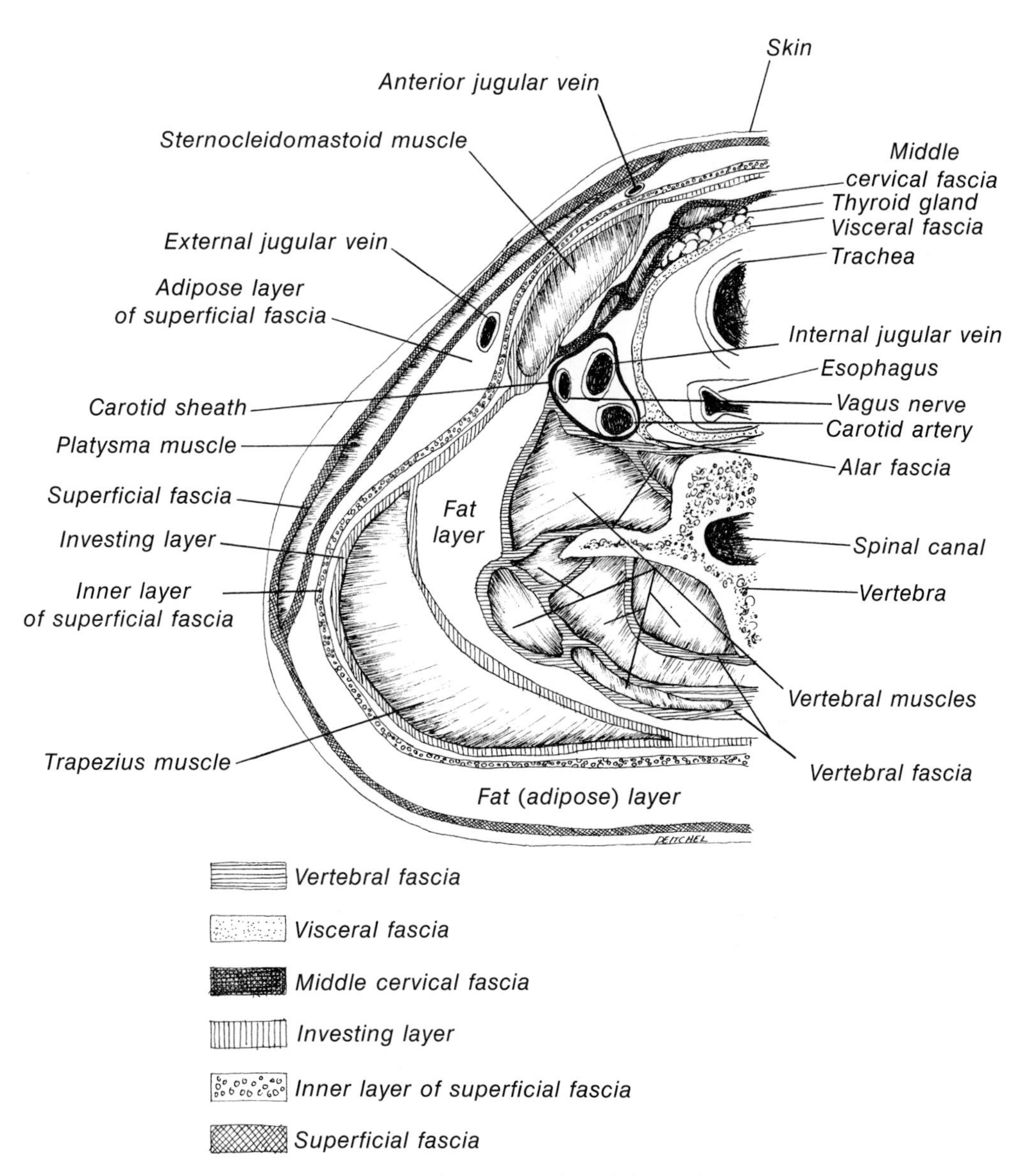

FIG. 8-4. Diagram of fascial layers as seen in cross section of the neck.

The geography of the deep fascia is quite complex and intricate. Once it is learned and understood, however, it is both interesting and fascinating. What makes the topic confusing, occasionally, is the use of contradictory nomenclature. One author may define a term in one manner, and another may use a different definition for the same term. An attempt will be made in the following discussion to keep the material well organized and to maintain a continuity of nomenclature.

The key to understanding is organization. With a knowledge of the deep fascia of the head and neck, the astute clinician has a clear understanding of the relations of the various anatomic structures therein.

Deep Fascia of the Jaws

Although the deep fascia is absent in the face and scalp, it is considered to be represented by the periosteum of the skull and facial bones. More posteriorly, deep fascia is present, and it invests the muscles of mastication.

A sheet of deep fascia, the *temporal fascia,* covers the temporalis muscle (Fig. 8-5). This fascia attaches to the superior temporal line, covers the temporalis muscle, and passes inferiorly to the zygomatic arch. Superiorly, the temporal fascia and fibers of origin of the temporalis muscle blend into a firm *aponeurosis.* An aponeurosis is a fibrous connective tissue which is extremely dense and firm; it appears as a flat tendon.

Inferiorly, the temporal fascia leaves the temporalis muscle and splits over the zygomatic arch. Below the arch, the deep fascia covers the lateral surface of the masseter muscle. It splits to enclose the parotid gland posteriorly; therefore, it is known as the *parotideomasseteric fascia.* The parotideomasseteric fascia is continuous with the deep cervical fascia below the jaw. It terminates anteriorly at the anterior border of the mandible. That portion of the parotideomasseteric fascia overlying the masseter muscle is known as the *masseteric fascia.* Over the parotid gland, it is *parotid fascia.* Posteriorly, as the masseteric fascia becomes the parotid fascia, it splits and envelops the parotid gland. In summary, the temporal fascia, the masseteric fascia, and the parotid fascia are continuous with one another, and they are all continuous with the deep cervical fascia of the neck.

From the pterygoid process, a sheet of deep fascia passes down on the *deep* surface of the medial pterygoid muscle and attaches

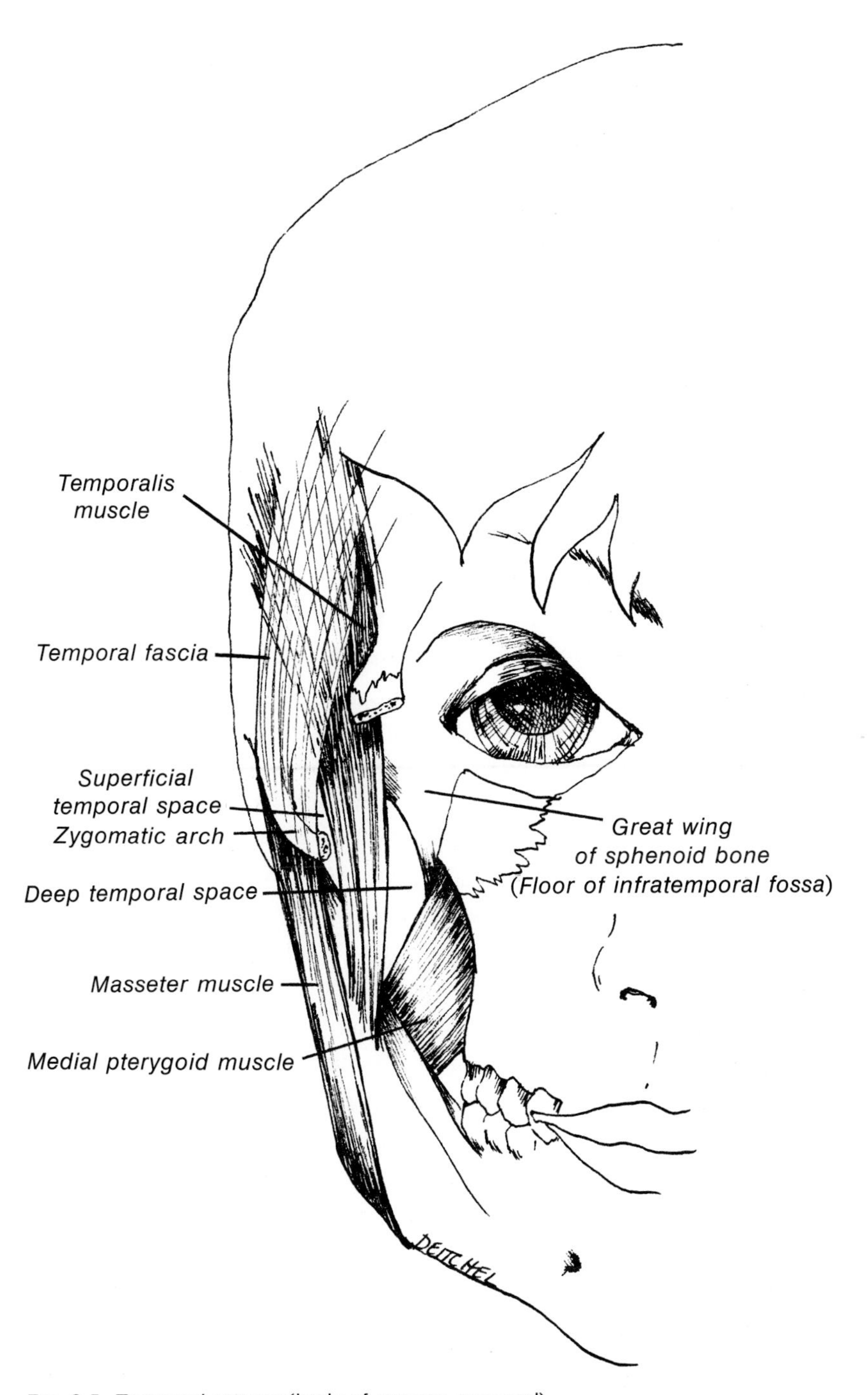

FIG. 8-5. Temporal spaces (body of zygoma removed).

to the inferior border and angle of the mandible. This is the *pterygoid fascia.* This layer of deep fascia is continuous, therefore, with the parotideomasseteric fascia on the lateral surface of the masseter muscle because both attach to the inferior border of the mandible. The pterygoid fascia also covers the medial surface of the medial pterygoid muscle and encircles the lateral pterygoid muscle.

From the above descriptions, we can expect some potential spaces to exist in the area of the muscles of mastication and parotid region. These will be described in the section on the spaces of the face and jaws.

Fascia of the Neck

The fascia below the mandible is referred to as the *cervical fascia.* It also has a superficial and a deep component. The *superficial cervical fascia* envelops the platysma muscle. It is continuous with the superficial fascia of the face, which invests the muscles of facial expression.

The *deep cervical fascia* invests the deeper muscles and organs of the neck. Superiorly it attaches to and splits at the inferior border of the mandible and becomes the parotideomasseteric fascia laterally and the pterygoid fascia medially.

Lower in the neck, the deep cervical fascia has many divisions based on relative anatomic positions. The most superficial layer of the deep cervical fascia is referred to as the *investing layer.* It surrounds the neck and invests the sternocleidomastoideus and the trapezius muscles. This layer is continuous with the deep fascia of the parotid gland and the muscles of mastication. The investing layer is the layer of deep cervical fascia that attaches to and splits at the inferior border of the mandible to become the parotideomasseteric fascia and the pterygoid fascia. Also, it encloses the hyoid bone in the anterior region of the neck.

As noted above, the investing fascia encircles the entire neck. It is attached to the whole length of the inferior border of the mandible. Anteriorly, it blends with the periosteum of the facial bones under the muscles of facial expression. Under the jaw, anteriorly, the investing fascia covers the anterior belly of the digastricus muscle. Also, it encompasses the submandibular salivary gland. On the deep surface of the gland, the investing fascia covers the stylohyoid muscle and posterior belly of the digastricus muscle. Anteriorly, the deep layers of investing fascia split to encircle the suprahyoid muscles and attach to the mandible. Posteriorly, it is attached to the styloid process and tongue. The *stylomandibular ligament* (Fig. 6-3) is merely a deep, dense band of investing fascia extending from the styloid process to the angle of the mandible. The investing fascia fuses with the ligamentum nuchae on the posterior aspect of the neck. Inferiorly, it attaches to the shoulder girdle and sternum. At the superior edge of the sternum in the midline, the investing layer splits and encloses a small amount of loose areolar tissue, the *suprasternal space* (space of burns). A prolific anastomosis between the right and left anterior jugular veins occurs in this small space. As in the anterosuperior region of the neck, where the investing layer splits to enclose the suprahyoid muscles below the hyoid bone, the investing fascia envelops the infrahyoid muscles. Here, however, it is given a separate name, the *middle cervical fascia.*

Deep and somewhat parallel to the sternocleidomastoideus muscle, the cervical fascia forms a tubular fascial compartment. This is the *carotid sheath* (Fig. 8-4). It extends from the base of the skull to the root of the neck. The carotid sheath envelops the internal carotid artery superiorly and the common carotid artery inferiorly. Throughout its entire length, the carotid sheath contains the internal jugular vein and the vagus nerve.

The trachea, esophagus, and thyroid gland are enclosed deep to the carotid sheath by another large tube of fascia, the *visceral* fascia (Fig. 8-4). This deep component of the deep cervical fascia is attached above to the thyroid and cricoid cartilages and pharyngeal tubercle of the occipital bone. It envelops the pharyngeal muscles and also attaches to the pterygomandibular raphe, mandible, and pharyngeal aponeurosis. Inferiorly, it continues into the thorax and blends with the fibrous covering of the heart, the pericardium.

A ribbon of fascia attaches the carotid sheath to the visceral fascia. This is the *alar fascia* (Fig. 8-4). It extends all the way from the skull to the seventh cervical vertebra.

Posterior to the visceral fascia and deep to the investing fascia and carotid sheath is the *vertebral fascia* (Fig. 8-4). This is another "tube" of the deep cervical fascia. It encloses the vertebral column and associated muscles. Anterolaterally, it blends with the carotid sheath and alar fascia bilaterally.

Fascial Spaces and Clefts

Understanding of the spread of infection is impossible without a clear understanding of fascial clefts or fascial spaces. A *fascial cleft* may be defined as a place of possible cleavage between two fascial surfaces. It is rich in tissue fluid and poor in fibers. The cleft between the superficial and deep fasciae is a good example. In contrast, a *fascial space* or *compartment* is a portion of anatomy that is partially or completely walled off by fascial membranes. The compartment housing the sublingual salivary gland is an example. This compartment is walled off by the mucous membrane and superficial fascia of the floor of the mouth above. Below, it is walled off by the deep fascia over the mylohyoid muscle. Even though the term *space* is ambiguous, it is in common usage, and it will be used here to imply *cleft* or *compartment.* However, it is more important to remember that these "spaces" are not empty voids. They are only potential spaces.

Spaces of the Neck

From the description of the fascia of the neck, we can predict that there are several potential spaces in the neck region. Moreover, the fasciae of the face and jaws are contiguous with those of the neck. The spaces of the face and jaws communicate with the spaces of the neck.

Lateral Pharyngeal Space (Fig. 8-6)

The lateral pharyngeal space lies high and deep in the neck. Its medial boundary is the visceral fascia overlying the superior constrictor of the pharynx. Since this space is so high, the lateral boundary is the pterygoid fascia and muscles, together with the fascia on the deep surface of the parotid gland. It communicates with the masticator space. The posterior boundary is the apposition of the vertebral and visceral fascia and the alar fascia. Inferiorly, the lateral pharyngeal space is bounded by the investing fascia of the digastricus muscle, posterior belly, and styloid process. The superior boundary is the petrous portion of the temporal bone. Thus, the lateral pharyngeal space extends from the base of the skull to the posterior belly of the digastric muscle. It contains a section of the carotid sheath. Communications include the submandibular space, masticator space, and retropharyngeal space.

Retropharyngeal Space (Figs. 8-4; 8-6)

The retropharyngeal space, another potential space, is a midline space extending from the base of the skull to the mediastinum (interior of thorax). It is a potential cleft between the visceral fascia and the vertebral fascia. The anterior boundary is the pharynx, and the posterior boundary is the vertebral column and alar fascia. It may communicate through the alar fascia to the lateral pharyngeal space superiorly or laterally.

Pretracheal Space (Fig. 8-4)

The pretracheal space lies directly in front of the trachea and is bounded anteriorly by the "strap muscles." It is somewhat below the lateral pharyngeal space, anterior and medial to it, and is limited superiorly by the attachment of the strap muscles to the thyroid cartilage and hyoid bone. Inferiorly, it communicates with the anterior mediastinum.

Spaces of the Face and Jaws

It is important to reemphasize that the spaces of the face and jaws communicate directly or indirectly with those of the neck.

Canine Space

The canine space, a potential cleft, lies deep to the skin and superficial muscles of the upper lip. The floor is the canine fossa and anterior surface of the maxilla. It is limited superiorly by the origin of the quadratus labii superioris muscle.

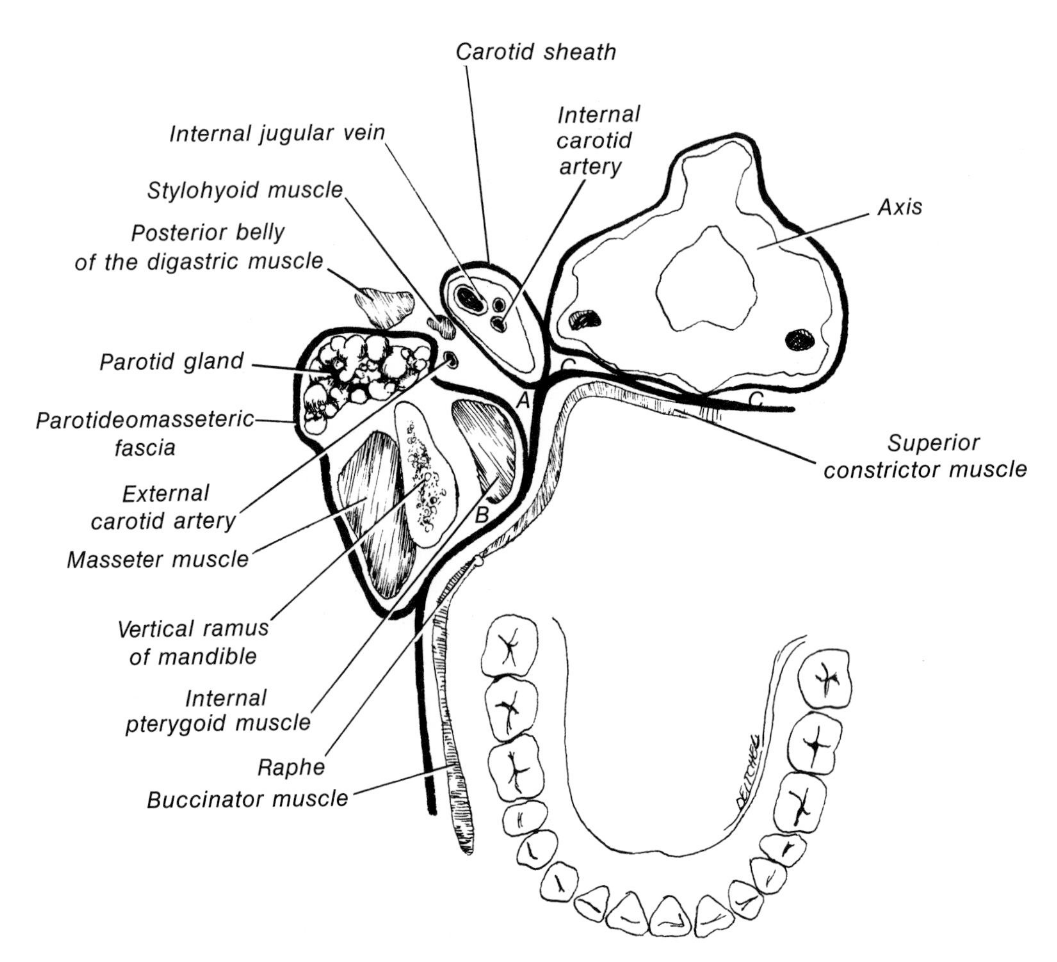

FIG. 8-6. Cross section of masticator space at the level of the occlusal plane. A. Lateral pharyngeal space. B. Pterygomandibular space. C. Retropharyngeal space.

Buccal Space

The buccal space, the "space" of the cheek, usually contains the buccal fat pad. The buccinator muscle bounds the space medially, and the skin limits it laterally. Superiorly, it is bounded by the zygomatic arch and the origin of the masseter muscle. The masseter muscle also serves as the posterior boundary of this cleft. Inferiorly, the buccal space is bounded by the mandible. The anterior boundary is the zygomaticus major and depressor anguli oris muscles.

Masticator Space (Fig. 8-7)

The masticator space is a potential cleft enclosed by the parotideomasseteric fascia laterally and the pterygoid fascia medially. Anteriorly and posteriorly the fascial walls are closed. At the anterior border of the masseter muscle, the masticator space is limited by the fusion of the masseteric fascia with the periosteum of the mandible, the superficial fascia, and the pterygoid fascia. Posterolaterally, the masticator space is bounded by the deep component of the parotid fascia. Posteromedially, the space is in relation to the lateral pharyngeal space. Superiorly, the temporal pouches are encountered. Inferiorly, the fascial walls are closed, since they fuse to the inferior border of the mandible and become the investing layer of deep cervical fascia.

Contents of this space include the masseter muscle, pterygoid muscles, pterygomandiular space, and space of the body of the mandible. Moreover, the parotid and lateral pharyngeal spaces communicate with the posterior area of the masticator space (Fig. 8-6).

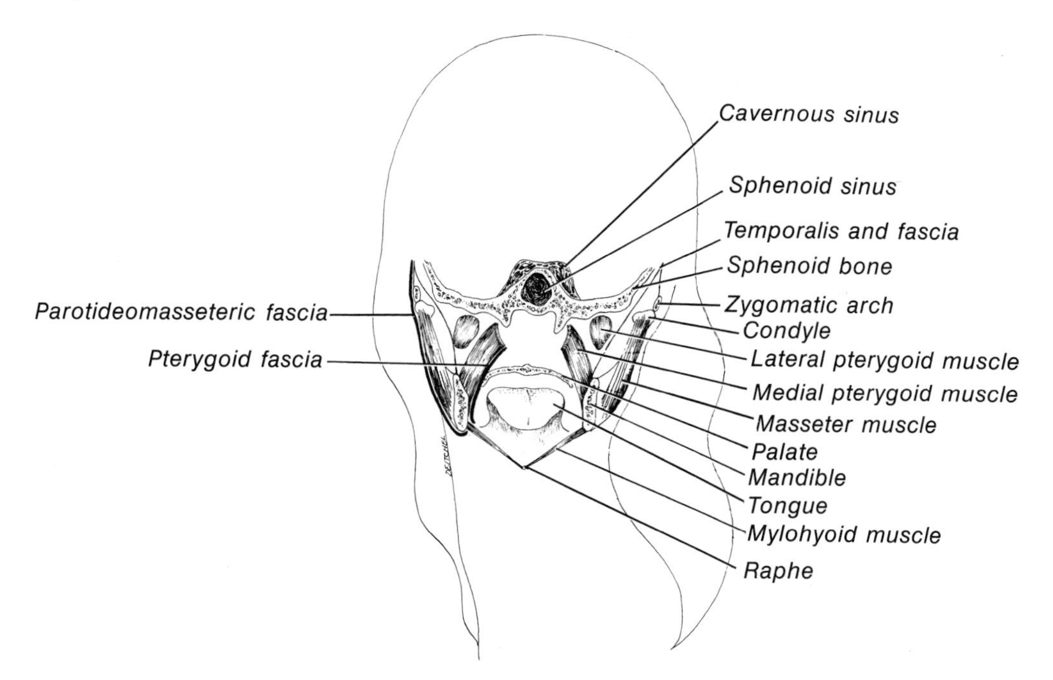

FIG. 8-7. Frontal section showing masticator space.

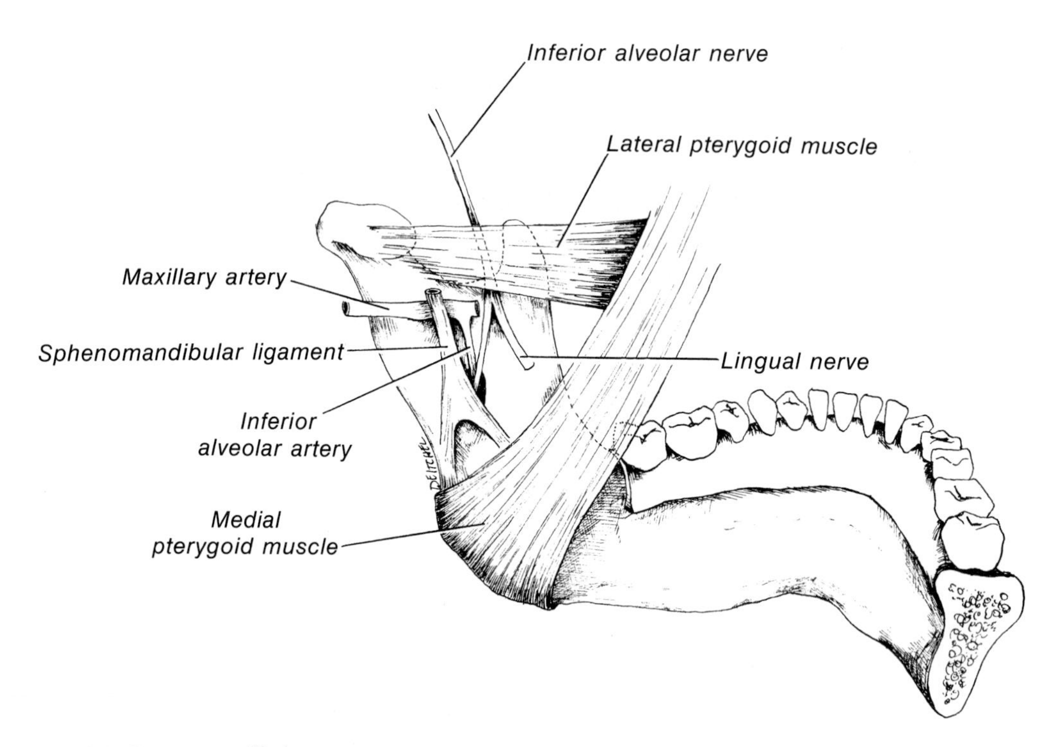

FIG. 8-8. Pterygomandibular space.

Temporal Spaces (Pouches) (Fig. 8-5)

The temporal spaces (pouches) are two in number, superficial and deep. The superficial temporal space is between the temporal fascia and the temporalis muscle. Between the temporalis muscle and the lateral surface of the skull is the deep temporal space. These spaces are filled with loose areolar tissue. Below the zygomatic arch, both the superficial and deep pouches communicate with the infratemporal region and pterygopalatine fossa.

Pterygomandibular Space (Figs. 8-7; 8-8)

The pterygomandibular space is a deep space containing the inferior alveolar nerve, artery, and vein and confined *within* the masticator space. It is filled with areolar tissue and is the site of injection for dental nerve blocks involving the lower teeth. Posteriorly, it communicates with the lateral pharyngeal space and the parotid gland. The medial boundary is the medial pterygoid muscle, and the lateral boundary is the vertical ramus of the mandible. The *bucco-pharyngeal curtain* is composed of the fascia, buccinator and superior constrictor muscles, and pterygomandibular raphe; which contribute to the origin of the buccinator and insertion of the superior constrictor muscles. This "curtain" forms the anterior boundary of the pterygomandibular space. Superiorly, the pterygomandibular space communicates with the temporal pouches, but the lateral pterygoid muscle is often considered the superior boundary.

Space of the Body of the Mandible

As the investing layer of deep cervical fascia splits at the inferior border of the mandible to form the masticator space, it also forms the space of the body of the mandible. The fascia at the inferior border fuses with bone and periosteum. The *periosteum enveloping the mandible* forms the potential cleft known as the space of the body of the mandible. The space is limited posteriorly by the insertions of the medial pterygoid and masseter muscles. Anteriorly, it is bounded by the attachment of the anterior belly of the digastric muscle. The fusion of fascia and periosteum encloses the space superiorly.

Parotid Space

The parotid space is the potential fascial cleft occupied by the parotid gland (Fig. 8-6). It is formed by the splitting of the parotideomasseteric fascia at the posterior border of the masseter muscle. Medially, there is communication with the lateral pharyngeal space. The masticator and pterygomandibular spaces are contiguous with the parotid space.

Submaxillary Spaces (Fig. 8-9)

There are five potential fascial clefts in the submaxillary group: the right and left sublingual spaces, right and left submandibular spaces, and the submental space.

The Submental Space. The submental space lies in the midline between the symphysis and the hyoid bone. The right and left anterior digastric muscles bound the space laterally. The floor is the

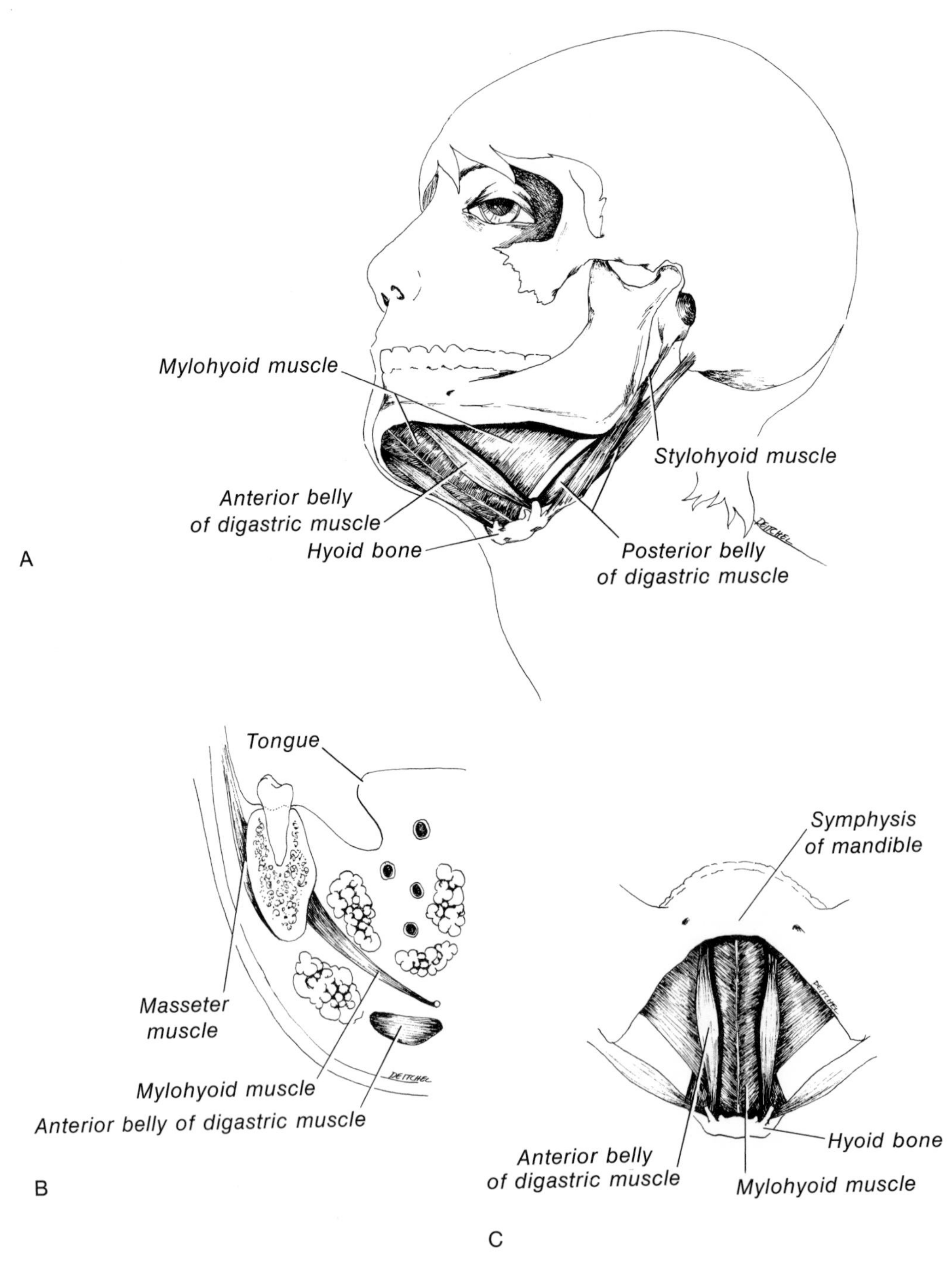

FIG. 8-9. Submaxillary spaces. A. Submandibular space. B. Sublingual space. C. Submental space.

mylohyoid muscle, and the roof is the suprahyoid portion of the investing layer of deep cervical fascia. This space contains submental lymph nodes and the origin of the anterior jugular veins.

SUBMANDIBULAR SPACE. Some anatomists refer to this space as the digastric space. It lies lateral and posterior to the submental space. Anteriorly and medially, it is bounded by the anterior belly of the digastric muscle. The posterior belly of the Digastricus, together with the stylohyoid muscle, bounds the submandibular space posteriorly and medially. The floor is formed by the mylohyoid and hyoglossus muscles. The roof is skin, platysma muscle, and the superficial layer of investing deep cervical fascia. As noted earlier, the investing layer of deep cervical fascia encloses the submandibular gland and the digastricus muscles. The inferior border of the mandible contributes to the lateral and superior boundaries.

The major contents of this space are the submandibular lymph nodes and salivary gland, the facial artery and vein, and the submental artery.

SUBLINGUAL SPACE. Above the submandibular spaces are the right and left sublingual spaces. The mylohyoid muscle separates the sublingual space from the submandibular space. However, since the posterior border of the mylohyoid muscle is a free edge, there is easy communication from the sublingual space to the submandibular space.

The roof of the sublingual space is the mucosa of the floor of the mouth. Laterally, it is bounded by the medial surface of the body of the mandible above the mylohyoid line. The geniohyoid and genioglossus muscles form the medial boundary.

The contents of this space include the sublingual salivary gland, the lingual nerve, the hypoglossal nerve, the lingual artery, and the submandibular (Wharton's) duct.

Clinical Notes

It is common for infection to occur in the spaces of the face and jaws. The teeth, mouth, nasal cavities, and sinuses all harbor millions of bacteria of many varieties. In the healthy state, these bacteria are in balance. Usually, therefore uncontrolled infection does not occur.

However, dental decay, lacerations, trauma, or surgery can cause a disruption in the delicate balance. Infection may result from one or more of the various types of bacteria becoming more active and productive than the rest. Since the spaces of the face and jaws communicate with the spaces of the neck, infection of the face and jaws has great potential for extension to the deep spaces of the neck. This may pose a definite threat to life.

Inflammation of muscle often results in spasm of the muscle. Dental infection, particularly of the molars, often spreads to the masticator space, which envelops the pterygoid muscles (particularly the medial pterygoid) and the masseter muscle. Since this space communicates with the temporalis muscle, it is easy to understand that patients with infected molars often cannot open their jaws. The elevators are in spasm because of the inflammation involving the fascial spaces enclosing them. This spasm of the masticatory muscles with resultant inability to open the jaw is referred to as *trismus.* Trismus, secondary to infected molars, typifies the extension of infection to the masticator space.

Swelling and infection involving the masticator, parotid, or laryngeal spaces are not uncommon. Unchecked, the infection can spread rapidly to contiguous spaces, such as the temporal spaces, and cause a rather severe clinical problem. Infection in the masticator space is usually of dental origin.

A swollen upper lip is typical of infection from maxillary anterior teeth, primarily the cuspids. Such infection usually involves the canine spaces and often spreads to the infraorbital region.

Infections of the lower molars frequently extend to the submandibular space or the sublingual space; those of the upper molars, to the buccal space. Infections of the lower incisors may spread to the submental space.

A knowledge of the various potential spaces of the face and jaws is mandatory to those whose profession centers in the maintenance of health in these areas. The origin and spread of infection can be predicted, prevented, or treated only with such knowledge. Table 8-1 is a consideration of the boundaries and relations of the various potential spaces of the neck, face, and jaws through which infection may spread.

TABLE 8-1. *The Potential Spaces of the Neck, Face, and Jaws*

SPACE	BOUNDARIES	COMMUNICATIONS
1. Superficial temporal space	Lateral—Temporal fascia Medial—Temporalis muscle Superior—Superior temporal line Inferior—Zygomatic arch	1. Masticator space 2. Pterygomandibular space 3. Parotid space 4. Infratemporal fossa
2. Deep temporal space	Lateral—Temporalis muscle Medial—Temporal squama Superior—Superior temporal line Inferior—Zygomatic arch	1. Masticator space 2. Pterygomandibular space 3. Parotid space 4. Infratemporal fossa
3. Masticator space	Lateral—Parotideomasseteric fascia Medial—Pterygoid fascia Superior—Zygomatic arch Inferior—Fusion of fascia Posterior—Fusion of fascia Anterior—Fusion of fascia	1. Temporal pouches 2. Lateral pharyngeal space 3. Parotid space 4. Sublingual space 5. Submandibular space
4. Pterygomandibular space	Lateral—Vertical ramus of mandible Medial—Medial pterygoid muscle Superior—Lateral pterygoid muscle Posterior—Parotid gland Anterior—Buccopharyngeal curtain	1. Temporal pouches 2. Lateral pharyngeal space
5. Space of mandible	Lateral—Periosteum Medial—Periosteum Superior—Fusion of fascia Inferior—Fusion of fascia Anterior—Anterior digastric Posterior—Insertions of masseter and medial pterygoid muscles	1. Masticator space 2. Submandibular space 3. Sublingual space

TABLE 8-1. *The Potential Spaces of the Neck, Face, and Jaws (Continued)*

SPACE	BOUNDARIES	COMMUNICATIONS
6. Parotid space	Lateral—Parotid fascia Medial—Parotid fascia	1. Lateral pharyngeal space 2. Masticator space 3. Pterygomandibular space
7. Canine space	Lateral (roof)—Skin and superficial fascia Medial (floor)—Canine fossa Superior—Quadratus labii superioris Inferior—Orblicularis oris	1. Buccal space 2. Infraorbital region
8. Buccal space	Lateral—Skin and superficial fascia Medial—Buccinator muscle Superior—Zygomatic arch Inferior—Mandible Anterior—Zygomaticus major and depressor anguli oris muscles Posterior—Masseter muscle	1. Masticator space 2. Canine space 3. Space of mandible
9. Submental space	Lateral—Right and left anterior digastric muscles Anterior—Symphysis Floor—Mylohyoid muscle Roof—Skin and superficial fascia Posterior (inferior)—Hyoid bone	1. Sublingual space 2. Submandibular space
10. Submandibular space	Lateral—Horizontal ramus of mandible Medial—Anterior and posterior digastric muscles and stylohyoid muscle Floor—Mylohyoid and hyoglossus muscles Roof—Skin and superficial fascia	1. Masticator space 2. Space of mandible 3. Sublingual space 4. Submental space

TABLE 8-1. *The Potential Spaces of the Neck, Face, and Jaws (Continued)*

SPACE	BOUNDARIES	COMMUNICATIONS
11. Sublingual space	Lateral—Body of mandible Medial—Geniohyoid and genioglossus muscles Floor—Mylohyoid muscle Roof—Mucosa of the floor of the mouth	1. Masticator space 2. Space of mandible 3. Submental space 4. Submandibular space
12. Lateral pharyngeal space	Lateral—Pterygoid fascia and muscles Medial—Visceral fascia and pharyngeal constrictors Posterior—Apposition of vertebral, visceral, and alar fascia Anterior—Pharyngeal constrictor Inferior—Fascia of digastric muscle Superior—Base of skull	1. Masticator space 2. Submandibular space 3. Retropharyngeal space 4. Carotid sheath 5. Parotid space
13. Retropharyngeal space	Anterior—Pharynx Posterior—Vertebral column Lateral—Fusion of fascia Superior—Base of skull Inferior—Mediastinum	1. Lateral pharyngeal space 2. Carotid sheath 3. Interior of thorax
14. Pretracheal space	Anterior—"Strap muscles" Posterior—Visceral fascia Superior—Thyroid cartilege and hyoid bone Inferior—Mediastinum	1. Lateral pharyngeal space 2. Interior of thorax

Infection usually spreads from one space to another through lines of least resistance. Therefore, infection of the sublingual space will spread to the submandibular space by taking a path around the free border of the mylohyoid muscle rather than by attempting to perforate through the mylohyoid muscle. Similarly, infection of a maxillary cuspid will perforate the labial bone and involve the canine space rather than perforate the more dense palatal bone and involve the palate.

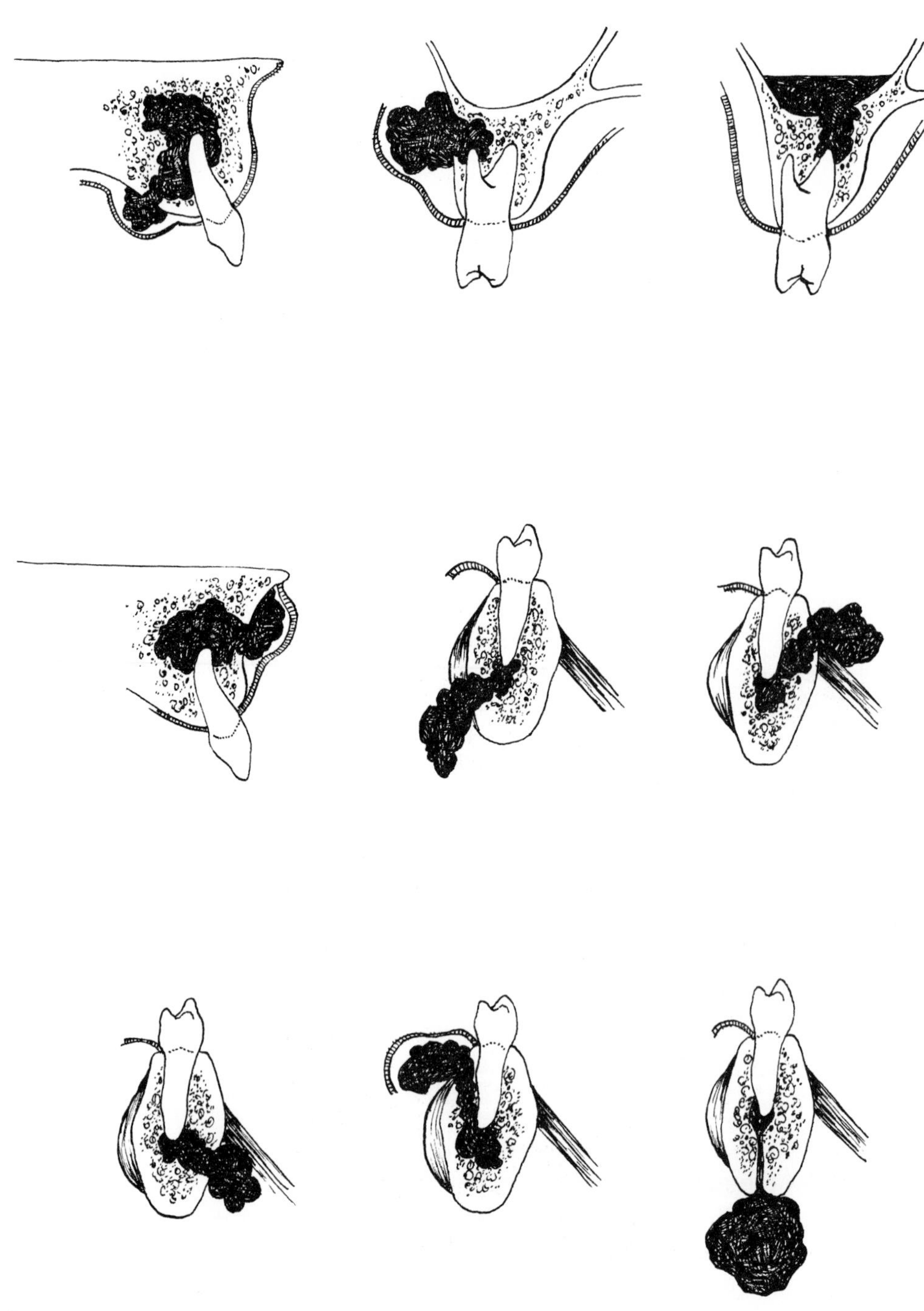

FIG. 8-10. Locations of potential regions of purulent perforation.

Infection may not, indeed it often *does not,* take the line of least resistance. The clinician must be aware of the variability of the course an infectious process may take (Fig. 8-10).

Treatment of infections of the face, jaws, and neck is aimed at control and elimination. Control involves procedures to prevent spread of the infection. Antibiotics and moist heat to the area are common successful therapeutic measures. Elimination involves removing the causative factors and evacuation of pus, if present. Many times control and elimination can be accomplished concurrently. Other times, control must be established before elimination. In still other cases, control is impossible without first eliminating the cause. The approach to treatment of an infection is a clinical judgment, and each case must be treated according to need.

Infection is often complicated by systemic disease, debilitation, and other general factors. Emphasis is made here that it is a patient who is infected—a total human being, not a tooth, a jaw, or a gland. The *patient* must be treated, not just his jaw. Extraction of an in-

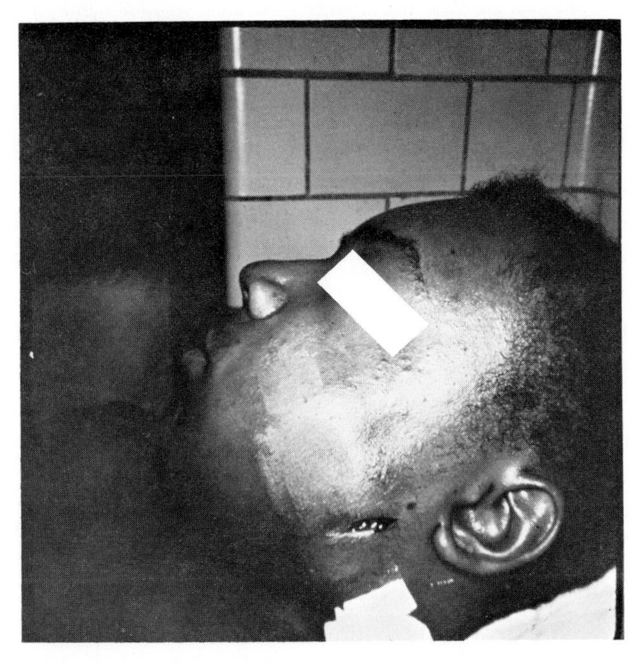

FIG. 8-11. Cellulitis. The infection was not localized and involved the masticator and temporal spaces.

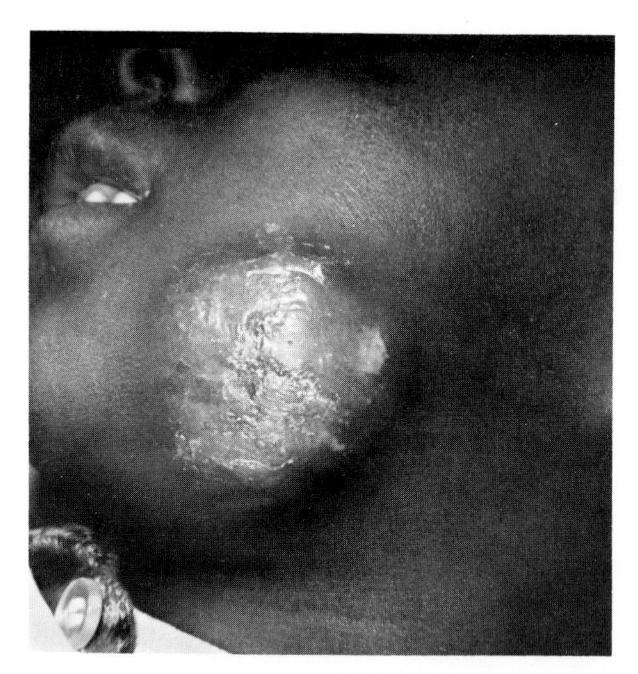

FIG. 8-12. Large facial abscess in a nine-year-old child with infection from a carious tooth. Note the well-defined borders of this abscess located in the masticator space.

fected tooth from a debilitated, elderly, poorly controlled diabetic is pointless without the concurrent use of fluid and electrolyte therapy, control of diabetes, and reestablishment of nutrition.

Table 8-2 shows the usual sites of localization of dental infections. It should be noted that the relation of the muscles to the apex of the root of the involved tooth is important in predicting where the infection will localize. But, as Figure 8-10 illustrates, variations are definitely possible.

Swelling may not necessarily mean that pus is present. Cellulitis and extensive edema may exist without pus being formed (Fig. 8-11). An attempt to drain such an area is fruitless. Treatment of such a condition by elimination of the causative factors (in this case, extraction of an infected tooth) and antibiotic therapy may result in resolution of the cellulitis without formation of pus.

If pus does accumulate (Fig. 8-12), however, it must be evacuated. This procedure is known as incision and drainage. Within 24 to 48 hours, the patient will improve. There are specific areas for

TABLE 8-2. *Potential Sites of Localization of Dental Infections*

INVOLVED TEETH	EXIT FROM BONE	POSITION OF MUSCLE ATTACHMENT IN RELATION TO ROOT APEXES	SITE OF LOCALIZATION
Upper central incisor	Labial	Above	Oral vestibule, canine space
Upper lateral incisor	Labial	Above	Oral vestibule, canine space
	Palatal		Palate
Upper cuspid	Labial	Above	Oral vestibule
	Labial	Below	Canine space
	Palatal		Palate
Upper bicuspids	Buccal	Above	Oral vestibule
	Palatal		Palate
Upper molars	Buccal	Above	Oral vestibule
	Buccal	Below	Buccal space
	Palatal		Palate
Lower incisors	Labial	Above	Submental space
	Labial	Below	Oral vestibule
	Lingual	Below	Sublingual space
Lower cuspid	Labial	Below	Oral vestibule
	Lingual	Below	Sublingual space
Lower bicuspids	Buccal	Below	Oral vestibule
	Lingual	Above	Submandibular space
Lower first molar	Buccal	Above	Masticator space
	Buccal	Below	Oral vestibule
	Buccal	Above	Buccal space
	Lingual	Above	Submandibular space
	Lingual	Below	Sublingual space
Lower second molar	Buccal	Above	Masticator space
	Buccal	Below	Oral vestibule
	Buccal	Above	Buccal space
	Lingual	Below	Sublingual space
	Lingual	Above	Submandibular space
Lower third molar	Lingual	Above	Submandibular or pterygomandibular space
	Buccal	Above	Masticator space

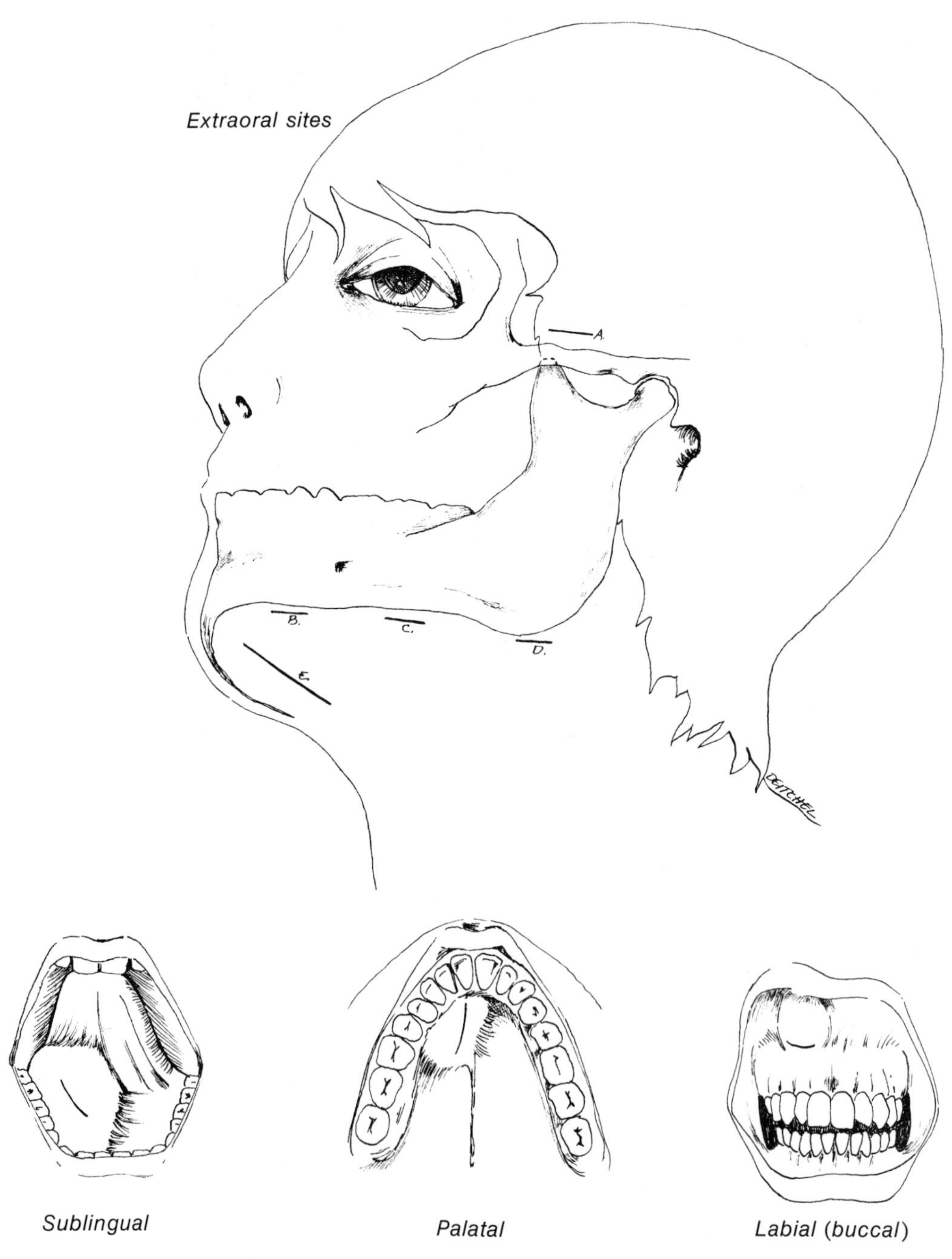

FIG. 8-13. Extraoral and intraoral sites of incisions for draining designated spaces: A, temporal pouches; B, sublingual space; C, submandibular space; D, masticator space; E, submental space.

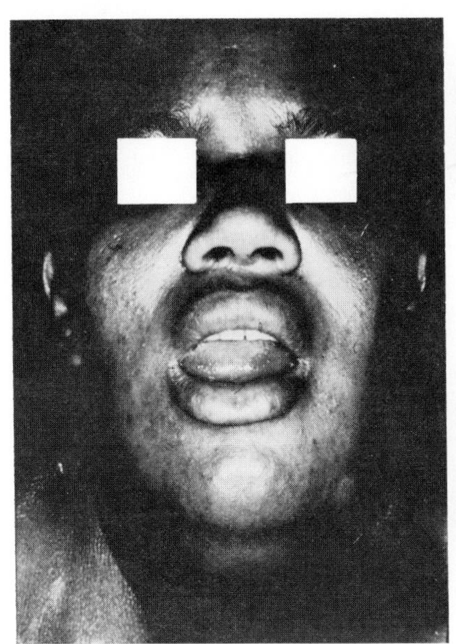
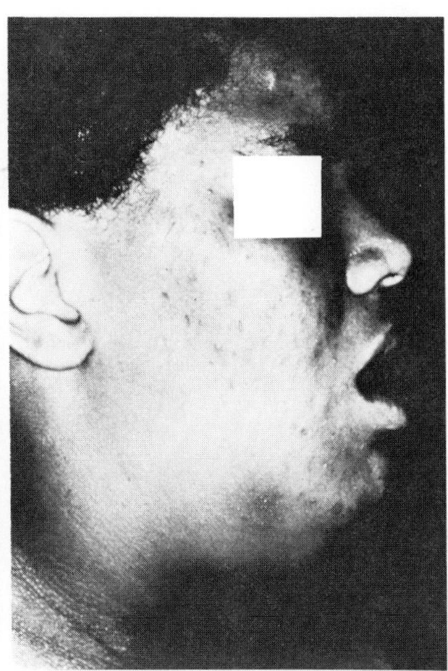

FIG. 8-14. Ludwig's angina. Note obstruction of the airway by the tongue.

incision and drainage, depending on the area of localization of the infection (Fig. 8-13).

It cannot be overemphasized that in the treatment of infection, a thorough knowledge of fascial planes and tissue spaces, together with a complete understanding of therapeutic measures, is mandatory. Without such wisdom, infections of the face and jaws can rapidly become a threat to life.

Ludwig's Angina (Fig. 8-14)

Ludwig's angina is a diffuse, uncontrolled infection of all five submaxillary spaces. Prior to the age of antibiotics, this infection often resulted in death, due to asphyxiation. The clinical picture is one of extensive edema of the oral and pharyngeal region. The floor of the mouth is elevated and with it, the tongue. Spread of the infection to the lateral pharyngeal spaces is common. Treatment is aimed at relief of respiratory obstruction and control of infection. Tracheostomy is often necessary. Aggressive antibiotic therapy, together with drainage of any purulence, is mandatory.

Nasal Cavity and Paranasal Sinuses

The nasal cavities and sinuses are located in the middle third of the facial skeleton. These hollowed-out areas provide a protective barrier against bacteria. They also warm the inspired air before it reaches the pulmonary tree. Many authorities believe another purpose of these cavities is to reduce the weight of the head.

Bony Architecture

The right and left nasal cavities are separated by the median *septum.* This bony septum is comprised of the vomer and the ethmoid bone (Fig. 9-1). Laterally, each nasal cavity is bounded by the *maxillary sinus.* The posterior boundary is the paired openings into the nasopharynx, known as the *choanae.* The floor of the nasal cavity is the palatine process of the maxillae and the horizontal process of the palatine bone. Thus, the palate separates the nasal and oral cavities. Projecting backward from the center of the posterior border of the hard palate is a small point of bone, the *posterior*

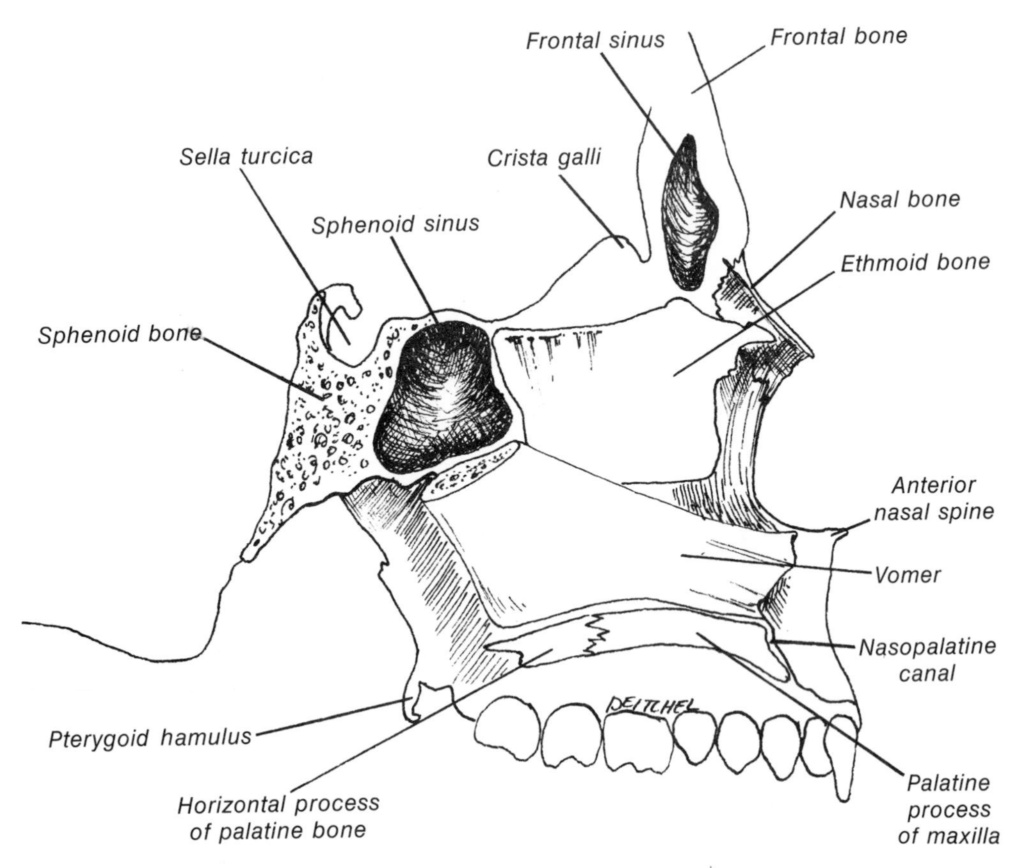

FIG. 9-1. Sagittal section to display medial wall of the nasal cavity (septum).

nasal spine. The *anterior nasal spine* is the pointed process seen at the anterior midline of the anterior nasal aperture (Figs. 9-1; 9-2). It is formed by a junction of the right and left maxillae. The roof of the nasal cavity is formed by the nasal bones, the cribriform plate of the ethmoid bone and the sphenoid bone. Anteriorly, the boundary is the *anterior nasal aperture* (Fig. 1-4B). This is also known as the piriform aperture and is heart-shaped. The rim of this opening is formed by the nasal bones and the maxillae.

Several anatomic landmarks can be identified within the nasal cavity. It should be noted that the roof is perforated by multiple small foramina, which transmit the olfactory nerves. Posteriorly on the roof, the opening into the sphenoid sinus can be seen. The large triangular notch noted on the anterior border of the septum is the area that receives the cartilagenous portion of the nasal septum. Also, the septum displays several furrows that mark the locations of vessels and nerves.

The lateral wall of the nasal cavity is most interesting (Fig. 9-2). The lacrimal bone, maxillae, ethmoid bone, sphenoid bone, and even the palatine bone contribute to this area. The inferior nasal concha is a separate bone, not a process of the ethmoid bone. The superior and middle conchae are processes of the ethmoid bone. The three chonchae are bulbous, banana-shaped structures attached lengthwise to the lateral wall. Below each choncha is a depression. These depressions are the three *meatus: superior, middle, and inferior meatus.* The posterior ethmoid air cells open into the superior meatus. Also, the sphenopalatine foramen opens here. The maxillary

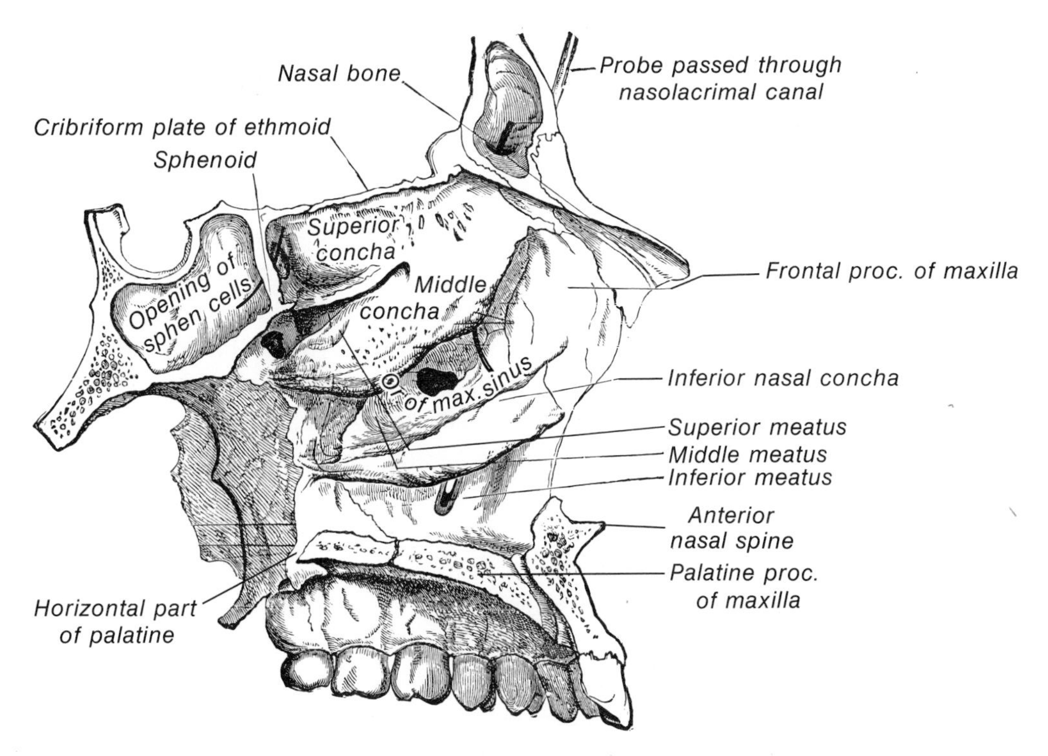

FIG. 9-2. Roof, floor, and lateral wall of left nasal cavity. (From Gray's Anatomy of the Human Body, 29th ed. C. M. Goss, editor. Philadelphia, Lea & Febiger, 1973.)

sinus opens into the middle meatus as do the middle and anterior ethmoid air cells. Moreover, the frontal sinus drains into the middle meatus. The *nasolacrimal canal* opens into the inferior meatus. The nasolacrimal canal drains tears from the eyes into the nose. Hence, excessive tearing, as in crying, produces a "runny nose."

Soft Tissues

The external nose is primarily cartilage, skin, and mucosa. The two openings into the nose, the *nares,* are guarded by a number of stiff hairs, *vibrissae.* These hairs remove foreign material carried in the current of inspired air. The cartilage of the external nose consists of the cartilage of the septum and the lateral and alar cartilages (Fig. 9-3). The outside of the nose is covered with skin and the inside of the nose and nasal cavity is lined with mucous membrane. The skin surrounds the nares and even passes into the nose for 2 to 3 millimeters to line the *vestibule.*

Further back into the nose on the lateral wall, just after entering the nasopharynx, the opening of the auditory tube can be seen. This opening leads into the middle ear via the *eustachian tube.* Removal of the three nasal conchae exposes numerous openings through the mucosa in the lateral wall of the nasal cavity. These openings correspond to the bony openings previously discussed. Note in Figure 9-4 the various openings and the complexity of the lateral wall of the nasal cavity. The *hiatus semilunaris* is merely a curved cleft at the depths of which lies the opening into the maxillary sinus. The *bulla ethmoidalis* is a prominence on the lateral wall just above the hiatus. It represents a bulging of the middle ethmoid air cells which open here.

Blood Supply

Details of arterial and venous anatomy are described in Chapter 3. In brief, the ophthalmic artery supplies blood to the nasal cavity via its ethmoidal branches. The sphenopalatine branch of the internal maxillary artery supplies the septum, meatus, and conchae. Venous drainage is via the ethmoid, sphenopalatine, and anterior facial veins.

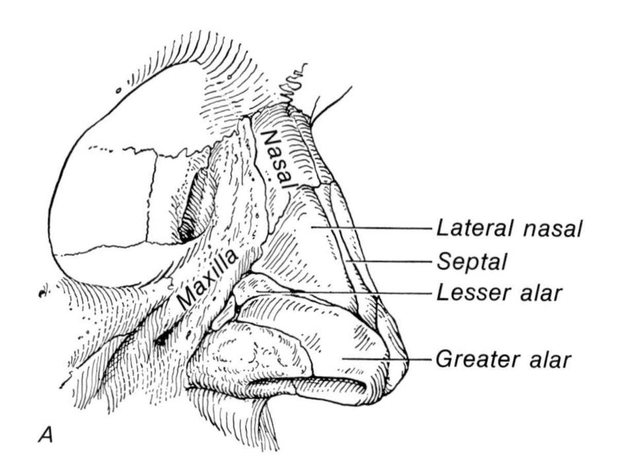

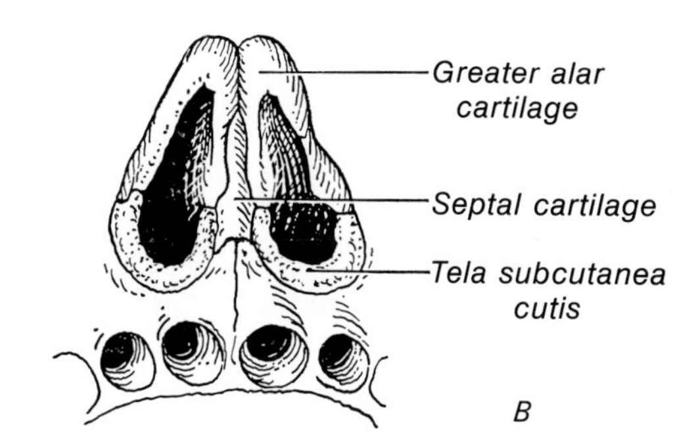

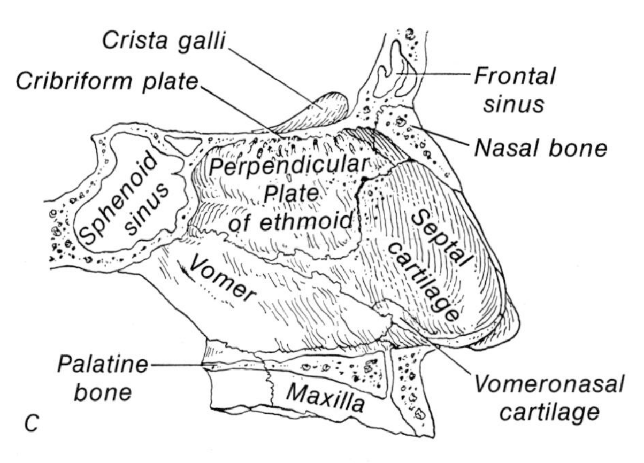

FIG. 9-3. Cartilages of the nose. A. Side view. B. Inferior view. C. Septum, right side. (Bones also depicted.) (From Gray's Anatomy of the Human Body, 29th ed. C. M. Goss, editor. Philadelphia, Lea & Febiger, 1973.)

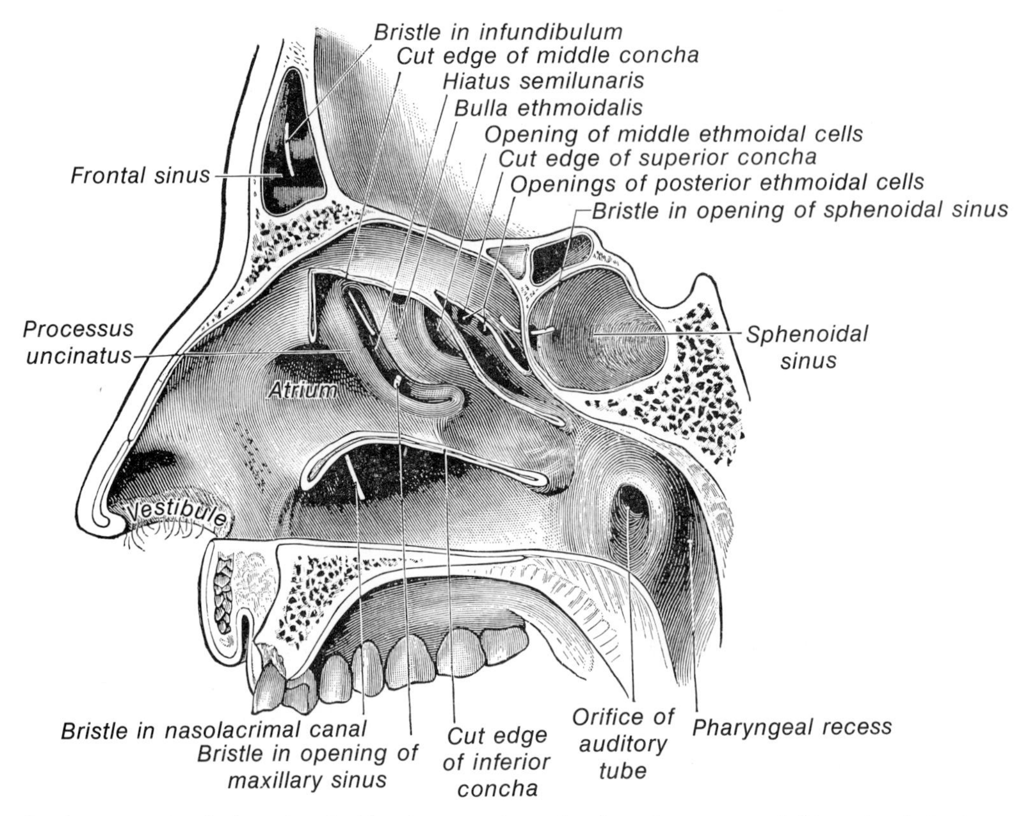

FIG. 9-4. Lateral wall of nasal cavity. The three nasal conchae have been removed. (From Gray's Anatomy of the Human Body, 29th ed. C. M. Goss, editor. Philadelphia, Pa., Lea & Febiger, 1973.)

Nerves

The nerves of ordinary sense—pain, touch, temperature—are carried via the nasociliary branch of the ophthalmic nerve, anterior superior alveolar branch of the maxillary nerve, nasopalatine nerve, and nasal branches of the sphenopalatine ganglion. Parasympathetic fibers for the glands of the nasal mucosa are also carried in branches of the pterygopalatine ganglion.

The olfactory nerve distributes special sensory fibers for smell via a plexus of nerves passing through the cribriform plate of the ethmoid bone (Fig. 4-15).

Paranasal Sinuses

The skull and facial skeleton are hollowed out into various compartments. The cranial cavity, oral cavity, nasal cavity, and orbits are the largest. However, in the region of the orbits and nasal cavity, even more voids are located in the bony skeleton. Within the substance of the maxillae, frontal bone, zygomas, ethmoid bone, and sphenoid bone are the paranasal sinuses. These sinuses are air-containing spaces, lined by mucous membrane, that communicate with the nasal cavity via the openings described earlier. They develop by the invagination of mucous membrane of the nose into the peripheral bone. In the young, these sinuses are much smaller than in adults. They slowly increase in size with age. This process is known as *pneumatization.*

Frontal Sinuses (Figs. 9-1; 9-5)

The frontal sinuses are located in the frontal bone just above the nasal cavity. The right and left frontal sinuses are divided by a septum which is somewhat perpendicular but rarely divides the two into a symmetrical pair. They drain into the ipsilateral middle meatus through a short canal, the *frontonasal duct.* Each at full size is about 2 to 3 centimeters in height, width, and depth.

Sphenoid Sinuses (Fig. 9-4)

Within the body of sphenoid bone, the air chambers known as the sphenoid sinuses are rarely symmetrical because of the lateral position of the intervening sinus septum. They communicate with the nasal cavity through an aperture above the superior concha. The size of each sinus is approximately 1.5 to 2.5 centimeters in diameter.

Ethmoid Air Cells

Numerous thin-walled cavities lie between the upper parts of the nasal cavities and orbits (Fig. 9-6). They are formed between the

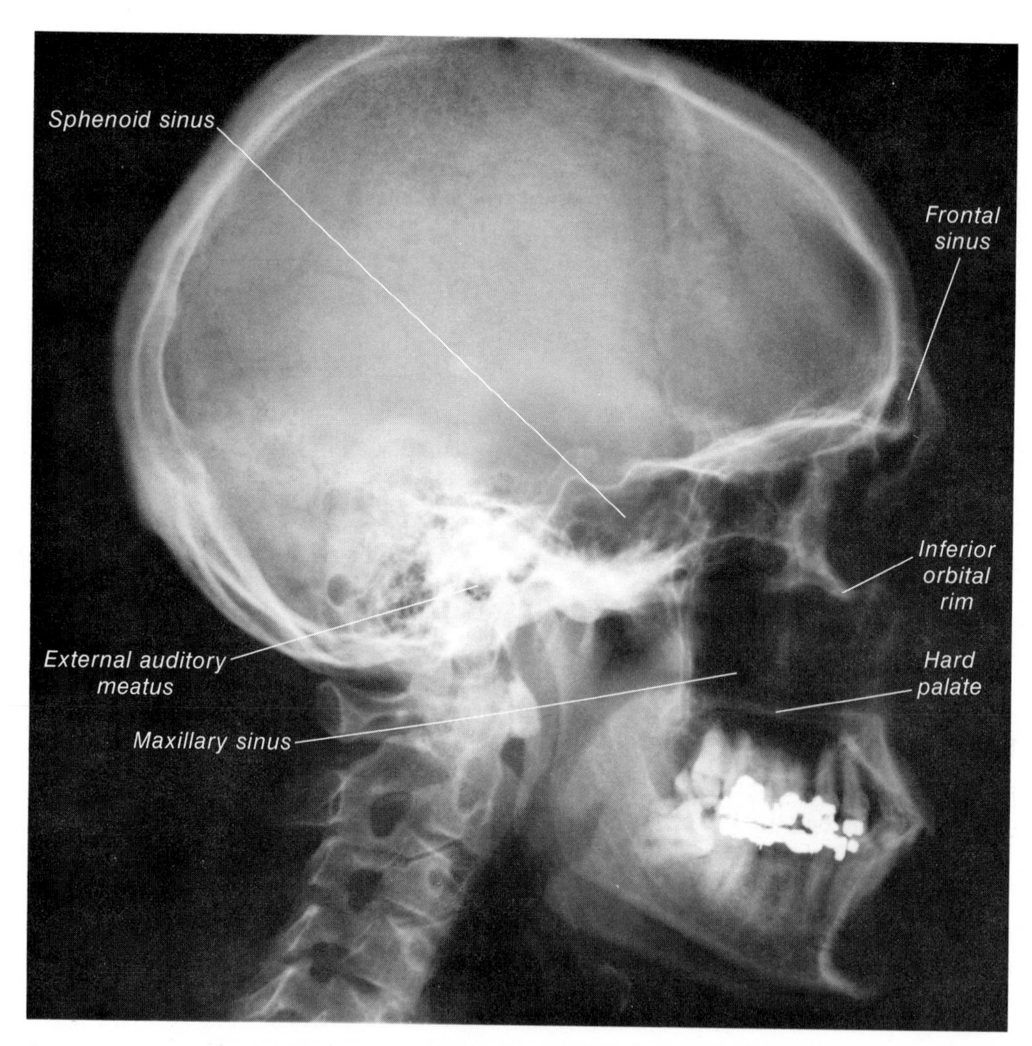

FIG. 9-5. Radiographs of skull and facial bones. A. Lateral.

lamina orbitalis (lamina papyracea) and the *superior* and *middle conchae* (Fig. 1-13) in skull. These ethmoid air cells open into the superior and middle meatus of the nasal cavity.

Maxillary Sinuses

The maxillary sinuses are the largest of the paranasal sinuses and occupy the body of the maxilla (Fig. 9-6; 9-7). Their shape is roughly

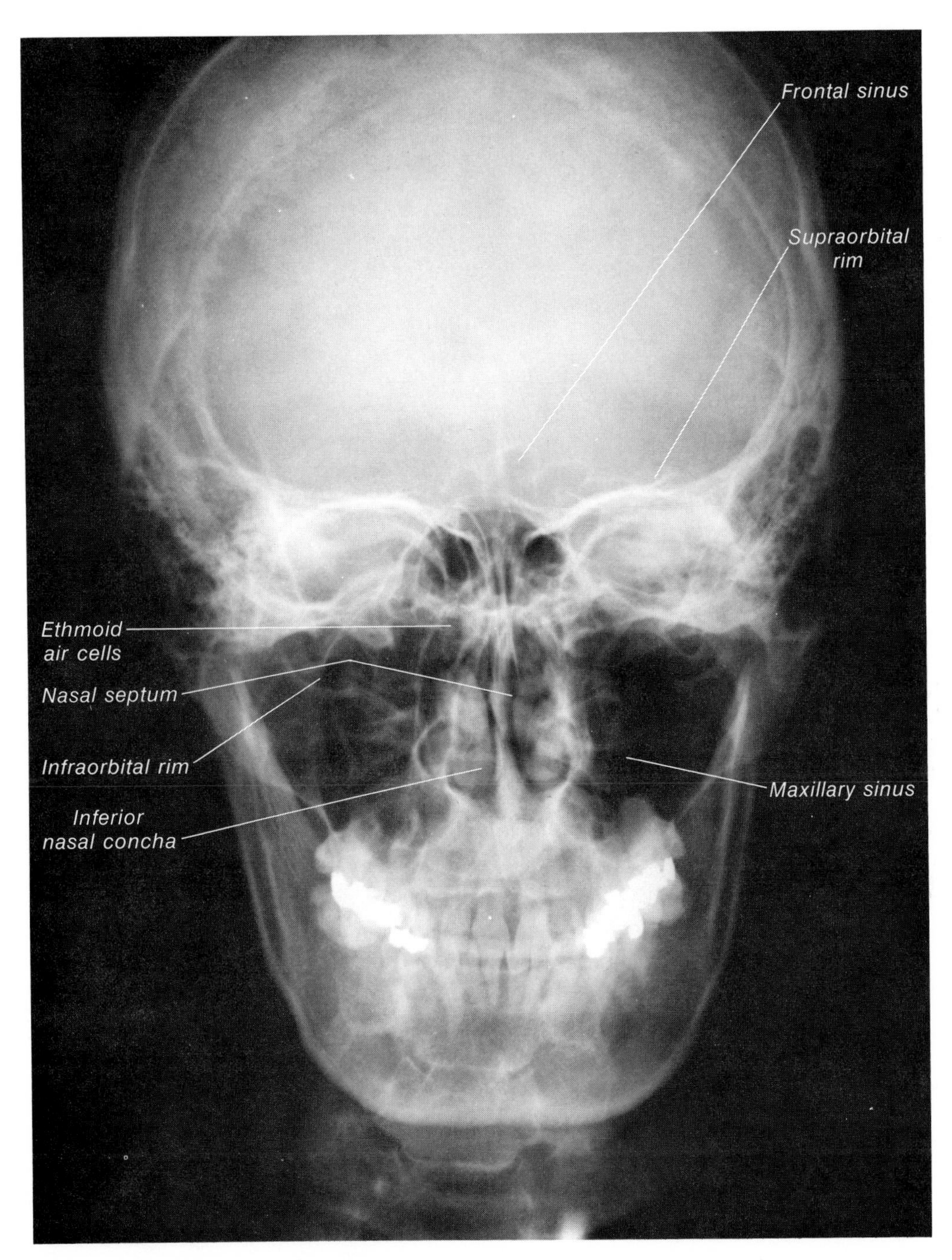

FIG. 9-5 (continued). B. Posterioranterior.

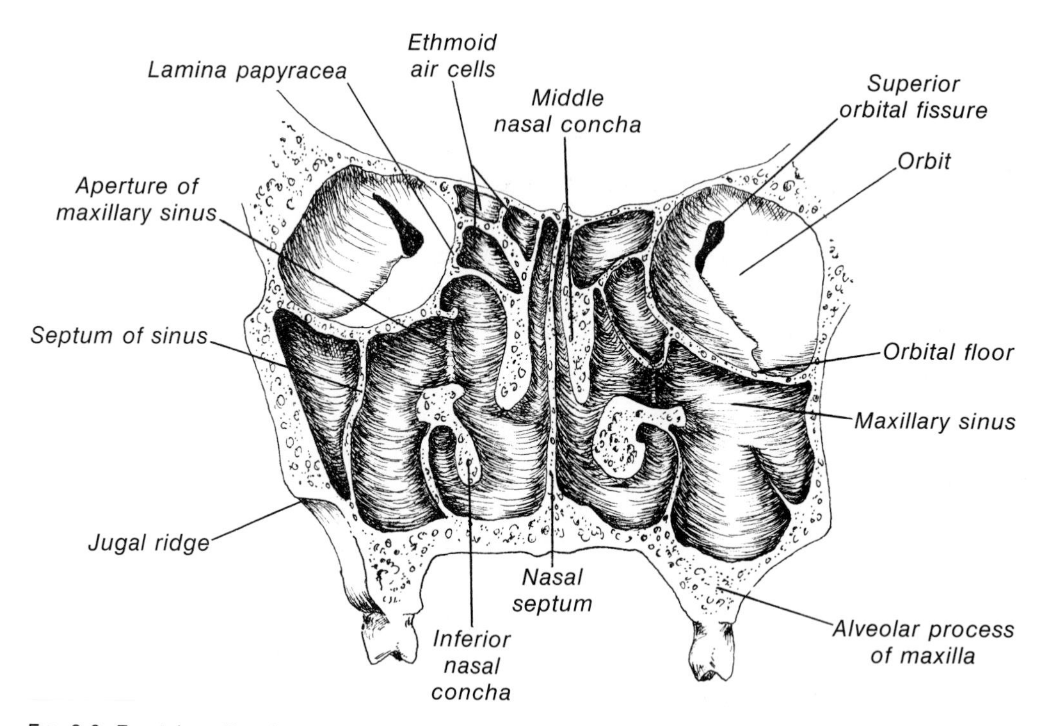

FIG. 9-6. Frontal section through orbits, nasal cavity, ethmoid and maxillary sinuses (region of the first molar).

pyramidal, with the base of the pyramid at the lateral wall of the nose and the apex pointing into the zygomatic process. The roof of the sinus is the orbital floor, and the alveolar process of the maxilla forms the floor of the sinus. The size varies from patient to patient, but on the average, the adult capacity is about 15 cubic centimeters. The vertical height is 4 centimeters; the anteroposterior depth, 3 centimeters; and the transverse breadth, 2 centimeters.

The maxillary sinus opens into the nasal cavity through an opening in the depth of the *hiatus semilunaris.* This is found under the middle concha.

The maxillary sinus is divided into various communicating compartments by bony walls and spicules; *septae.* As the sinus pneumatizes with age, it surrounds the roots of the teeth and even progresses into the body of the zygoma.

Vessels and Nerves

The ethmoid branches of the ophthalmic artery supply the ethmoid air cells and frontal sinuses. The sphenoid sinus is supplied by the sphenopalatine branch of the maxillary artery, and the maxillary sinuses are supplied by the infraorbital and the alveolar branches of the maxillary artery. The venous drainage is via ophthalmic veins and pterygoid plexuses. The nerve supply is via various branches of the first two divisions of the trigeminal nerve. Of note is that the alveolar branches of the maxillary nerve ramify in the maxillary sinus before entering the roots of the maxillary teeth. For this reason, a sign of the clinical entity of maxillary sinusitis is often dental pain.

Lymphatic drainage of the nose and paranasal sinuses ends in the submandibular nodes, retropharyngeal nodes, and superior deep cervical nodes.

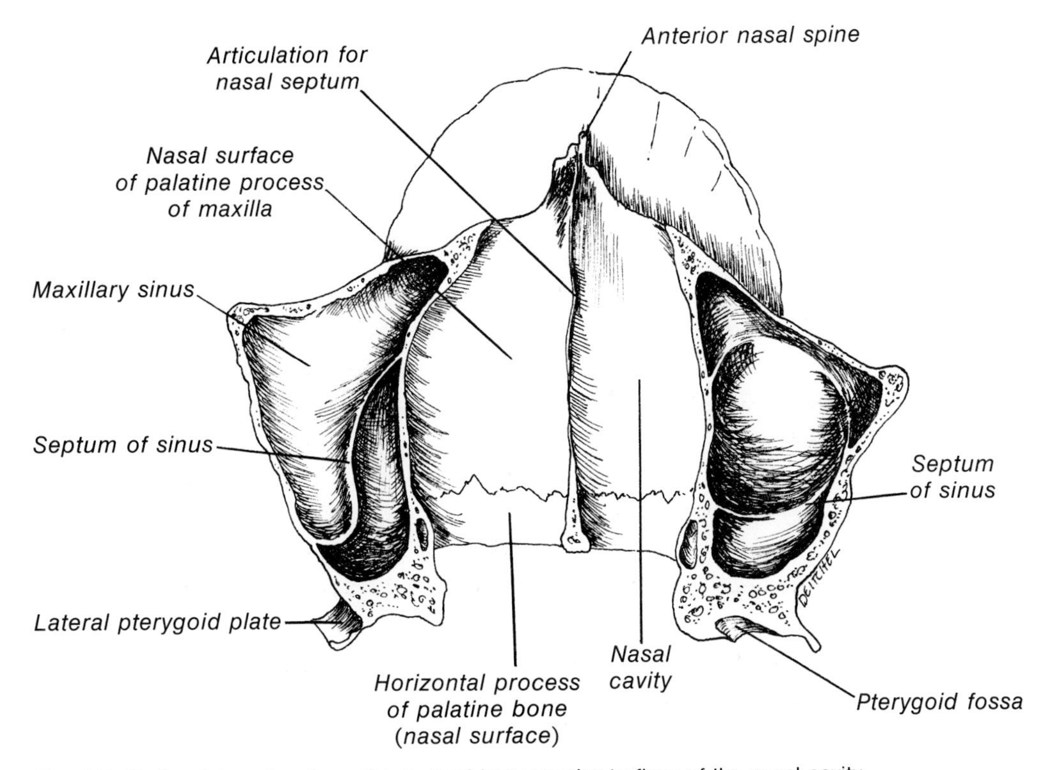

FIG. 9-7. Horizontal section from above, level just superior to floor of the nasal cavity.

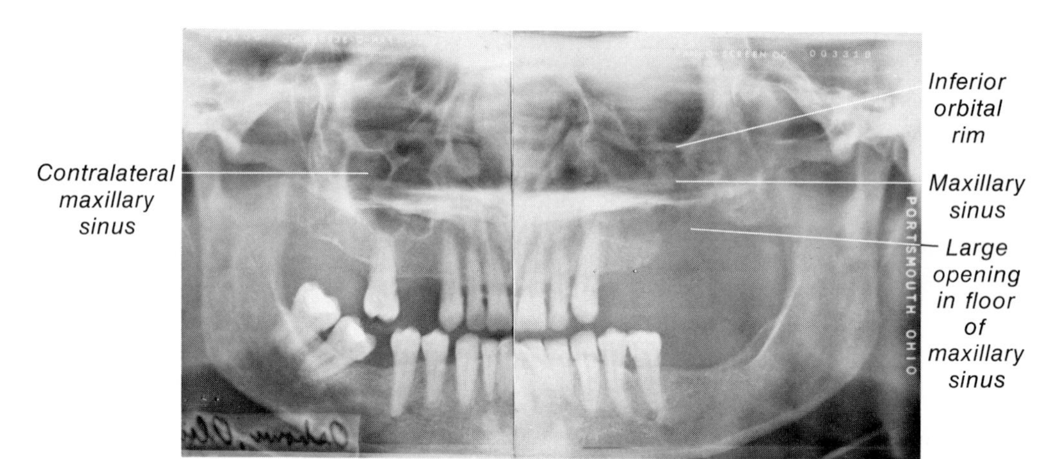

FIG. 9-8. Panoramic radiograph. Note the cloudiness of the sinus adjacent to the oral opening. This indicates chronic infection secondary to entrapment of food debris from the oral cavity. Contrast this with the dark (clear) contralateral maxillary sinus.

Clinical Notes

To the dental professional, the maxillary sinus is the most clinically significant. Inflammation of this sinus often causes pain in the maxillary teeth. Also, removal of the root tips of maxillary teeth can be hazardous in that these roots may be accidentally pushed into the sinus during the procedure. When these sinuses become inflamed, their mucous membranes produce large amounts of mucus via the various intramucosal glands. The maxillary sinus drains poorly, since its ostium is above the floor. Collection of material in the sinus sometimes produces considerable discomfort. Infection of a maxillary molar or a bicuspid commonly perforates and extends into the sinus to produce a secondary sinusitis.

Once an opening has been established through the floor of the sinus into the oral cavity, this opening must be closed to prevent influx of oral material into the sinus with resultant chronic sinusitis. Such an opening is termed an *oral-antral fistula* (Fig. 9-8).

Trauma to the facial skeleton, particularly the orbital region, may cause disruption of the sinus roof and herniation of the orbital contents down into the sinus. This may produce double vision, *diplopia,* or other difficulties in function of the eye.

10

Oral Cavity and Pharynx

Persons who specialize in the care and treatment of the oral cavity have a great responsibility. The oral cavity participates very actively in respiration, nutrition, and excretion—the three essentials of life itself. Indeed, unless the oral cavity is functioning normally, these three essentials are hampered. Proper health and function of the oral cavity are mandatory for sustained, physiologic life. Ill health of the mouth results in decreased function of the organism in general. A simple toothache can lead to dehydration, weight loss, infection, sepsis, and even death. Trauma to the face and jaws commonly results in severe general decline of the individual's health. Before treating abnormalities or pathologic conditions, the clinician must be knowledgeable of normalcy.

The Oral Cavity

The oral cavity is divided into two major portions: the *vestibule* and the *cavum oris.*

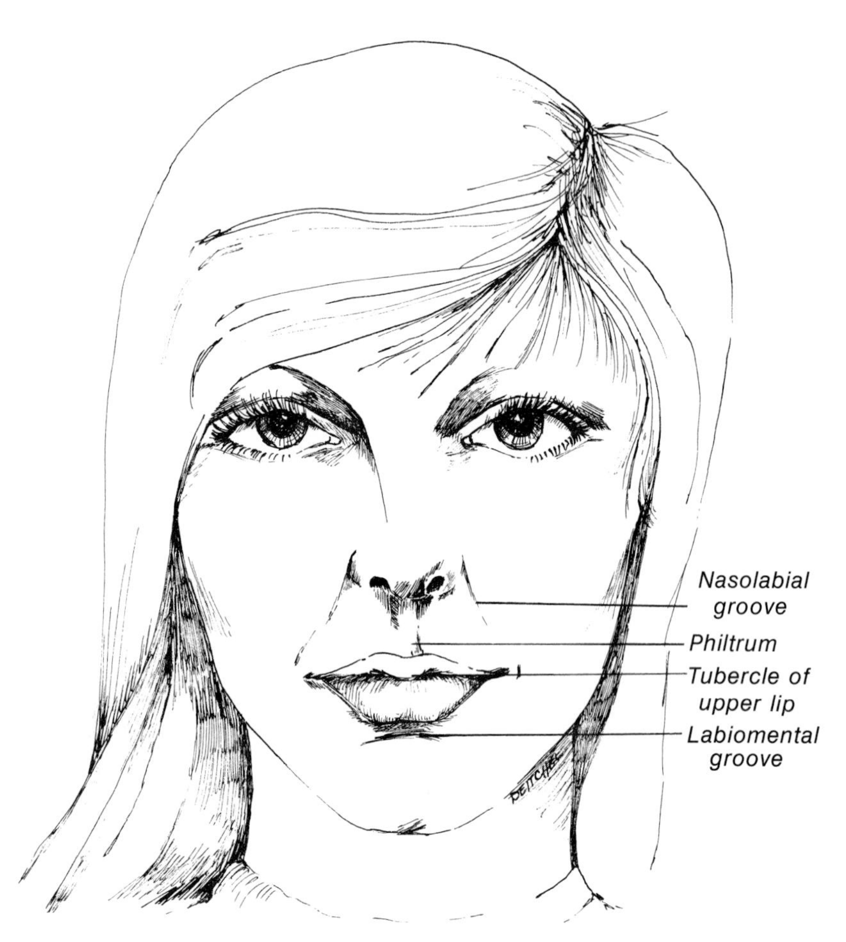

FIG. 10-1. Face of adult.

The Vestibule

The vestibule is bounded externally by the lips and cheeks and internally by the alveolar processes and teeth. Above and below, it is bounded by the reflection of the mucosa from the cheeks and lips onto the alveolar process. Laterally, it is known as the *buccal vestibule,* and anteriorly it is referred to as the *labial vestibule.* The parotid salivary gland duct empties into the buccal vestibule opposite the maxillary molars. Palpation of this vestibule reveals a smooth moist trough. No masses or lesions are noted in the normal state. The mucosa should be uniformly pink, throughout.

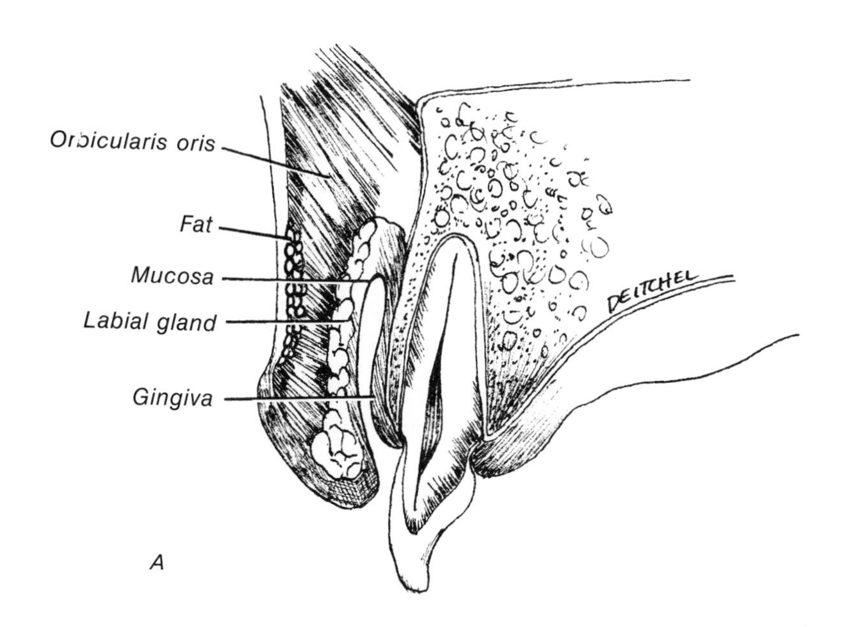

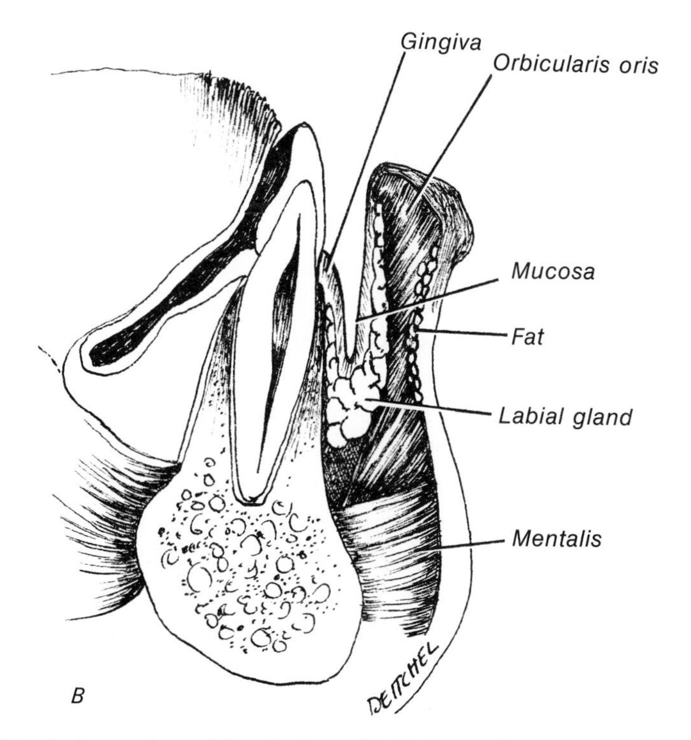

FIG. 10-2. Incisor regions of lips. A. Upper lip. B. Lower lip.

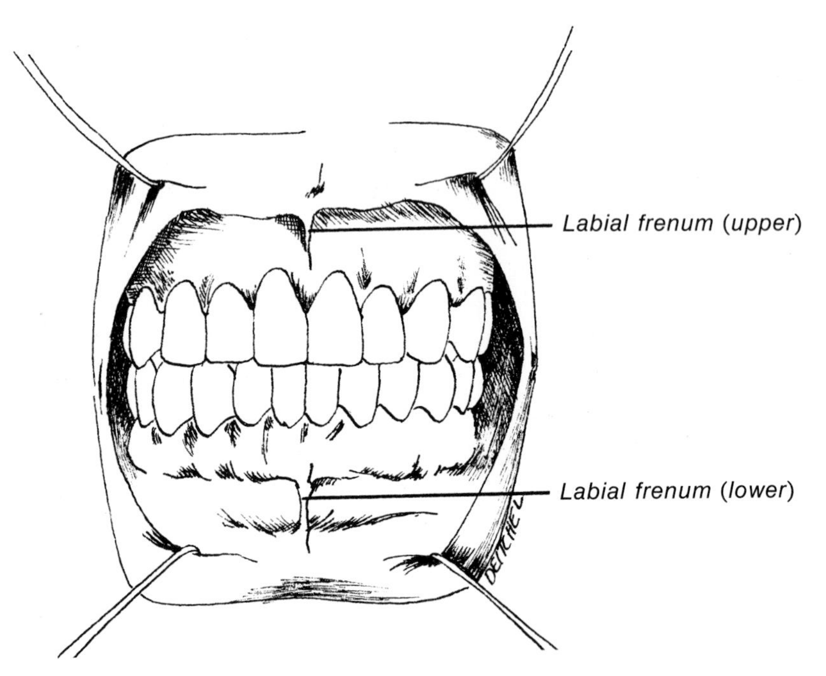

FIG. 10-3. Labial frenum.

The lips are the fleshy folds of tissue surrounding the orifice of the mouth (Fig. 10-1). They are lined by moist mucous membrane intraorally. Extraorally, they are covered by skin. Where the skin and mucous membrane join, a distinct margin, the *vermilion border,* is seen. From this line to the wet mucousa inside the lip is the *vermilion zone.* This vermilion zone is characteristic of man only. It is pink because the epithelium is thin and the rich network of underlying vessels shows through.

In the midline of the upper lip, a small protrusion, the *tubercle* is noted. Just above this is the *philtrum,* a depression leading to the columella of the nose. The lower lip has a small, midline depression corresponding to the tubercle.

The substance of the lips is primarily composed of fatty tissue laterally and salivary glands medially, with the orbicularis oris muscle sandwiched between. Connective tissue binds the various structures into a functioning unit. Moreover, the skin and mucous membrane are tightly fixed to this connective tissue by multiple fibrous bands that transverse the entire thickness (Fig. 10-2).

The upper and lower lips are attached to their respective alveolar processes by a web of fibrous tissue in the midline known as the *labial frenum.* The frenum is covered with mucosa (Fig. 10-3).

The cheeks, which form the lateral boundary of the vestibule, are similar in structure to the lips. Fatty tissue lies under the skin, and glands lie under the mucosa. Sandwiched between the fat and glands is the buccinator muscle.

Intraorally, the mucosa is bound to the deeper structures of the cheek by fibrous tissue. This aids the mucosa in conforming to the shape of the Buccinator during contraction and helps to prevent cheek biting during chewing.

Small yellow dots may be seen on the cheek mucosa just inside the corner of the mouth. They are not abnormal and represent rudimentary sebaceous glands. They are referred to as *Fordyce's spots.*

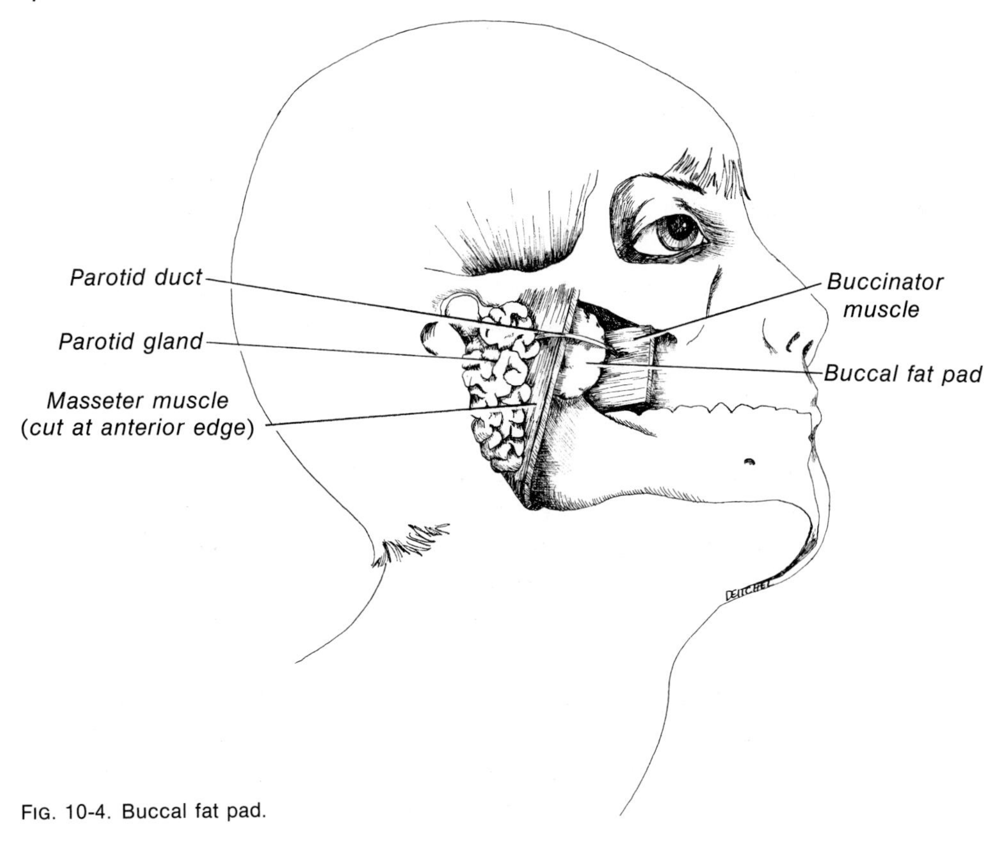

FIG. 10-4. Buccal fat pad.

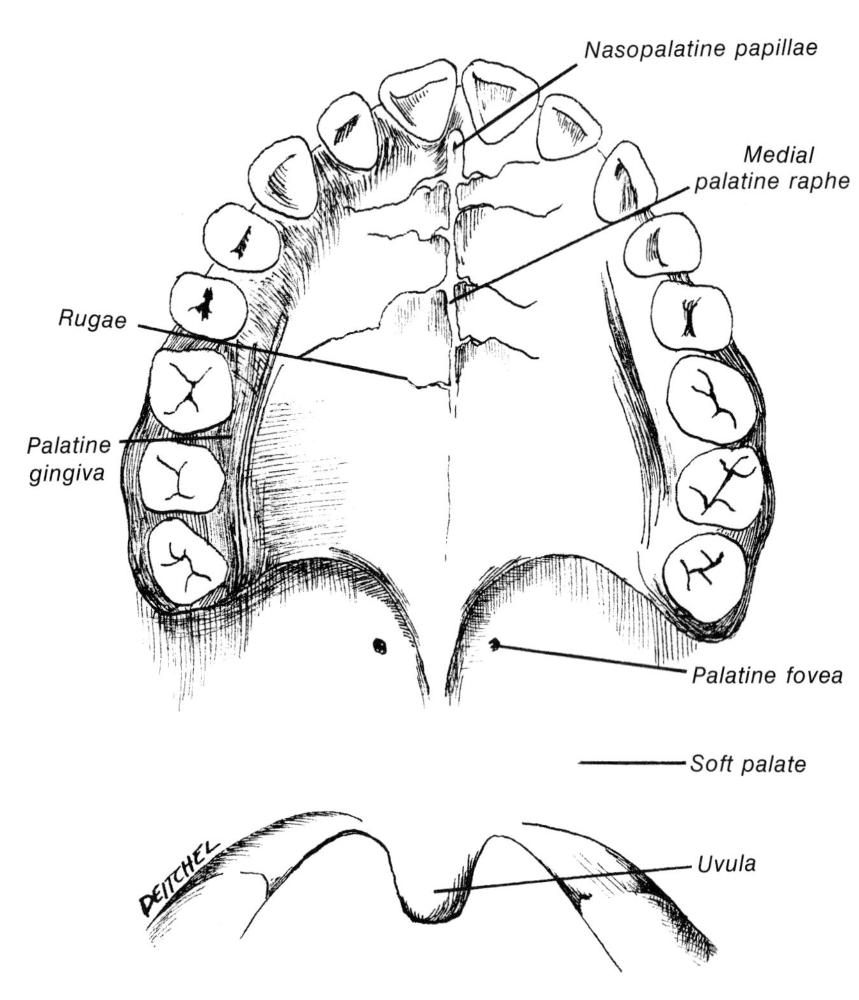

FIG. 10-5. Hard and soft palates.

The *buccal fat pad* is a lobular convex structure occupying a small area between the buccinator muscle and the masseter muscle. Some of this fat extends back into the space between the temporalis and the pterygoid muscles. Moreover, the fat may also reach superiorly into the temporal fossa and inferiorly into the pterygomandibular space. The mass of fat lying between the buccinator and masseter muscles is the buccal fat pad, often referred to as the "fat pad of Bichat" (Fig. 10-4). It is most interesting that even in patients suffering from severe malnutrition, the buccal fat pad persists much longer than does subcutaneous fat in other portions of the body.

Cavum Oris

Within the confines of the alveolar arches is the *cavum oris.* Superiorly, the boundary is the hard and soft palate. The mylohyoid muscle forms the floor. The anterior tonsillar pillar acts as the posterior boundary, leading into the oropharynx.

The *palate* or roof of the oral cavity is divided into the *hard palate* and the *soft palate* (Fig. 10-5). The maxillary palatine processes and horizontal processes of the palatine bone contribute to the skeleton of the hard palate. A thick layer of soft tissue covers the hard palate. Peripherally, the mucosa surrounding the teeth is known as the *palatine gingiva.* The soft tissues over the posterior two thirds of the hard palate are more spongy than at the anterior one third, owing to the fact that in the mucosa of the posterior two thirds lie most of the palatine salivary glands. The mucosa of the anterior one third of the palate reveals radiating ridges, the *palatine rugae.* Just behind the central incisors is the incisive (nasopalatine) *papilla,* which marks the nasopalatine foramen. Extending posteriorly from the papilla in the midline is the *median palatine raphe.* It is a narrow, sharp ridge of fibrous tissue covered by mucosa, and it marks the median palatine suture.

The mucosa covering the hard palate extends posteriorly to envelop the palatine muscles. This area is the *soft palate.*

The muscles in the soft palate are the *tensor veli palatini* supplied by the trigeminal nerve, the *levator veli palatini* supplied by the pharyngeal plexus, and the *musculus uvulae* also supplied by the pharyngeal plexus of nerves. The mucosa of the soft palate is more red than that of the hard palate because of the greater keratinization of the hard palate and the rich and densely arranged vessels of the soft palate.

Just behind the boundary between the hard and soft palates is a pair of small pits on either side of the midline, the *palatine foveae.* Some of the ducts of the palatine glands empty here. When the mouth is wide open, a fold of mucous membrane can be seen extending from the lateral root of the soft palate to the retromolar pad of the mandible (Fig. 10-6). Beneath this fold is the *pterygomandibular raphe,* which is interposed between the buccinator muscle and the pharyngeal constrictor muscle. Further posteriorly, a

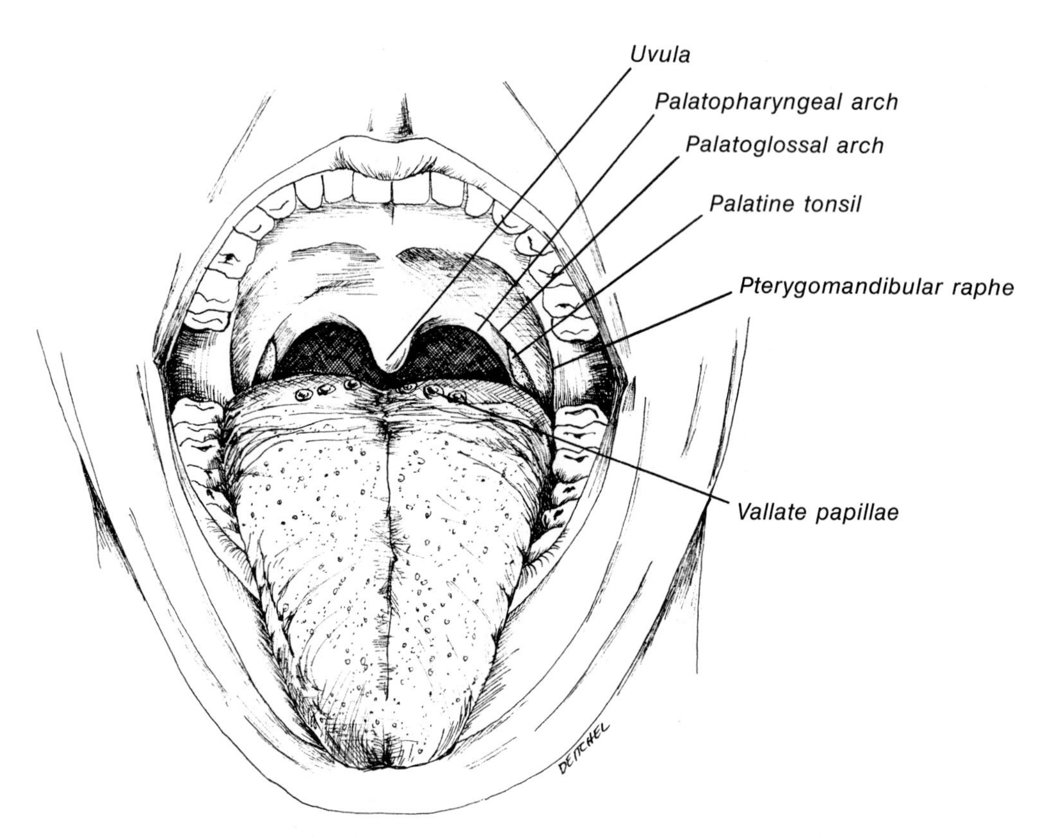

FIG. 10-6. Oral cavity (cavum oria).

second fold can be seen extending to the base of the tongue. This is the *palatoglossal arch,* or *anterior tonsillar pillar,* and represents the mucosal covered palatoglossus muscle. A third fold can be seen more medial and posterior, the *palatopharyngeal arch* or *posterior tonsillar pillar.* This arch represents the palatopharyngeus muscle. Between the anterior and posterior tonsillar pillars a depression, the *tonsillar fossa,* is evident. Herein lies the palatine tonsil.

The posterior border of the soft palate is free and doubly concave. The mucosa continues onto the nasal mucosa. A small projection of muscle and mucosa, the *uvula,* is located in the midline of the free posterior border.

The *alveolar arches* form the lateral and anterior boundaries of the cavum oris, separating it from the buccal and labial vestibules. They are bony processes of the maxilla and mandible and are composed of cancellous bone covered with a thin plate of cortical bone. The roots of the teeth are anchored in the alveolar processes in sockets lined with cortical bone known as *lamina dura.* The mucosa covering the alveolar processes is the *gingiva.* Triangular projections of gingiva between the teeth are the *interdental papillae.* As the gingiva reflects into the vestibule, it is smoother, more mobile, and darker in color.

Behind the last mandibular molar, a fleshy area, *the retromolar pad,* is located. It contains glands and fibrous tissue. As the alveolus atrophies following loss of teeth, this pad becomes more prominent. Clinically, it is continuous with the pterygomandibular fold.

The tongue occupies the space between the lower alveolus and the floor of the mouth. It is a muscular organ covered with specialized mucosa and is attached at its base and central portion to the floor of the mouth. The muscles of the tongue are described in Chapter 2.

The tongue may be divided into an anterior two thirds and a posterior one third (Fig. 10-7). The body and the tip are the anterior one third. The posterior one third is the base and the root.

A V-shaped trough, the *terminal sulcus,* separates the two divisions. Embryologically, the anterior two thirds of the tongue develops from the first branchial arch. It receives nerve innervation from the chorda tympani and lingual nerves. The posterior one third develops from the second and third branchial arches and is innervated by the glossopharyngeal nerve. At the apex of the terminal sulcus is a blind pit, the *foramen cecum,* which marks the point of embryologic development of the thyroid gland.

The dorsal surface of the anterior two thirds of the tongue faces the palate. Four different types of papillae are located anterior to the terminal sulcus (Figs. 10-7; 10-8). The mushroom-shaped prominences directly in front of the terminal sulcus are the *circumvallate* papillae. They have a trough surrounding them. The walls of this trough contain taste buds and the depths contain the *serous salivary glands of Von Ebner. Filiform papillae* cover the entire palatine surface of the tongue and give the tongue its velvety appearance. Interspersed among the filiform papillae are the *fungiform papillae,* which appear as red dots and contain taste buds. Along the poste-

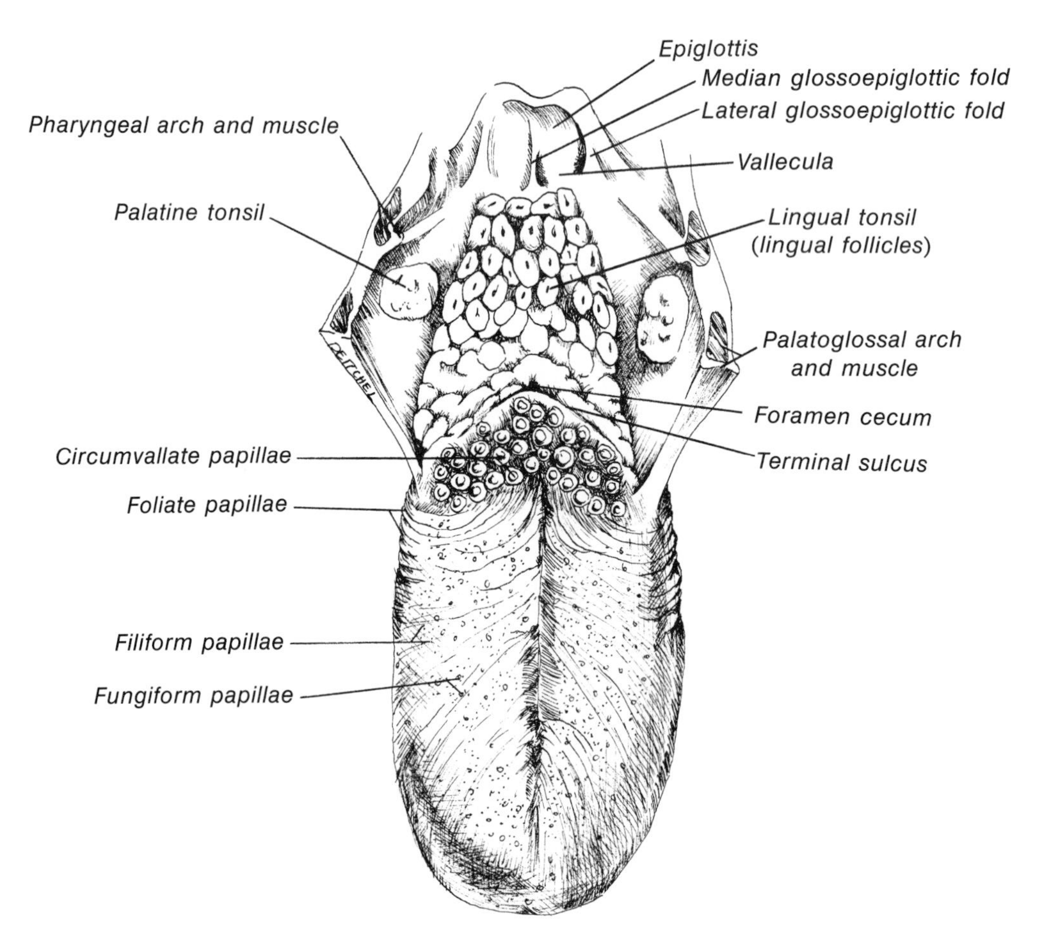

FIG. 10-7. Dorsum of tongue.

rior lateral area of the tongue are parallel folds of tissue, *the foliate papillae.* These also contain taste buds.

Posterior to the terminal sulcus, the dorsal surface of the tongue contains oval prominences surrounded by shallow furrows. This is lymphatic tissue and is often termed the *lingual tonsil.*

The root of the tongue is attached to the epiglottis by a medial fold, the *glossoepiglottic* fold. Laterally, the *glossopharyngeal* folds attach the root of the tongue to the pharyngeal walls. On either side of the *glossoepiglottic* fold, between it and the glossopharyngeal folds, are the right and left *valleculae.* The valleculae are deep blind pouches. The palatoglossal arch or fold attaches the root of the tongue to the palate.

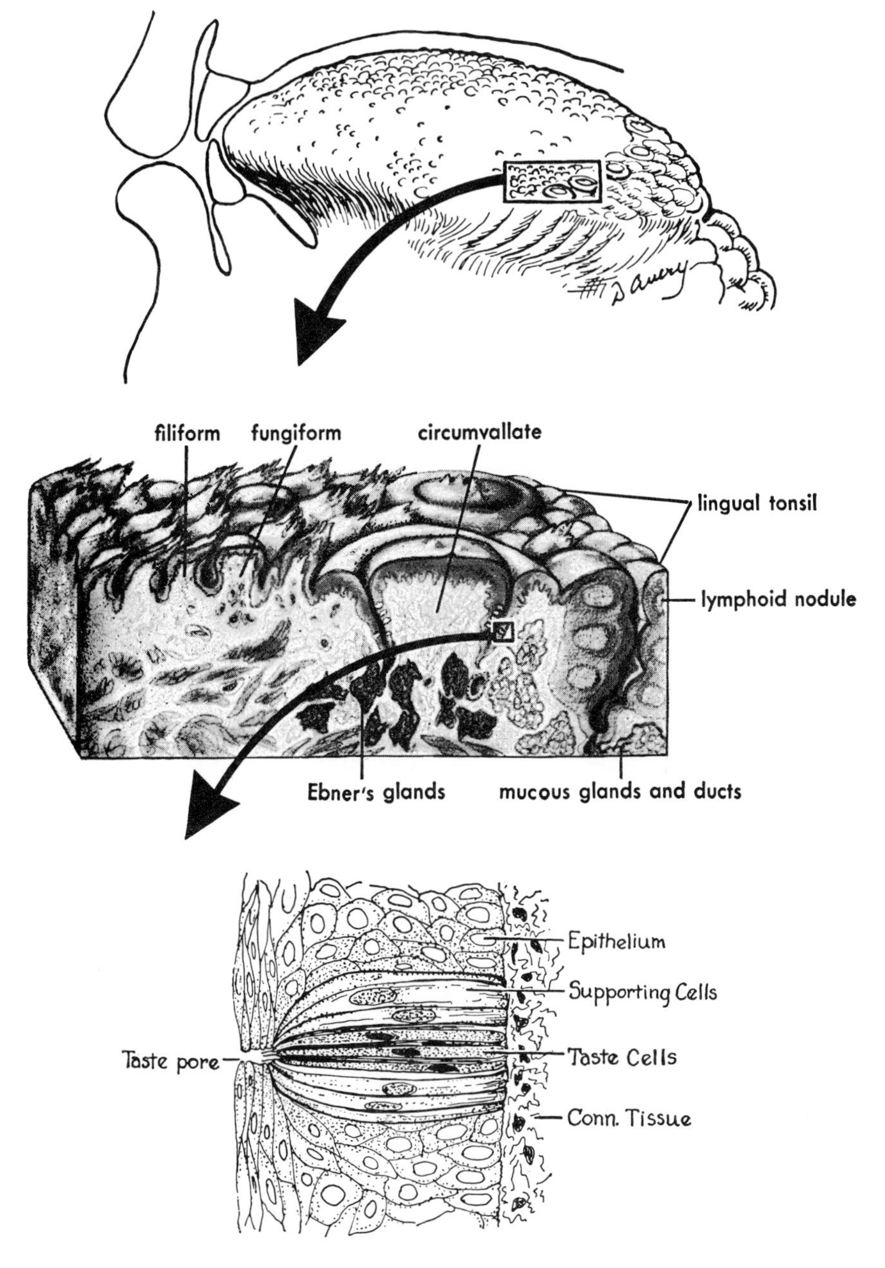

FIG. 10-8. Above, diagram of the body of the tongue. Center, a diagram showing the microscopic anatomy of the tongue. Below, a taste bud with a taste pore opening on the surface of the epithelium. (From Steele, P. F.: Dimensions of Dental Hygiene, 2nd ed. Philadelphia, Lea & Febiger, 1975.)

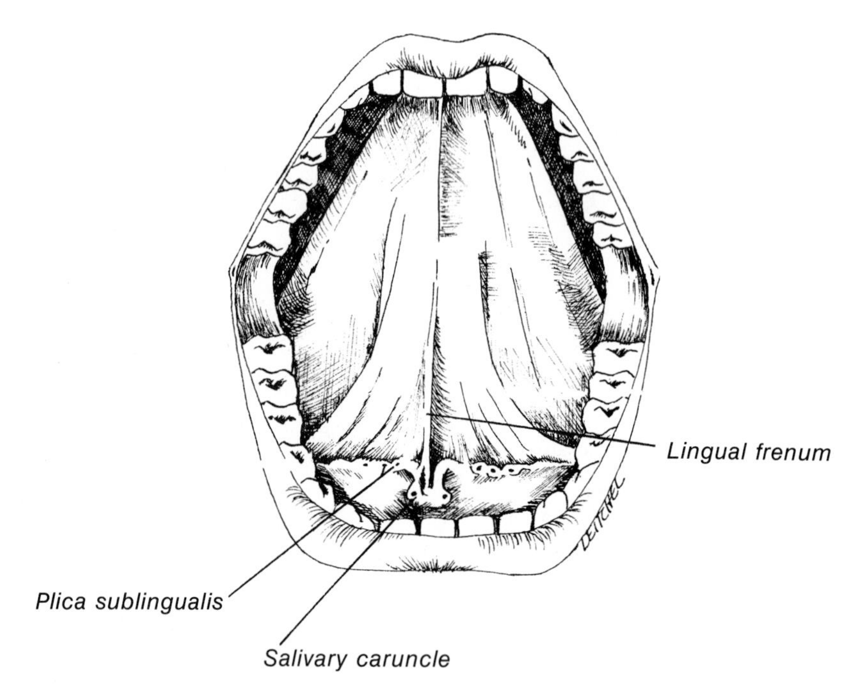

FIG. 10-9. Sublingual region of the tongue.

The ventral or inferior surface of the tongue rests on the mucosa of the floor of the mouth. It is covered by smooth mucosa. In the midline is the *lingual frenum,* which is a sickle-shaped fold attaching the underside of the tongue to the floor of the mouth (Fig. 10-9). Anteriorly, on either side of the frenum, can be seen the *caruncles,* openings for the right and left submandibular ducts. An irregular fold, the *plica sublingualis,* begins here and runs posteriorly along the floor of the mouth, marking the openings of the ducts of Rivinus of the sublingual gland.

Muscles, nerves, and blood supply to the lips, cheeks, oral cavity, and tongue are described in detail in Chapters 2, 3, and 4.

The Pharynx

That part of the respiratory-digestive tube placed behind and below the nasal cavities and mouth is the *pharynx.* It is a conical, muscular tube with the base pointed upward. The boundaries are

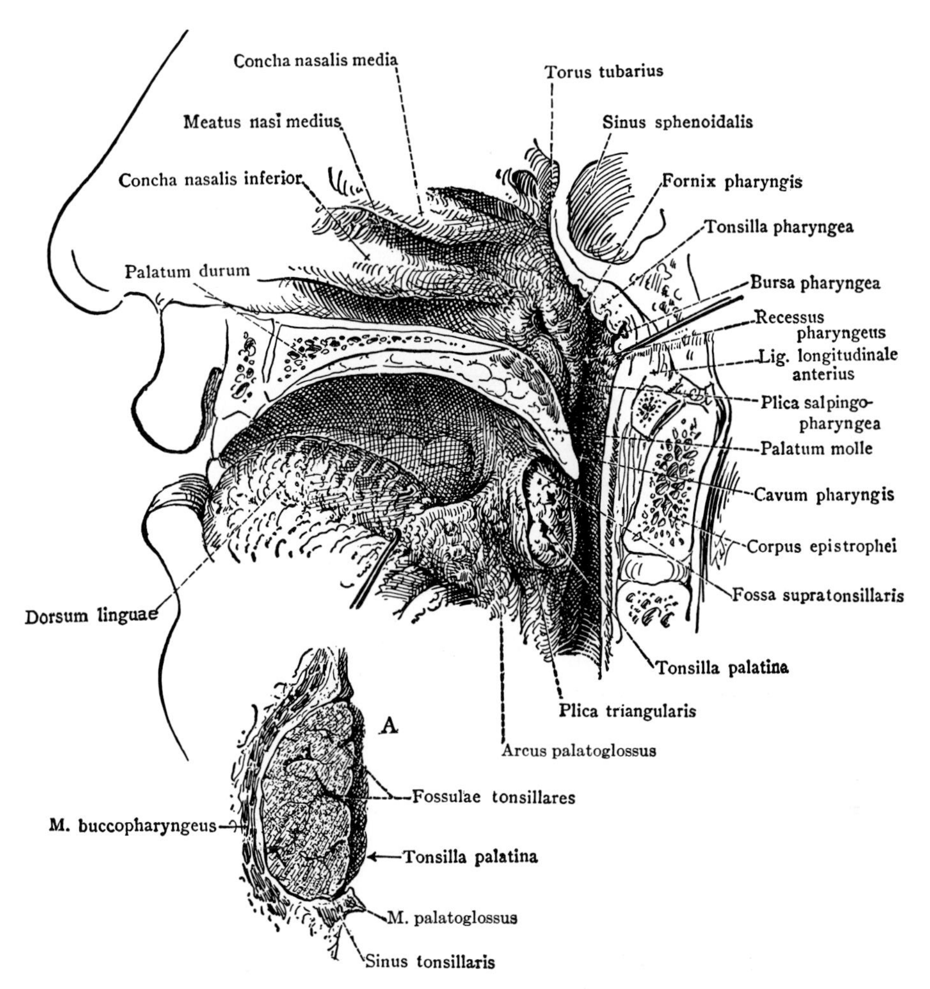

FIG. 10-10. The oral and nasal pharynx in median sagittal section, showing palatine and pharyngeal tonsils. A. Detail of palatine tonsil in frontal section. (After Eycleshymer and Jones in Gray's Anatomy of the Human Body, 29th ed. C. M. Goss, editor. Philadelphia. Lea & Febiger, 1973.)

from the base of the skull to the sixth cervical vertebra. The narrowest point is at its termination in the esophagus.

The pharynx may be divided into three parts: *nasopharynx, oropharynx* and *laryngopharynx.* As with the oral cavity, nerves, blood vessels, and muscles are described in detail in Chapters 2, 3, and 4.

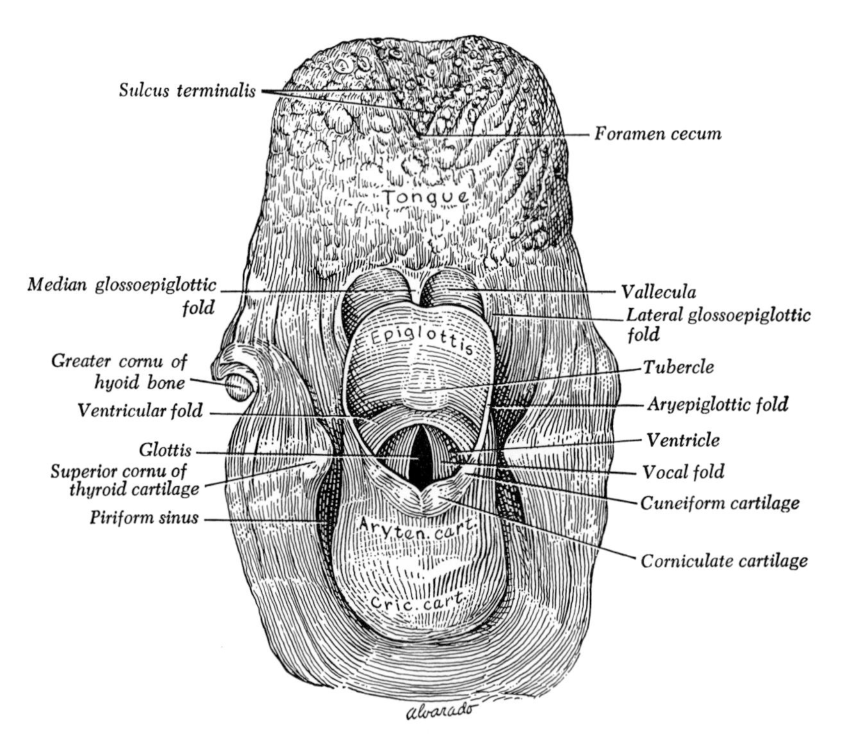

FIG. 10-11. Entrance of the larynx, posterior view. (From Gray's Anatomy of the Human Body, 29th ed. C. M. Goss, editor. Philadelphia, Lea & Febiger, 1973.)

Nasopharynx

The nasopharynx is about one inch in length and lies behind the nasal cavity, above the oral cavity and in front of the first cervical vertebra (Fig. 10-10). On the lateral wall is seen the opening for the auditive (eustachian) tube. This opening is at the level of the inferior concha. Superiorly and posteriorly guarding the opening is a crescent-shaped bulge, the *torus tubarius.* The posterior-inferior end of the torus tubarius passes down as a fold, the *salpingopalatine fold,* to attach to the palate. The *salpingopharyngeal fold,* marking the salpingopharyngeal muscle, is more prominent and arises in the same area to pass down and attach to the lateral wall of the lower pharynx. Behind the torus is a blind pouch, the *pharyngeal recess.* Lymphatic tissue, known as the pharyngeal tonsil (adenoids), lies in this area.

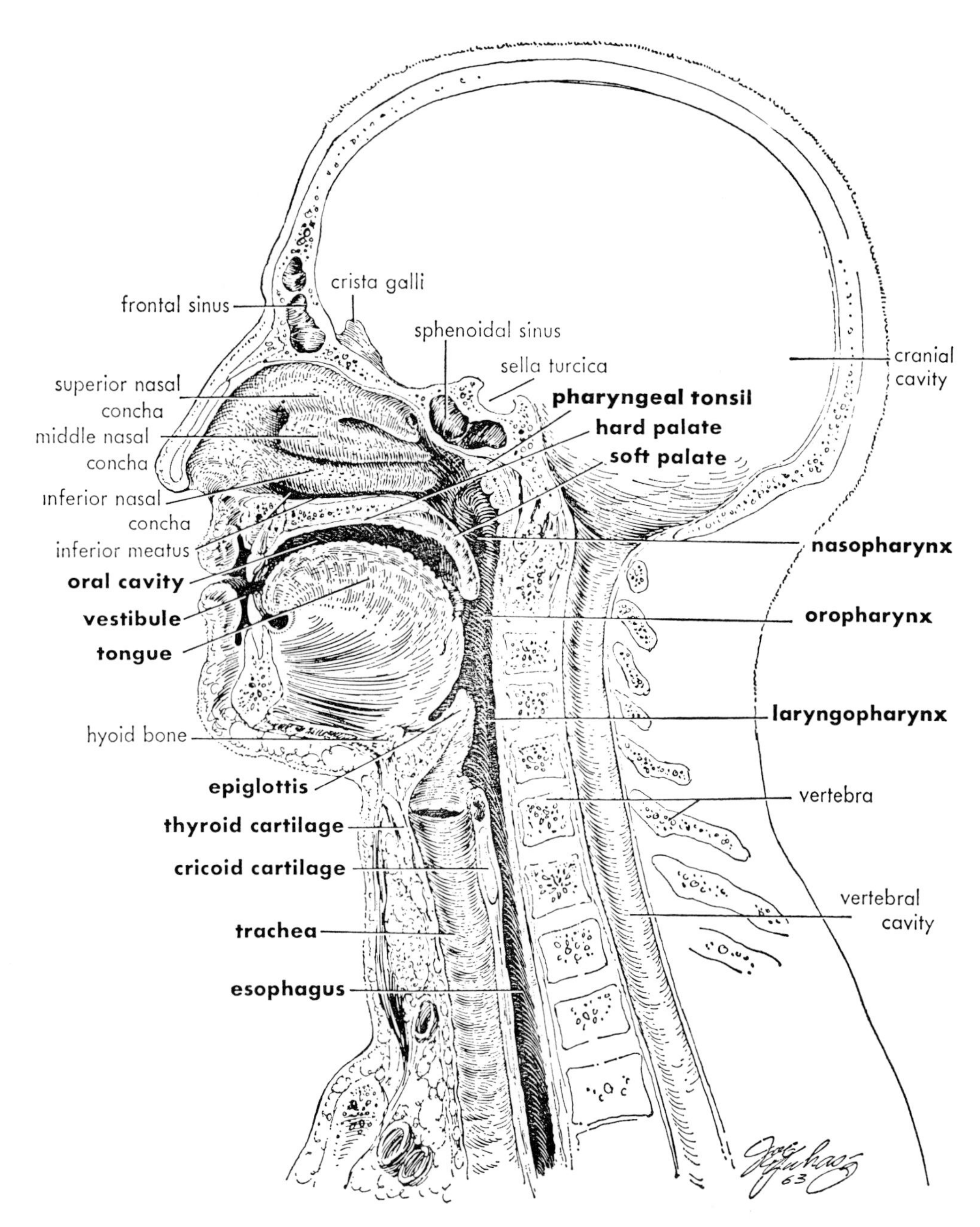

FIG. 10-12. Median section of the head and neck showing relationships of upper digestive and respiratory systems. (From Crouch, J. E.: Functional Human Anatomy, 2nd ed. Philadelphia, Lea & Febiger, 1972.)

Oropharynx

The oropharynx is about 2 inches in length, lies in front of the second and third cervical vertebrae, and extends from the soft palate to the hyoid bone. The anterior tonsillar pillar (*glossopalatine fold*) is considered the anterior boundary of the oropharynx. In the oropharynx are seen the medial *glossoepiglottic fold* and the two lateral *pharyngoepiglottic* folds, which attach the epiglottis to the tongue and the wall of the pharynx. The depressions between them are the *valleculae* (Fig. 10-11). The epiglottis is a flap of cartilage covered with mucosa, shaped like a leaf, and projecting obliquely upward behind the root of the tongue. The free end is capable of flapping over the entrance to the trachea during swallowing. It keeps material from entering the trachea and directs swallowed material backward so that the material will enter the esophagus (Fig. 10-12).

Laryngopharynx

The laryngopharynx is that portion of the pharynx that extends from the level of the hyoid bone to the lower border of the cricoid cartilage. It lies in front of the fourth, fifth, and sixth cervical vertebrae and is approximately 3 inches in length.

The *larynx,* located in the laryngopharynx, is the voice box. It projects up into the laryngopharynx and is the entrance to the trachea which leads into the lungs (Fig. 10-12). On either side of the laryngeal opening, between it and the walls of the pharynx, is a blind pouch, the *piriform sinus* (Fig. 10-11).

Three major cartilagenous forms make up the bulk of the framework of the larynx: the *thyroid cartilage,* the *epiglottis,* and the *cricoid cartilage* (Fig. 10-13). The thyroid cartilage is shaped like a shield. The cricoid cartilage is shaped like a signet ring and the epiglottis is leaf-shaped. They are held together by ligaments and muscles (Fig. 10-14). When the muscles and ligaments are removed, it is easier to see how the parts fit together (Fig. 10-15). The arytenoid, corniculate, and cuneiform cartilages are small, paired sesamoid structures that complete the framework.

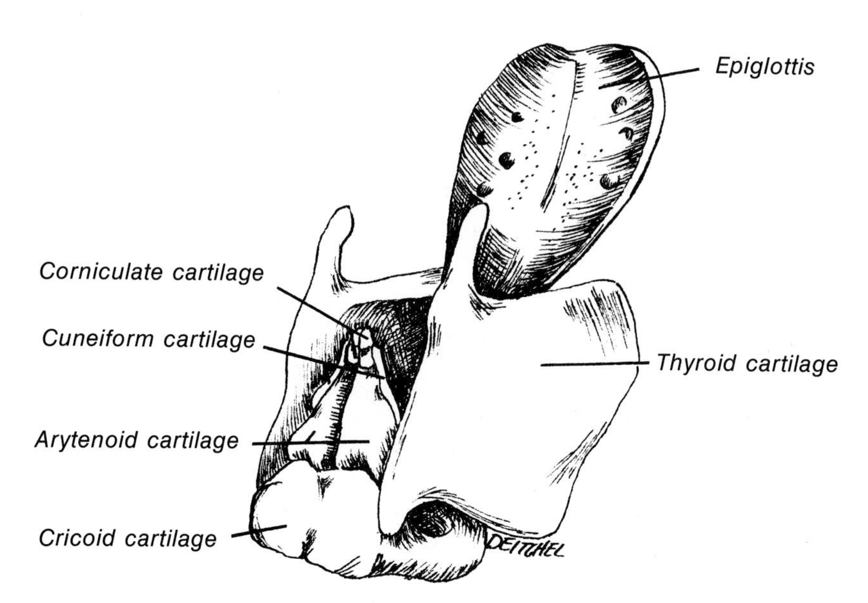

FIG. 10-13. Framework of the larynx (together).

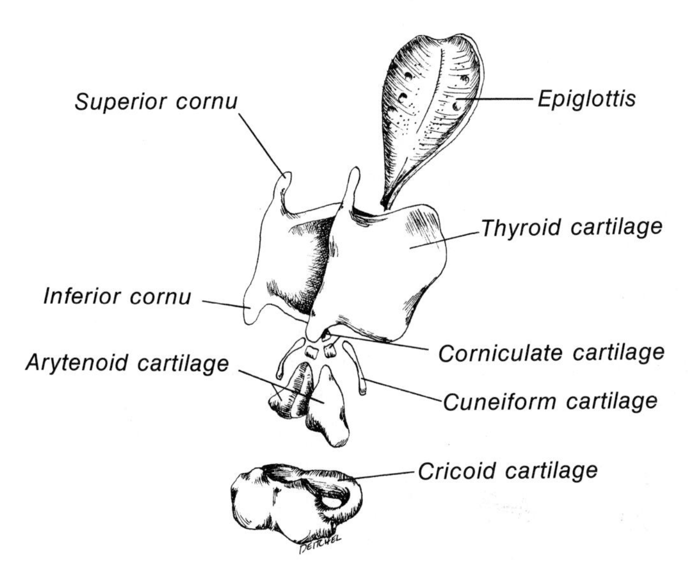

FIG. 10-14. Framework of larynx (apart).

Looking down into the larynx from above, we can see the vocal cords crossing the interior (Fig. 10-16). As air passes over them, the cords vibrate and sound is produced. The sounds vary, depending on the tightness of the cords. This tension is controlled by several muscles all working in a very coordinated fashion. The larynx is, indeed, a highly sophisticated and specialized organ.

The interior of the larynx is covered with mucosa and clinically may be examined with a laryngoscope (Fig. 10-17). This instrument retracts the root of the tongue downward and pulls the epiglottis forward, opening the larynx. The view seen by the clinician is exemplified in Figure 10-16.

Below and posterior to the laryngopharynx is the esophagus leading to the stomach.

The muscles of the pharynx are the superior, medial, and inferior constrictors, the stylopharyngeus, salpingopharyngeus, and the palatopharyngeus. The nerves and blood supply are discussed in Chapters 3 and 4. The constrictors and salpingopharyngeus are supplied by the pharyngeal plexus and glossopharyngeal nerve. The blood supply is via the ascending pharyngeal artery, superior thyroid artery, and thyrocervical branches of the subclavian artery.

Clinical Notes

Debris or a foreign body such as a fish bone is occasionally trapped in the valleculae or in the piriform sinus. Removal of such an item can be quite difficult, particularly if the patient is a child.

When the lingual frenum attaches close to the top of the tongue, a condition known as *ankyloglossia* (tongue-tie) occurs (Fig. 10-18). This limits the mobility of the tongue and may impair speech. If so, a *frenectomy* may be performed to increase tongue function.

If the mucosa of the larynx is stimulated by a foreign body, such as filling material, saliva, or food, the muscles controlling the vocal cords will cause the cords to close the opening into the trachea. This is known as *laryngospasm.* It protects the tracheobronchial tree from debris. However, it also prevents air from entering the lungs and, therefore, often causes an emergency. Intravenous injection of a rapidly acting muscle relaxant will paralyze the cords, opening the

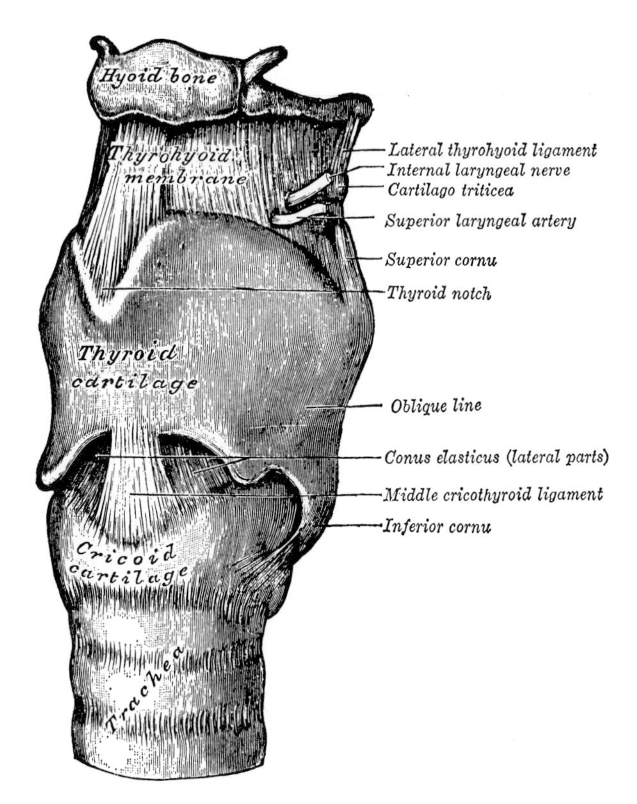

FIG. 10-15. The ligaments of the larynx. Anterior and slightly lateral view. (From Gray's Anatomy of the Human Body, 29th ed. C. M. Goss, editor. Philadelphia, Lea & Febiger, 1973.)

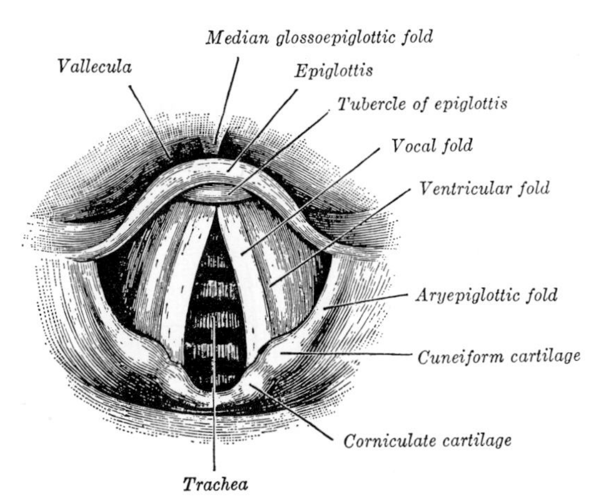

FIG. 10-16. Laryngoscopic view of interior of larynx. (From Gray's Anatomy of the Human Body, 29th ed. C. M. Goss, editor. Philadelphia, Lea & Febiger, 1973.)

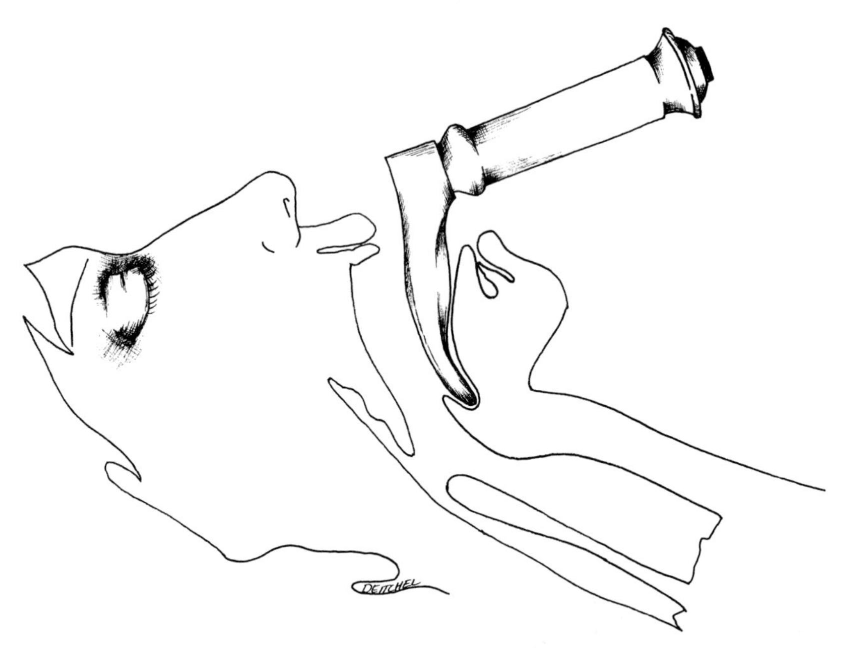

FIG. 10-17. Use of laryngoscope to expose glottis.

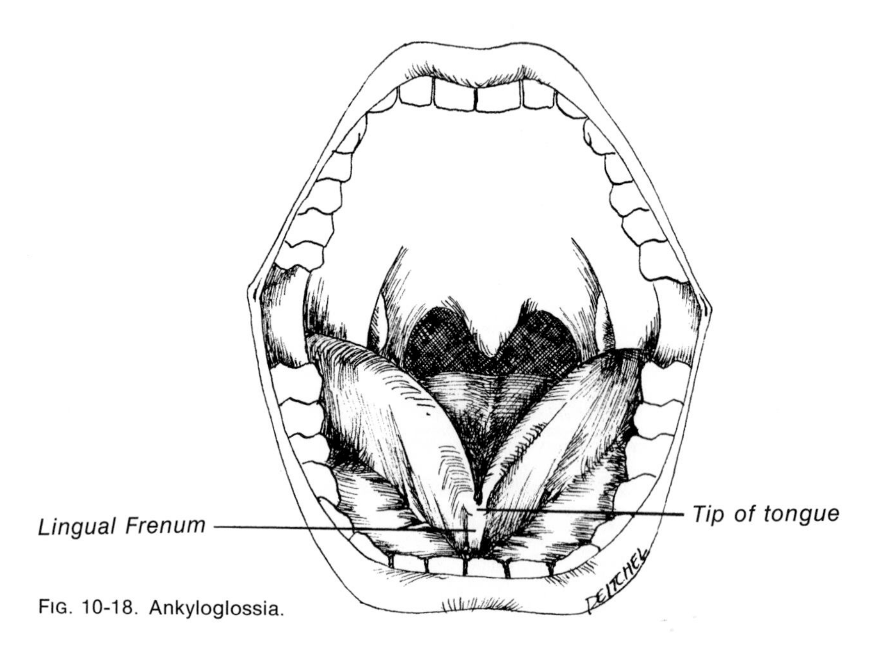

FIG. 10-18. Ankyloglossia.

passage into the trachea. Once the cords are relaxed, air can be introduced into the trachea and lungs. However, one must be careful not to project the causative foreign body into the trachea.

References

ARCHER, W. H.: Oral Surgery, 4th ed. Philadelphia, W. B. Saunders Co., 1966.

BOUCHER, C. O.: Current Clinical Dental Terminology. St. Louis, MO, C. V. Mosby, 1974.

CROUCH, J. E.: Functional Human Anatomy, 2nd ed. Philadelphia, Lea & Febiger, 1972.

DEWEESE, D. D., and SAUNDERS, W. H.: Textbook of Otolaryngology, 4th ed. St. Louis, MO, C. V. Mosby, 1973.

Dorland's Illustrated Medical Dictionary, 25th ed. Philadelphia, W. B. Saunders Co., 1974.

DRIPPS, R. E., et al.: Introduction to Anesthesia: The Principles of Safe Practice. Philadelphia, W. B. Saunders Co., 1972.

EDWARDS, L. F.: Concise Anatomy, 3rd ed., New York, McGraw-Hill, 1971.

GRANT, J. B.: An Atlas of Anatomy, 6th ed. Baltimore, Williams & Wilkins, 1972.

GOSS, C. M. (editor): Gray's Anatomy of the Human Body, 29th ed. Philadelphia, Lea & Febiger, 1973.

GUYTON, A. C. Textbook of Medical Physiology, 4th ed. Philadelphia, W. B. Saunders Co., 1971.

HAMM, A. W.: Histology, 3rd ed. Philadelphia, J. B. Lippincott Co., 1957.

HOUSSAY, B. A.: Human Physiology, 2nd ed. New York, McGraw-Hill, 1955.

KRUGER, G. O.: Textbook of Oral Surgery, 3rd ed. St. Louis, MO, C. V. Mosby, 1968.

ORBAN, B. J.: Oral Histology and Embryology, 5th ed. St. Louis, MO, C. V. Mosby, 1962.

PERNKOPF, E.: Atlas of Topographical and Applied Human Anatomy. Philadelphia, W. B. Saunders Co., 1963.

SHAFER, W. G.: HINE, M. K., and LEVY, B. M.: A Textbook of Oral Pathology, 3rd ed. Philadelphia, W. B. Saunders Co., 1974.

SICHER, H., and DUBRUL, E. L.: Oral Anatomy, 6th ed. St. Louis, MO, C. V. Mosby, 1975.

SNELL, R. S.: Clinical Anatomy for Medical Students. Boston, Little, Brown & Co., 1973.

STEELE, P. F.: Dimensions of Dental Hygiene, 2nd ed. Philadelphia, Lea & Febiger, 1975.

THOMA, K. H.: Oral Surgery, 5th ed. St. Louis, MO, C. V. Mosby, 1969.

WARFEL, J. H.: The Head, Neck, and Trunk, 4th ed. Philadelphia, Lea & Febiger, 1973.

WHEELER, R. C.: A Textbook of Dental Anatomy and Physiology, 4th ed. Philadelphia, W. B. Saunders Co., 1965.

Index